THE AMERICAN WHALEMAN

Sperm Whaling—"The Chase"

From a lithograph by Endicott & Co., New York, 1859, after drawings by A. Van Best and R. S. Gifford, corrected by Benjamin Russell

THE AMERICAN WHALEMAN

A STUDY OF LIFE AND LABOR
IN THE WHALING INDUSTRY

BY

ELMO P. HOHMAN

[1928]

AUGUSTUS M. KELLEY • PUBLISHERS
CLIFTON 1972

First Edition 1928
(New York: Longmans, Green & Company, 1928)

Reprinted 1972 by
Augustus M. Kelley Publishers
REPRINTS OF ECONOMIC CLASSICS
Clifton New Jersey 07012

I S B N 0-678-00834-5
L C N 68-55737

PRINTED IN THE UNITED STATES OF AMERICA
by SENTRY PRESS, NEW YORK, N. Y. 10013

PREFACE

THE American sailing-ship has vanished from the seas, but not from memory. The twentieth century is witnessing a growing revival of interest in the California clippers, the Liverpool packets, the East Indiamen and China clippers, the African slave-traders, and even the lowly coasting schooners. Something there was about the clouds of towering canvas and the thoroughbred lift and dip of a Donald McKay clipper which gripped the memory of those who knew them and which to-day challenges the imagination of those who are less fortunate. So alluring, so imperative is this challenge that it is beginning to penetrate through the loose ranks of sea yarns and reminiscences into the guarded studies of objective scholarship; and in consequence large stores of documentary treasure, buried for the most part in Atlantic Coast libraries and customs-houses, are gradually being salvaged and interpreted in the light of modern research methods.

By no means the least interesting member of this departed fleet was the whaler — ship, bark, or mere "plum-pud'ner." As a vessel she was signally lacking in the grace, speed, and slender feminine beauty of the clipper; but in staunch seaworthiness, in bulldog battling with wind, wave, and whale, and in the range and seeming endlessness of her voyages she was the peer of any craft afloat. In spite of her slowness, her breadth of beam, and her broad-ended wallowing, the typical whaler was the vehicle of a life which merged adventure and exploitation, courage and brutality, abandon and niggardliness, as fully as any occupation in America, past or present.

Recent emphasis upon the social sciences and a growing humanitarianism are partly responsible, no doubt, for an interest in the sailing-ship era which extends to the men as well as to the vessels. What manner of men occupied the forecastles and cabins, and how did they live while at sea? Where did they hail from, and what were the tricks of their trade? What of hours, wages, and working conditions? And what of dangers and discipline?

The present work seeks primarily to answer such questions, in the case of the whalemen, by holding a mirror up to reality. It is designed to be a study of labor conditions and of labor problems in a much-neglected corner of American economic history — an excursion into the field of whaling man-power rather than of whaling tonnage. A considerable proportion of the material presented has been taken directly from original manuscripts — log-books, Consular Letters, crew-lists, and, most valuable of all, crew-accounts and other whaling account-books — which have scarcely been touched, heretofore, for research purposes. And yet the entries in scores of crew-accounts are more accurately eloquent of whalemen's earnings and treatment than hundreds of pages of general description!

Seen in retrospect, the whaleman assuredly did not merit the scorn which was heaped upon him so generously by his cousin, the merchant seaman. On the contrary, "whether we consider the stupendous object of his pursuit, or the vast extent of waters over which he roams to secure his prey, or the dangers and perils peculiar to his avocation," the whaleman is a figure to conjure with. And that without the aid of a single spark of fancy! For the unadorned facts of his life became at times so strange or so harrowing that they had no need of the pale aid of imagination. Here was daring, languorous love in the South Seas, exotic adventure, and exploring wanderlust; and here also was danger, monotony, exploitation, shipwreck, "ugly" and "gallied" whales, vice, cruel hardship, and "Nantucket sleigh-rides!" Small wonder that such ingredients, mixed in constantly varying proportions by romantic fortune, made for experiences far stranger than those undergone by many heroes of fiction!

Part II is meant to present a picture of whaling life at the time when the fishery was in full career. This is the backbone of the work, containing the bulk of whatever contribution these researches may have to offer. But to present this material without considering the remainder of the industry seemed like tearing a passage from its context, or presenting a play without stage settings. Parts I and III, therefore, are intended to provide the scenery against which the action in Part II takes place. They do not pretend to form an adequate history of American whaling: rather do they provide the main elements of an absorbing story of economic development and devolution for those who have neither time nor inclination to delve into much fugitive historical material. The footnotes are road-signs for those who care to read them; the appendices contain samples of the raw material out of which the book was constructed; and the bibliography may serve as chart and compass to those who wish to explore the less familiar waters of whaling literature.

It is a privilege to acknowledge my indebtedness to Professors W. Z. Ripley, S. E. Morison, and F. W. Taussig, of Harvard University, for encouragement and invaluable counsel; to Messrs. George H. Tripp and Frank Wood for ready access to the wealth of material gathered by the New Bedford Free Public Library and by the Old Dartmouth Historical Society; to Mr. George Francis Dow and Mr. Harry Neyland for assistance in reproducing certain illustrations; and, most of all, to Helen Fisher Hohman for the discriminating judgment and appraising criticism of a fellow-craftsman.

ELMO P. HOHMAN

Evanston, Illinois
June, 1928

CONTENTS

ILLUSTRATIONS

THE AMERICAN WHALEMAN

THE

AMERICAN WHALEMAN

CHAPTER I

BEARINGS

AMERICAN whaling as a significant industry vanished with the coming of the twentieth century. In the record of national economic and maritime development it is a closed chapter. The typical "blubber-hunter" — broad on the beam, bluff-bowed, square-rigged, and resembling, on dark nights when the try-pot fires were burning, a miniature floating inferno — has disappeared from the seas. With look-outs perched precariously at her mast-heads and with three or four whaleboats swinging conspicuously from her davits, she no longer excites the curiosity of ocean travelers or the derision of merchant seamen. The ports which sent such whalers to the ends of the earth have only fading memories of voyages which brought much of the colorful and of the exotic to the shores of provincial New England. Whaling products in part have outlived the uses to which they were put, and in part have been supplanted by satisfactory substitutes. And the whaleman is a figure forgotten or remembered only in retrospect, along with the peculiar factors of his onerous and ill-starred occupation.

But such a characterization applies only to the present and to the recent past. For as the eighteenth century witnessed the gradual rise and hesitant expansion of American whaling, and as the twentieth century has seen its extinction, so the intervening period marked its greatest growth and highest development. Throughout the greater part of the nineteenth century Yankee whaling products supplied many vital needs both at home and abroad; and in so doing they carried the industry to such impressive proportions that the combined whaling fleets of all other nations were hopelessly outdistanced.

Sperm oil constituted the best standard illuminant, and was used extensively in light-house beacons and wherever a bright, clean light was desired. Spermaceti, a spongy, oil-containing substance found in the head of the sperm whale, formed the basis for the better grades of candles. Whale oil was employed in the cheaper types of illumination and for a variety of lubricating purposes. Whalebone, utilized in the manufacture of stays, corsets, riding and carriage whips, umbrellas, and other objects requiring both strength and flexibility, served a wider range of functions. Ambergris, extremely scarce and correspondingly valuable, was much sought both in the manufacture of perfumery and in various Mohammedan countries, where it was regarded as an infallible aphrodisiac. Other minor and incidental products included blackfish oil; sea-elephant oil; and the teeth and jawbone of the sperm whale, patiently transformed during long hours of shipboard monotony into countless specimens of the scrimshawer's art. There was little commercial value in the last two commodities: they served rather for the making of curiously-carved gifts for the women at home, or for the casual and licentious companions so readily found in foreign ports.

This list of products gave whaling an economic and industrial position significant to the entire nation, and vital to the New England seaboard. For some time before the Civil War only the manufacturers of shoes and of cottons stood between whaling and the front rank amongst all the industries of Massachusetts. Lieutenant Wilkes, writing the last chapter of his monumental "Narrative of the United States Exploring Expedition" during the early forties, was able to state: "Our whaling fleet may be said at this very date to whiten the Pacific Ocean with its canvas. . . . The ramifications of the business extend to all branches of trade . . ." and ". . . are spread through the whole Union."

In 1838 the value of the oil and bone brought into New

Bedford alone was $2,490,051; and during the ensuing forty years, with six annual exceptions, it remained between $2,000,000 and $6,000,000.[1] The whaling fleet attained its greatest size as early as 1846, when 735 vessels, aggregating 233,189 tons, were engaged in the industry. The years of greatest production, however, extended over the decade 1843–1853. Sperm oil, with 166,985 barrels, reached its peak in 1843; whale oil, with 328,483 barrels, in 1851; and whalebone, with the enormous catch of 5,652,300 pounds, in 1853.[2]

The size and scope of the industry stand out clearly in two cross-section views. The first is a statistical photograph taken during the year 1833, when the whaling merchants were just entering upon their greatest era. The second portrays the business as it appeared on January 1, 1844, while gathering resources to send out its greatest fleet two years later.

In 1833 there were some 392 American whaling vessels, displacing about 132,000 tons and carrying approximately 10,000 men. The value of vessels and outfits was placed at $12,000,000; and the annual income for each of the three preceding years was estimated to have been about $4,500,000. In addition, there were numerous other occupations which were wholly or partly dependent upon the industry. Such were the manufacture of whaling equipment and of whaling products; shipbuilding and repairing; ship-chandlery; the conduct of sail-lofts and of rope-walks; the furnishing of outfits and of board and lodging for prospective and returning crews; and the provision, for whalemen on shore leave, of facilities for recreation and dissipation. Including these subsidiary callings, it was held that about $70,000,000 worth of property and some 70,000 persons were involved, directly and indirectly, in the business of whale-hunting.[3]

Within the next decade, however, even these generous figures were far outdistanced. By January 1, 1844, the fleet had

[1] Pease, Z. W., and Hough, G. A., "New Bedford, Mass.: Its History, Industry, Institutions, and Attractions," p. 36. Published by the New Bedford Board of Trade in 1889.

[2] Figures taken from the *Whalemen's Shipping List,* the authoritative weekly organ of the whaling industry. Figures for a given year are found in a January issue of the ensuing year.

[3] Williams, J. R., in *North American Review,* XXXVIII, pp. 84–116.

increased to 644 vessels, and its value had risen to $19,430,000, — a rate of growth which averaged $675,000 per year. Together with the catchings at sea, the value of the whole was placed at $27,784,000. These 644 whalers, of 200,484 tons' burden, were manned by 17,500 officers and men, who consumed annually $3,845,500 worth of commodities. The annual yields of oil and whalebone were sold, in the crude state, for about $7,000,000; and when manufactured, for $8,000,000 to $9,000,000.[4] In order to supply whaling products in such quantities it was necessary to kill some 10,000 whales each year. About eight to ten per cent of this total number were lost, even after having been captured and killed, as the result of rough weather, sinkings, and other exigencies of the chase. The remaining carcasses, having been stripped of their blubber and bone, were cast adrift to form huge and bloody feasts for sharks and for birds of prey.[5]

But nineteenth-century whaling was not only significant in the economic life of New England and of the United States: it occupied a predominant position in the field of world whaling as well. By 1850 the supremacy of American whaling had been established beyond question. In 1847 it was estimated that the whaling fleet of the entire world consisted of about 900 vessels; and of these no less than 722 belonged to the United States.[6] Since the remainder sailed under a variety of flags, no single nation could be regarded as more than a nominal rival. Thus the American whaleman attained an undisputed mastery of the world's whaling industry at the time of its greatest development, and held it until the close of the century.

This world-wide reach of New England whaling inevitably gave it certain international and cosmopolitan aspects. The

[4] Grinnell, Joseph, "Speech on the Tariff, with Statistical Tables of the Whale Fishery," a 16-page pamphlet published in 1844. See Appendix B for further details regarding this material.

[5] Estimate made by Wilkes, Charles, in his "Narrative of the U. S. Exploring Expedition," V, Ch. XII.

[6] The size of the world's whaling fleet was given in a speech made by William H. Seward in the U. S. Senate on July 29, 1852, and reported in the *Congressional Globe* for the 32nd Congress, 1st Session, XXIV, Part III, pp. 1973–1976. The second figure is taken from the annual statistics for 1847, published by the *Whalemen's Shipping List* in January, 1848.

long voyages which carried vessels and crews into remote harbors and over uncharted waters turned the whaleman into a globe-trotter who often flaunted his sophistication in the faces of provincial neighbors in his home port. The boarding-houses and brothels of New Bedford and of Nantucket were clamorous with men who had followed a compass from Boston Light to Cape Horn, Honolulu, and the Fiji Islands, or from Zanzibar and Java Head to Shanghai, Spitzbergen, and the Sea of Okhotsk.

Whaling contributed more than its share, too, to the work of maritime discovery. The constant effort to open up fresh whaling grounds in unfrequented waters led to much matter-of-fact exploration, unheralded but effective. American whalers worked their curious, pioneering path into the North and South Pacific, in particular, always alert for new fields for operations; and certain islands in the South Seas still bear the hardly tropical names of the Nantucket or New Bedford captains who discovered them. A report transmitted to the Secretary of the Navy on September 24, 1828, gave a long and impressive list of the islands, reefs, and shoals which had been discovered by American whalemen or were scarcely known outside of their ranks. The data used in drawing up this report, which covered the Pacific, Indian, and "Chinese Oceans," resulted wholly from examinations of whaling log-books, journals, and charts, and from first-hand interviews with the masters and mates who had been responsible for the original entries.[7]

The world-traversing activity of whaling found its counterpart in the character of the crews. Here were to be found representatives of practically every race, nationality, class, type, and temperament. Never in this country have there been more thoroughly cosmopolitan, polyglot mixtures than crowded into the teeming forecastles of mid-century whaling vessels. Specimens of every nationality of Europe and of dozens of South Sea Islands; ne'er-do-well sons of wealthy American families, so popular in period novels; Cape Verde

[7] This "Report on Islands Discovered by Whalers in the Pacific" was made by J. N. Reynolds. It is contained in House Executive Documents III, No. 105, 2nd Session, 23rd Congress.

and Azores Islanders; scions of great whaling houses; farm boys and sophisticated city youths; escaped criminals and men with endless varieties of "pasts"; conscientious and capable youngsters who intended to "get ahead" in the industry; coal-black negroes and half-breeds with the blood of aristocrats and of men of wealth in their veins; soft-voiced, clean-limbed natives of the Sandwich Islands; bold and brutal mutineers with the hearts of pirates, — all these and many others responded to the shouts and imprecations of the mates during a spell of "dirty weather," or patiently carved out masterpieces of scrimshaw work during long hours of monotonous ill luck on the whaling grounds.

But if the United States imported whalemen, she repaid the debt by exporting a goodly share of the products which they brought home. As early as 1844 the annual whaling exports were valued at $2,000,000, or about one-fourth of the entire catch.[8] In 1850 New Bedford and Boston alone sent abroad 72,338 gallons of sperm oil, 136,465 gallons of whale oil, and 32,475 pounds of whalebone.[9] The great bulk of the exports found a market in Europe. Holland and the German States were amongst the best customers, though England and France also made large purchases after the decay of the British whaling fleet.

In strange contrast to the international distribution of products and to the colorful, cosmopolitan crews was the extreme localization of the industry in a few provincial ports. For American whaling was synonymous with New England whaling. And not even the whole of maritime New England was included; for the center of the trade was at Nantucket and in Buzzards Bay, with a number of outposts in Narragansett Bay and on Long Island Sound to the west and a solitary sentry at Provincetown to the east and north. In 1843, for instance, Nantucket and the four largest Buzzards Bay ports of New Bedford, Fairhaven, Westport, and Mattapoisett sent out 366, or 57%, of the 642 American whaling ves-

[8] Grinnell, Joseph, "Speech on the Tariff, With Statistical Tables of the Whale Fishery," — a sixteen-page reprint, (1844) of a speech made in Congress. See Appendix B.

[9] *Whalemen's Shipping List,* annual statistics for 1850, published in January, 1851.

sels.[10] And thereafter, as the prestige and prosperity of Nantucket declined, the degree of concentration in Buzzards Bay showed a steady increase.

But the localization of whaling was not so much a matter of regions as of ports. The merchants of only five seaport towns — Nantucket, New Bedford, Fairhaven, New London, and Sag Harbor — dominated the industry throughout its period of importance. In 1843 New Bedford fitted out one-third of all American whalers; the remaining members of the quintet contributed another one-third; and all other ports were forced to combine in order to muster the last one-third.[11] By 1857 New Bedford's fleet had risen from one-third to one-half of the total; and as the smaller ports dropped out with the waning fortunes of the business after the Civil War, this one city rapidly absorbed a growing percentage of the diminishing total tonnage.

It is notable, too, that the leading whaling ports, with the single exception of New Bedford, were relatively small towns, even when judged according to contemporary standards of population. Boston, New York, Baltimore, and the other great Atlantic cities were content to retain the merchant shipping and the carrying trades for which they were best fitted. Now and again, when they made half-hearted attempts to engage in whaling, their ventures were uniformly unsuccessful.

Rooted in both the land and the people were the forces which thus threw the world's whaling trade into the hands of a few small ports situated along a short stretch of New England seacoast. The unfruitful and back-breaking soil which early drove all of tide-water New England to seek a livelihood from the sea existed here as elsewhere; and, together with the broken and indented coastline which afforded numerous safe harbors, it sufficed to explain why men went down to the sea. But why did the choice fall upon whaling rather than upon fishing or the merchant trade? It was true that the villages to the north of Cape Cod were nearer to the great

[10] These figures were derived from a list of whaling ports and their vessels which was published in the *Whalemen's Shipping List,* Vol. I, p. 32, on April 4, 1843.

[11] *Ibid.*

fishing banks, and that the towns of Massachusetts Bay possessed a larger and more accessible hinterland which gave them marked advantages in the mercantile pursuits. But neither Nantucket nor New Bedford was able to boast of any marked natural features which were peculiarly adapted to the pursuit of whales, unless the absence of any easier means of livelihood might have been so construed.

The niggardliness of nature, however, was offset by the character of the people. Whaling demanded an unusual combination of qualities on the part of those followers who would succeed in it; and the Quaker-Puritan-Yankee stock of southern New England, through training, temperament, and cultural environment, possessed these traits in rare degree. The catalogue was a long and mixed one. Courage, hardihood, skill, thrift carried to the point of parsimony, shrewdness, stubborn perseverance, ingenuity, sturdy independence, a cold lack of squeamishness in driving bargains, and a righteous scorn for luxuries, — these were characteristics, partly admirable, partly unpleasant, which rendered yeoman service to their possessors in the whaling industry. And these were amongst the qualities, inculcated by the philosophy of Quaker and Puritan forbears and modified by several generations of struggle with a niggardly and often hostile New World environment, which characterized the Nantucket and New Bedford whalemen who carried whaling to its highest point.

Beyond doubt it was true, of course, that the rise and fall of whaling was due to the operation of fundamental economic forces which created or withheld a demand for its products. When this demand was heavy and increasing, the industry waxed and grew prosperous: when the demand fell off permanently, the industry waned and died. But the fact that leadership in supplying the required products passed into the hands of Nantucket and New Bedford, rather than of New York or Charleston or New Orleans or even Boston, must be attributed in no small degree to the character and background of the inhabitants.

The further specialization which took place within the general field of whaling was probably due to minor causes rooted in the realm of historical accident. Whatever the original reasons, however, certain ports clung tenaciously to various specialized branches of the industry. Throughout her history Nantucket maintained a stubborn preference for sperm oil, and unwisely assisted in shortening her whaling career by so doing. New London and Stonington found their greatest successes in the Antarctic, where they hunted the sea-elephant as well as the right whale. Sag Harbor remained true to northern right whaling, in spite of the higher price of sperm oil. Provincetown, with her small, light vessels, known in the jargon of the trade as "plum-pud'ners," never ventured beyond the confines of the Atlantic. New Bedford, on the other hand, laid the foundations for her prosperity through a canny discernment of the shifting possibilities of the industry, and a willingness to send her vessels wherever the greatest profits were to be had.

With the exception of the small vessels and short cruises of the Provincetown fleet, however, the important activities in the various branches of the industry were essentially similar. All were characterized by long voyages of ten to fifty months, made in ships or barks ranging in size from 200 to 500 tons. Generally the sperm whalers, like their prey, roamed all the seas of the world, while right whalers were more likely to restrict their cruising radius to the North Atlantic or the North Pacific. But there were many exceptions in which the latter sailed as many nautical miles as the former.

There was, too, a superb seamanship and a consummate skill in the handling of whaling vessels which caused them to suffer far less than merchantmen from stress of weather. Navigating expertness was to be expected, of course, on the part of men who spent year after year in their floating prisons with only brief and infrequent furloughs on land. But the relative freedom from serious loss is to be explained also by the construction of the vessels and the manner of sailing them. Speed was of little importance. In bad weather sail was shortened in ample time, and on the whaling grounds was taken in every night. The whaling functions proper, separate and distinct from the ordinary duties of seamen, required that all whalers be heavily manned; and this surplus man-power proved a great asset in all times of emergency. Whaling vessels had to be staunch craft of heavy construction and broad on the beam in order to withstand the terrific strain of the huge carcasses of many tons lashed alongside while the blubber was being cut-in and hoisted on board. And maximum capacity, an essential consideration for ships which had to bring home their own catches and carry supplies for a period of several years, further demanded both strength and breadth.

The builders of whaling craft consistently emphasized strength and seaworthiness at the expense of speed and grace of line; and consequently a whaler was readily distinguishable from her more attractive half-sisters in the merchant marine. "It is impossible," wrote one contemporary observer, "to meet a whale-ship on the ocean without being struck by her mere appearance. The vessel under short sail, with lookouts at the masthead, eagerly scanning the wide expanse around them, has a totally different air from those engaged in a regular voyage."

But if speed and grace of line were missing in the larger vessels they were possessed in superlative degree by the whaleboats in which the game was pursued at close quarters. These perfected rowboats were shallow, double-ended craft, twenty-eight to thirty feet long, propelled by paddles, spritsail, or five long and heavy oars, with a still longer steering-oar at the stern. In spite of their loads of hundreds of fathoms of whale-line, of whaling craft and gear, and of four oarsmen, harpooner, and boatheader, they were admirably fast and seaworthy. Nor did they lose anything in the manner of their handling. For the American whaling crews yielded to no class of men known to history as skillful and daring boatmen. Sheer necessity, in fact, compelled them to handle their small craft with amazing skill. Whether towed on a "Nantucket sleigh-ride" by a harpooned victim that ran along the surface at such speed that the boat seemed to leap from one wave crest to the next; or pulled up to a "gallied" whale to attempt the final lance-thrusts; or maneuvred about an "ugly" animal where a single sluggish response meant destruction by jaws or flukes; or sent into the teeth of a gale of wind and a heavy sea,—under such threats a whaleboat in action placed an imperative premium upon clever and skillful handling that rapidly eliminated the unfit and the timid.

The dangers and vicissitudes of whaling were appallingly numerous. In addition to the risks of storm, shoal, and reef, common to all seafarers, there were many hazards peculiar to whalers alone. Long absences in unfrequented waters, far from the regular lanes of commerce; enforced contacts with strange peoples in new lands; liability to fire from the flames under the try-pots, as well as from the boiling contents; ice and cold in the Arctic or Antarctic; and the countless dangers of the chase itself, — each of these categories of mischance took its toll of life and property.

Accidents to the men in the boats were both frequent and severe. A "stove boat," smashed into bits by a blow of the flukes or the jaws of a whale too closely pursued, was a commonplace. Sometimes the six occupants were rescued, unhurt; sometimes only two or three survived, badly injured; and sometimes all were lost. Now and again a flying loop of a foul line snatched a man out of his boat and dragged him under water with the ease and speed of a malevolent magician. A boat's crew, losing sight of the mother ship during the twilight hours of a long and absorbing chase, might be discovered the next morning, after a night of terror; or the frail craft might go down with all hands; or, in a few extreme cases, two or three survivors might be picked up weeks later, having drifted hundreds of miles on the open sea and having been driven to subsist on the flesh and blood of their dead comrades.

But even if a whaling voyage proceeded without untoward incidents, the lot of the crew was not enviable. Both nature and man combined to create a situation which, at its best, was hard, and at its worst represented perhaps the lowest condition to which free American labor has ever fallen. The very nature of the industry made it impossible to escape from cramped and inadequate living quarters, a small variety of staple foods, drab discomforts, maddening monotony, long hours of gruelling and dangerous work, repeated exposure to hazards in many forms, and a rigid discipline.

Some generous owners and just and kindly masters did

what was possible, without undue effort or expense, to ameliorate these conditions. Others, however, (and they constituted a good percentage of the whole) aggravated the inevitable evils many-fold and then added others of their own contriving. Discipline was too often mistaken for tyrannical license by irascible officers who enforced bullying orders with heavy sea-boots and a steady flow of profanity and vituperation. In the effort to save money both food and quarters were made far worse than was necessary. Clothing and supplies were furnished through the slop-chest at more than generous prices. Deductions were made from the lay on every possible pretext, reasonable and unreasonable; and as a result it happened often that a man would return from a three or four years' voyage to find himself actually in debt to the firm for which he had been working!

Worst of all was the practice, fortunately not universal, of driving the crews to the limit of human endurance by systematic beating, hazing, flogging, "working up," and general maltreatment, not excluding various ingenious forms of cruel and abusive indignity. It should be stated, in partial extenuation, that the character of the men in the forecastle often suggested, if it did not necessitate, stern and rigorous treatment. The deck of a whaler at sea was no place for soft words or an appearance of timidity. Rigid and unquestioning discipline was imperative in order to insure both success and safety; and the human scourings which comprised a large percentage of the mid-century whaling crews were not amenable to subtle suggestion. The criminal and unscrupulous elements, in particular, could be impressed only by strong words backed by the threat of stronger fists or belaying-pins.

But in spite of these facts, extenuating under certain circumstances, it remained true that many officers bullied and abused their crews needlessly and shamelessly. Such treatment was rendered both safe and easy by the fact that the legal status of whalemen at sea, like that of seamen in general, was strongly reminiscent of mediæval serfdom. In part this unfree status resulted from the inevitable demands of discipline. But it was due also to political weakness on the part of seamen as a class, which made it difficult to secure protective legislation and to enforce those laws which were on the statute books.

And when, after enduring many months of such conditions, the whaleman finally returned to his home port, his lot only passed from bad to worse. For the average foremast hand on shore was helpless in the toils of an organized system of exploitation. Outfitters and infitters, shipping-agents, and boarding-house keepers all took heavy toll from his slender purse. And their satellites and confederates, the runners, grog-dispensers, and keepers and inmates of brothels, took equally heavy toll from his character — and completed the pilfering of his purse. In the end, after a few days or weeks of hectic and degrading dissipation, he was usually cajoled or bullied into signing on for another cruise of three to four years. And very soon thereafter, with an empty pocketbook and an outfit for which he was being charged an exorbitant price, plus interest at twenty-five per cent per voyage, he was dumped into the forecastle of an outward-bound whaler, commonly in a comatose state of drunkenness which only many hours and much profane abuse on the part of the mates could dispel.

What were the compensations for such a life of hardship, danger, and degradation? The financial remuneration took the form of a "lay," or a fractional share in the net proceeds of an entire voyage. A green hand commonly started with a lay of 1/200; an able seaman secured about 1/150; a boat-steerer often obtained 1/75; whereas a few favored captains sometimes bargained for as much as 1/10 or 1/12. Since the amount due under this system could not be calculated until the conclusion of a cruise, the ordinary whaleman was dependent upon the owners and master for the advance of any goods or cash which he might require during the long absence of one to four years. Consequently it was necessary to keep an individual account for each member of a crew; and the final settlement involved only the payment of the net difference between the totals of a rather elaborate set of debit and credit items.

Such a system provided excellent opportunities for extortion and financial chicanery — and these opportunities were by no means ignored by outfitters and by many merchants and masters. If a voyage had been unusually successful, the foremast hands often received appreciable amounts in cash; but after ordinary or poor cruises the gross lays, none too large at best, were greatly depleted by a series of heavy debit charges. In general, the average earnings of foremast hands as a whole were decidedly lower than the average wages paid in the merchant marine and in unskilled occupations on shore.

Other compensations were few and inadequate. Now and again a veteran whaleman, with a sentiment of his own, was wont to regard the actual pursuit of the game as a royal form of sport, providing thrills and a savage joy of the chase beside which other modes of hunting seemed colorless and insipid. Such a feeling, however, could hardly apply even to the bravest foremast hand, who rowed with his back to the game and blindly obeyed the orders of the mate without being able to see the course of the action. Scrimshawing, even though it be regarded as a fine art, furnished but a makeshift relief from the tedium of long months at sea in a vessel so small and bare as to make it seem more like a prison than a home.

Nor was there more than negative consolation in the long hours of inactivity that often accompanied periods of poor luck on the whaling grounds. Aside from the fact that such inactivity only prolonged the length of the entire voyage, there was woefully little with which to occupy the spare time which did accrue. Sleeping, mending, reading, scrimshawing, "skylarking," and spinning years, interminably repeated, practically exhausted the possibilities of amusement or recreation.

Only when a "gam" took place was there a welcome break in the monotonous round of life between sea and sky. For a chance meeting with another whaler was usually made the occasion for an exchange of visits which consumed several hours, or perhaps an entire day. If the other vessel was homeward bound, mail was despatched by her; or, better still, if outward bound, she might have letters to deliver. Aside from shore leave, a "gam" was perhaps the whaleman's greatest joy.

For the rest, however, the life of the American whaleman

was as hard and cheerless as that of any group of free workers in the history of the United States. And small wonder, when man did so much to aggravate, and very little to alleviate, the forbidding conditions already determined for the industry by nature!

PART I

LAUNCHING

CHAPTER II

THE EUROPEAN BACKGROUND

THE beginnings of the whale fishery are still shrouded in mist. Only at rare intervals does the modern chronicler catch glimpses of vague and shadowy forms moving across the foggy expanse of mediæval waters.[1] Possibly the Japanese and the Tartars were the first to engage in the practice of whale-hunting, making use of large boats which seldom ventured far from shore. The Northmen, however, were probably the first Europeans to molest the whale through deliberate pursuit, even though their hunting was but occasional, unsystematic, and non-commercial. So small was the scale of their operations, and so completely have the traces thereof been obliterated, that

[1] Authentic information regarding early European whaling is meager in the extreme. The main sources of material are to be found in the following works: Hakluyt, "Principall Navigations, Voiages, Traffiques, and Discoveries of the English Nation"; Anderson, A., "Historical and Chronological Deduction of the Origin of Commerce"; Moriniere, Noel de la, "Histoire Generale des Peches"; McCulloch, J. R., "A Dictionary, Practical, Theoretical, and Historical, of Commerce and Commercial Navigation." Amongst the works devoted entirely to whaling there are three, in particular, which contain important sections concerned with the early history of the industry. These are Scoresby, W., "An Account of the Arctic Regions, With a History and Description of the Northern Whale-Fishery"; Scammon, C. M., "Marine Mammals of the Northwestern Coast of North America"; and (the only recent one) Jenkins, J. T., "History of the Whale Fisheries." Virtually all of the material embodied in this chapter was taken from one or another of these publications; and further specific references will not be made to them.

little is now known of their methods or of the weapons and vessels which they employed.

At least one indirect and incidental reference, however, seems to justify the belief that some method of capturing the right whale was in vogue among the Northmen at least as early as 850 A. D. About 885 A. D. to 890 A. D. one Octher, a Northman, gave the English King Alfred an account of his discoveries along the coast of Norway. These explorations had been made while engaged in the search for whales; and, added the narrator proudly, he had "sailed along the Norway coast, so far north as commonly the whale-hunters used to travel." Such a statement can leave little doubt that whales had been hunted in Norwegian waters for several preceding decades, at least; but unfortunately King Alfred did not urge his traveler-guest to dwell further upon the subject of whales and their captors.

Whaling on a systematic, commercialized basis was first followed by the Basques along the shores of the Bay of Biscay, where it is still possible to trace the remains of their watch-towers and furnaces. Beginning well before 1000 A. D., these mediæval whalemen brought their industry to its apogee as early as the twelfth and thirteenth centuries — although they continued to capture whales, in gradually diminishing numbers, until the close of the seventeenth century.

During this long period the Basques were eminently successful both in the capture of their game and in the sale of their products. At first they remained cautiously near the shore. Then, with more experience, came voyages which grew courageously longer, until at length they were accustomed to cover great distances on the open sea. Long before Columbus discovered America they were making frequent trips to Iceland; and the middle of the sixteenth century saw them often in the vicinity of Newfoundland. Their captures became so tempting, in fact, that they soon fell under the sway of the ubiquitous feudal tax-gatherers. In 1261, for instance, they were paying a tithe on all whales' tongues sold in Bayonne; and in 1388 Edward III was collecting a duty of six pounds sterling for every whale brought into Biarritz.

The mediæval Basques seem to have built up, too, an important trade in whaling products. Whale meat, especially the tongue, was sold regularly in the market-places of Biarritz, Bayonne, Cibourre, and other less important towns on the coast, and the remaining commodities enjoyed a larger market. The oil was distributed over a wide area, considering the inadequacies of mediæval transportation and the complexities of the countless local tariffs; whalebone was used for so many different purposes that the demand insured a generous price; and the blubber was salted and sold inland, particularly in the east of France.

Coincident with the final decline of Basque whaling after 1600 came the rise of the fishery on a larger scale at Spitzbergen. This inhospitable region had been discovered by the Dutchman Barents in 1596. Not until fourteen years later did an English exploring expedition under Thomas Edge sail into the same teritory; but in spite of this tardiness the English Muscovy Company, armed with a royal charter from James I, claimed the exclusive right of whaling in all waters adjacent to Spitzbergen. The vigorous opposition to this claim made by the Dutch led to a decade of bitter rivalry which was characterized by constant friction and some fighting. By 1623, however, the Dutch had gained a clear predominance. Soon the English Muscovy Company ceased its efforts to compete; and by the end of the seventeenth century practically all English whalers had been driven out of these northern waters.

Other vessels, however, flying other flags, had taken their places. By 1721 the Greenland and Davis Straits fleets included 55 crews from Hamburg, 24 from Bremen, 20 from the Biscayan ports, and 5 from Bergen. But Holland was still so far in the lead that even Hamburg was completely overshadowed. During the decade 1699–1708 the Dutch equipped no less than 1652 vessels and captured between 8,000 and 9,000 whales, representing a gross income of some 26,000,000 florins.

Partly because of national jealousy, and partly because of the prevailing Mercantilist philosophy, the English merchants showed great concern over their lack of success in whaling. Finally the South Sea Company roused itself for an attempt to revive the fortunes of the fishery. Parliament was per-

THE CELEBRATION OF MASS ON THE BACK OF A WHALE

From an engraving in "Nova Typis Transacta Navigato," Manacho, 1621

THE MANNER OF CATCHING WHALES

From an engraving in Monck's "Account of a most dangerous Voyage to Greenland," in "Churchill's Voyages," Vol. I, London, 1744

suaded to provide that whale fins, oil, and blubber, if they had been taken in English vessels with an English master and a crew not more than two-thirds foreign, were to be admitted free of duty for seven years after Christmas, 1724. Fortified with this assistance, the Company began operations in 1725. But the long years of national inactivity proved to be a costly handicap. Expert high-priced harpooners had to be secured from Holstein, just as the English Muscovy Company had had to retain experienced Basque whalemen more than a century before. Inexperience in all phases of the industry caused inefficiency and inordinately heavy expenses. As a result, the enterprise was far from successful. Indeed, during the first eight seasons the Company suffered a total loss of £177,782/3/0.

After this lamentable record, it was deemed essential to resort to bounties. In 1733 Parliament authorized a bounty of twenty shillings per ton on all whaleships of more than 200 tons which had been fitted out in Great Britain. Even this encouragement, however, proved insufficient. Consequently in 1740 the bounty was raised to thirty shillings per ton; and in 1749 it was further increased to forty shillings, with the added proviso that under certain prescribed conditions this same sum was to be allowed on vessels built in the North American colonies.

As a result of such heroic subsidizing, the British whaling fleet commenced to grow in earnest. In 1759 the English Greenland fishery included a list of thirty-four vessels, sailing largely from the three ports of Whitby, Hull, and London. Even this number, however, was still hopelessly inferior to the large and efficiently-handled fleet which the Dutch had been maintaining in northern waters for more than a century. Hamburg, too, had doubled this record a full generation before; and Bremen and the Basque ports had almost equalled it.

But the English whalemen were finally emerging from their period of apprenticeship, and were soon to outdo their erstwhile masters. Their star was clearly in the ascendant, whereas that of their European rivals was declining. For three-quarters of a century Britain was to be the leading whaling nation of Europe; and only her own rebellious sons across the Atlantic were to offer a serious bid for supremacy. But that challenge, slowly gathering weight after the Revolution, was to be so successful that both Britain and all of her competitors were to be completely eclipsed. In the nineteenth century the center of gravity of the whaling industry was to shift heavily from Europe to hardy, youthful America.

CHAPTER III

WHALING IN THE COLONIES

THE first American whalemen were not white, but red. For an undetermined length of time before the white man came, the Indian tribes living along the southern coast of New England had pursued the whale in canoes from the shore. Their implements were crude and their captures relatively few, as shown by the fact that whale meat and blubber were considered great delicacies in all the Indian villages; but nevertheless they developed a definite technique and some degree of skill in the pursuit of their game.

When a whale was sighted, as many canoes as could be manned put out from shore in an attempt to surround the animal. If successful in this, numerous rude wooden harpoons, attached by short lines to wooden floats, or droges, were implanted in the blubber. The cumbersome droges were intended to impede the flight of the victim to such an extent that the pursuers, by dint of heroic paddling, might keep within striking distance. The death of the game, if finally achieved, then became the result of a lengthy harassing process or running fight which continued for many hours. Fastening directly to the whale by means of a long line attached to the sunken harpoon, as the white whalemen did later, was out of the question in such frail craft as the birch bark canoes of the Indians. Consequently they were forced to keep within range as best they might, and to be content with the delivery of such occasional lance thrusts as opportunity might allow. Both the strategy of the chase and the inadequacy of the weapons employed were more likely to weaken the victim through the loss of blood from many minor wounds than to result in any mortal blows. If the prey remained near the shore, and if the pursuers could paddle long enough and hard enough, they might perchance be permitted to witness the death throes of their harassed and stricken prize. But in the great majority of cases the whales must have escaped by sounding or by swimming out to sea, in spite of the harpoons which may have been imbedded in their flesh and the inadequate droges which dangled behind them.[1]

Despite their lack of real success, however, the Indian whalemen possessed both courage and skill; and the early colonists were not slow to recognize this fact. When the whaling knowledge and dexterity of the Indians were combined with the heavier boats and implements of the colonists, the percentage of captures rose materially. As a result it was common for the two parties to enter into partnerships which involved the joint use of native labor and of white capital. As early as 1650, for instance, the settlers at Southampton, Long Island, were employing members of the neighboring tribes to man their boats, and were allowing them a given percentage of the captured oil in lieu of wages. At times the terms of these whaling agreements were inscribed in the town books, as at Easthampton on April 2, 1688. On that day Jacobus Skallenger and others hired certain Indians to engage in whaling from November 1 to April 1 at three shillings per day apiece, with the whites furnishing all necessary equipment.[2]

There are references to whales and whaling from the very beginning of the colonies in New England and New York. Captain John Smith reported that he had sighted large numbers of whales while sailing along the New England coast in 1614; and Richard Mather wrote in 1635, that he had seen "many mighty whales spewing up water in the air, like the smoke of a chimney, and making the sea about them white and hoary." [3]

[1] Starbuck, A., "History of the American Whale Fishery," pp. 6–12, contains a number of references to early Indian whaling. The most detailed and most readable account of Indian whaling, however, is to be found in Spears, J. R., "Story of the New England Whalers."

[2] Starbuck, A., "History of the American Whale Fishery," pp. 6–12.

[3] See Scammon, C. M., "Marine Mammals of the Northwestern Coast," p. 202.

Unexpected prizes in the form of drift-whales constituted the settlers' first introduction to the whaling industry; and the disposition of these gifts of providence gave rise to many sets of canny regulations. In the early years of the Plymouth and Massachusetts Bay colonies it was provided that the sum realized from every drift-whale was to be divided into three parts — one-third to go to the colonial government, another portion to the town having jurisdiction, and the final third to the finder. In 1662 the town of Eastham, on Cape Cod, voted appreciatively that a certain portion of every whale cast ashore should be devoted to the support of the minister. And Southampton, on Long Island, developed an elaborate code for the proper handling of its whaling problems. In 1644 it was ordered that the town be divided into four wards of eleven persons each, and that two individuals from each ward, chosen in rotation, be employed to cut up every drift-whale. Each cutter was then to be allowed a double share in the proceeds realized from his particular whale; while every other male inhabitant was to receive only a single share. One year later it was decreed that anyone who reported the location of a stranded whale to a magistrate was to receive a reward of five shillings. But this munificent sum was forfeited if the carcass were discovered on Sunday! [4]

From such accidental and occasional contacts with whales the colonists soon passed to an organized prosecution of the fishery. During the third quarter of the seventeenth century the deliberate pursuit of whales was begun by communities on the eastern end of Long Island, on Martha's Vineyard, on Cape Cod, in the Bermudas, and on Nantucket. Probably Southampton was the first. Sometime between 1645 and 1655 the settlers of that village began to fit out boat expeditions which cruised speculatively along the coast during absences of one to two weeks, landing each night in order to sleep on shore. Martha's Vineyard adopted similar tactics in 1652; and about 1665 whale oil was being sought in the bays of the Bermudas. Cape Cod, too, must have entered the ranks before 1670; for during the years 1672–1690 the Nan-

[4] Starbuck, A., "History of the American Whale Fishery," pp. 6 ff.

tucketers were willing to admit that the people of Cape Cod had attained a "greater proficiency in the art of whale catching than themselves." [5]

Nantucket herself, the future queen of the industry, lagged somewhat in her beginnings. The first whale killed by the islanders was secured only because it injudiciously remained in the harbor for three days, thereby giving the inhabitants time to devise a crude harpoon and to press the attack. Soon thereafter, in 1672, a contract was made with a certain James Lopar, a Cape Cod whaleman, who agreed to move to Nantucket and to engage in whaling for one-third of the proceeds, while the remaining two-thirds went to the town. But eighteen years later the business had shown so little progress that another Cape Cod man, Ichabod Paddock, was engaged to give instruction in the latest whaling methods.

With the turn of the century, however, Nantucket rapidly took the lead in the system of boat-whaling from the shore which characterized the earliest American phase of the fishery. This method was essentially the same in all localities. Large spars, serving as lookout stations, were erected at prominent points along the coast; and the boat crews were quartered in small huts nearby, within easy earshot of the lookout man. When whales were sighted the alarm was given and the boats put out in pursuit. Very often the harpooner and one or two other members of each crew were Indians; and the Indian method of attack, employing long-continued harassing tactics, was followed with but few modifications. If the chase ended in success the great carcass was towed laboriously to the shore and hauled up on land near a rude try-works, where the blubber was stripped off and the oil tried out.

In spite of the short range and comparative crudity of this scheme of operations, it was effective. In 1726, for instance, no less than eighty-six captures were made off the coast of Nantucket alone. The danger, too, was far less than that encountered on the subsequent longer voyages; for Obed

[5] The best accounts of the early colonial whaling are to be found in Macy, Obed, "History of Nantucket"; Starbuck, A., "History of the American Whale Fishery"; Scammon, C. M., "Marine Mammals of the Northwestern Coast of North America"; and Jenkins, J. T., "History of the Whale Fisheries." The first two are particularly valuable.

Macy was able to make the surprising statement that from 1690 to 1760 not a single white person was killed in the boat-whaling carried on from the island. Nantucket whaling began to show possibilities of fulfilling the prophecy made on Folly House Hill in 1690. According to a well-worn tradition (which need not be taken too seriously) some man then remarked feelingly to a number of persons who were observing the spouting of whales at sea: "There is a green pasture where our children's grandchildren will go for bread."

In fact, the children usurped the occupation assigned to the great-grandchildren. In 1712, when Christopher Hussey was blown off shore in a gale and captured the first sperm whale far out at sea, the superior quality of the new oil was recognized at once. Sperm oil was lighter and purer than the whale oil of the right whales which had theretofore constituted the only victims of the harpoon. But the sperm whale was a haughty, elusive aristocrat of the high seas; and he would not deign to soil his flukes in the shallow coastal waters which were frequented by his more phlegmatic cousin, the right whale. So it was necessary to develop larger and more seaworthy craft which could venture into deeper water and pursue the cachalot in his own domain. And Nantucket took to sperm whaling with a native enthusiasm which continued throughout her history. By 1715 she had six small sloops of thirty to forty tons each which were making voyages of several weeks' duration, sometimes sailing as far as Newfoundland in search of game.

During the ensuing half-century the entire American fishery, with Nantucket in the lead, underwent a steady development in technique, in range of operations, and in size. The technique of whale-hunting was improved by fastening the boat to the whale by means of a long line attached to the harpoon; by working out the lines and dimensions of the graceful and versatile whaleboats; and by developing a series of ingenious contrivances whereby the blubber could be stripped off and the oil boiled out and stowed away at sea. Harpoons, lances, whale-line, cutting-spades, and other articles of whaling craft and gear evolved in a similar manner. Vessels, too, became not only larger and heavier, but underwent gradual changes in rig and hull which transformed them from tiny sloops to more pretentious schooners and brigs.

These improvements vastly increased the radius of activities. Freed from the necessity of returning to shore after each capture, the whalemen ventured on longer and longer voyages. By the time the Revolution broke out they had explored most of the whaling grounds of the North Atlantic and were preparing to invade the regions of the South Atlantic. Tradition has it, in fact, that the first American whaler to cross the equator — the brig *Amazon*, of Nantucket, Uriah Bunker, Master — reached home on the very day which witnessed the battles of Lexington and Concord. Before this, however, the ever-expanding activities of the New England whalers had taken them to Davis Straits between 1732 and 1746; to Baffin Bay by 1751; to the Gulf of St. Lawrence and Hudson Bay by 1761; to the coast of Guinea in 1763; to the Western and Cape Verde Islands, the West Indies, and the Caribbean Sea by 1765; and to the coast of Brazil in 1774.

In this widening search for new whaling grounds, as in other phases of the industry, the Nantucket fleet took the lead. And her initiative was rewarded with larger and larger catches. In 1730 her 25 small vessels, of 38 to 50 tons each, secured 3700 barrels of oil which sold for 3200 pounds sterling. In 1756 her fleet of 80 vessels, averaging about 75 tons per vessel, brought home some 12,000 barrels of oil which exchanged for 27,600 pounds sterling. And by 1775 she had 150 whalers, ranging from 90 to 180 tons each, which captured about 30,000 barrels of oil and enabled her to lay claim to 167,000 pounds sterling.[6] Such figures meant that whaling, to the islanders, was neither an adventure nor an experiment: it had become an industry.

But other regions also felt the stimulus and the lure of expanding activities. Southern New England and eastern Long Island, in particular, began to find in whaling a serious

[6] These figures are taken from an article written in 1785, but published in 1794, entitled "The Progress of the Whale Fishery at Nantucket." Massachusetts Historical Society Collections, Series 1, Volume III, p. 161.

means of livelihood. Boston and Salem engaged in boat-whaling on a small scale, though for the most part Boston remained content to act as a trading factor for the other ports. Cape Cod was represented in the fraternity of whaling villages by Provincetown, Eastham, Chatham, Barnstable, and Yarmouth. Edgartown, as the center of the fishery on Martha's Vineyard, became a town of some importance. The shores of Rhode Island and of Connecticut formed the land base for a growing fleet of small craft; and the inhabitants of the eastern end of Long Island, led by Sag Harbor, Southampton, and Easthampton, developed an unmistakable aptitude for the pursuit of this royal game. Even Williamsburgh, Virginia, far out of the orthodox whaling zone, fitted out a small sloop in 1751 and secured a good catch. And finally, about 1755, the town of Dartmouth, which under its later name of New Bedford was to become the greatest whaling port in the world, entered the industry in a manner which was humble enough. Joseph Russell, the founder of the city, was also the originator of its long whaling activity; but after ten years he and two associates owned only four diminutive sloops of about fifty tons each. New Bedford's preëminence in whaling, like many another form of success, was a flower of slow growth.[7]

In spite of its prevailing growth and prosperity, however, pre-Revolutionary whaling had its misfortunes and provincial vexations. During the early years of the century the Long Islanders were greatly troubled by the jealous exactions of the governors of New York, who not only demanded a tax of five per cent on the proceeds of every whale captured, but also imposed a series of captious regulations in an attempt to divert the trade of the whaling ports from Massachusetts and Connecticut to New York. The opposition to the governors was led by Samuel Mulford, a picturesque and sturdy figure who finally went directly to London (with the inner linings of his pockets protected against the famed metropolitan pickpockets by means of fishhooks), in order to seek redress. After a

[7] See the early chapters of Ricketson, D., "History of New Bedford," for a detailed account of the beginnings of the town and of its greatest industry.

long and persistent struggle the tax was at length repealed in 1720.

Nor did the colonial whalemen receive any better treatment at the hands of the English governors of the newly-conquered French provinces at the close of the French and Indian Wars; for here, too, they were hampered and hindered in many ways. In these cases, however, the governors insisted that the annoying regulations had been necessitated by the whalemen's brutal treatment of the natives along the shores of Newfoundland and of Labrador. At the same time the British Government discriminated against colonial whaling by requiring that oil and bone be exported only to English ports, where it was chargeable with a duty. This order seemed peculiarly discriminatory because it came at a time when British whalers, under rather lax stipulations, were entitled to a handsome bounty.

Other losses were caused both by man and by nature. For a decade or more French and Spanish privateers formed a constant menace to the small and unarmed whaling vessels, and an appreciable number were actually captured and despoiled. Nature, too, exacted her toll. With longer voyages and protracted periods of continuous cruising, often in unfamiliar waters, came a steadily mounting number of ships that never returned. In 1755, and again in 1756, Nantucket alone lost three sloops, with all hands, near the Grand Banks. And other ports were by no means immune.

Price fluctuations constituted another prolific source of uncertainty. The relatively small yearly fleets were subject to so many vicissitudes that it was impossible to calculate in advance the size of the season's catch. If good fortune presided over the whaling grounds, and if the losses to vessels and cargoes chanced to be small, the resulting large stock of oil depressed the price. If, on the other hand, a poor season was accompanied by many sinkings and by depredations of privateers, unsatisfied demand would place a heavy premium upon the small catch which was brought to port. In the later years of large and scattered fleets the law of probability served in some degree to equalize losses and bonanza voyages, so that the percentage of variation in the catches of succeeding seasons was not as great as the hazardous nature of the industry would have seemed to indicate. But the smaller fleets of the eighteenth century, restricting their activities to the North Atlantic, possessed neither the numbers nor the world-girdling range of activities required for the effective cancellation of gains and losses within a given season. Consequently the price of oil fluctuated widely and unpredictably, as is indicated by the random figures which chance to be available:[8]

Year	Average Price of Oil per Tun £	s.	Year	Average Price of Oil per Tun £	s.	Year	Average Price of Oil per Tun £	s.
1730	7		1748	14		1765	16	11
1742	18	13	1753	21		1767	18	
1743	14	8	1756	18		1768	18	
1744	10		1762	17	15	1770	40	

1772 to 1775, £44 to 45

The risks and difficulties which beset the fishery, however, were not sufficient to halt its steady growth; and progress and expansion were particularly marked during the decade preceding the outbreak of the Revolution. Under the Grenville Ministry England finally gave up the long and disappointing struggle to force the British whalers into successful competition with the Dutch. The duty on whale products from the colonies and the bounty on British whalers were removed at the same time, and every encouragement was given to the industry in North America. This action at once stimulated the technique and scope of colonial whaling operations. Improved methods and types of equipment were carried to the point of increased efficiency, and larger and more seaworthy vessels visited fresh and more distant whaling grounds to meet the steadily enlarging demand.

The best picture of the state of the industry just before

[8] No complete price statistics for this early period are available. These figures, coming at irregular intervals, have been taken from three independent sources, viz., Massachusetts Historical Society Collections, Series 1, Volume III, p. 161; Macy, Obed, "History of Nantucket," p. 54; and Scammon, C. M., "Marine Mammals of the Northwestern Coast," p. 206. The prices are in English terms of pounds sterling per tun of eight barrels, rather than the later American terms of dollars per barrel or cents per gallon.

the Revolution was given in a report on the whale and cod fisheries submitted to the House of Representatives by the then Secretary of State, Thomas Jefferson, on February 1, 1791. Fortunately the Secretary saw fit to add to the main body of the report an "Appendix on the State of the Whale Fishery in Massachusetts, 1771 to 1775," which contained this enlightening table:[9]

Home Ports	Vessels Annually Northern	Tonnage	Vessels Annually Southern	Tonnage	Seamen Employed	Annual Catch Sperm Oil (Bbls.)	Annual Catch Whale Oil (Bbls.)
Nantucket	65	4875	85	10,200	2,025	26,000	4,000
Wellfleet	20	1600	10	1,000	420	2,250	1,250
Dartmouth (New Bedford)	60	4500	20	2,000	1,040	7,200	1,400
Martha's Vineyard	12	720	—	—	156	900	300
Lynn	1	75	1	120	28	200	100
Barnstable	2	150	—	—	26	240	—
Boston	15	1300	5	700	260	1,800	600
Falmouth (Barnstable County)	4	300	—	—	52	400	—
Swanzy	4	300	—	—	52	400	—
Totals	183	13,820	121	14,020	4,059	39,390	7,650

Massachusetts then, as later, was far ahead of any other state in the size of her whaling fleet and the value of its products. The ports of all other states taken together, however, did raise the figures for the industry as a whole appreciably above those applying to Massachusetts alone. For in 1774 the entire colonial fleet consisted of 360 vessels, aggregating about 33,000 tons, and employing some 4,700 men.

[9] This report, including the appendix, was first printed separately at Philadelphia in 1791. Its title was "Report of the Sec. of State on the Subject of the Cod and Whale Fisheries, Feb. 1, 1791." Later it was joined with the "Report on the Principal Fisheries of the American Seas," made by Lorenzo Sabine, and both were printed in House Miscellaneous Documents No. 32, 2nd Session, 42nd Congress.

And these vessels brought into port during the course of a year approximately 45,000 barrels of sperm oil, 8,500 barrels of whale oil, and 75,000 pounds of whalebone.[10]

Such figures as these, together with the spirit of daring enterprise which they indicated, caused Edmund Burke to exclaim, during the course of his well-known speech on American affairs, delivered in Parliament in 1774: "Look at the manner in which the New England people carry on the whale fishery. . . . We learn, that while some of them draw the line or strike the harpoon on the coast of Africa, others run the longitude, and pursue their gigantic game along the coast of Brazil. No sea, but what is vexed with their fisheries — no climate that is not witness of their toils."

[10] Starbuck, A., "History of the American Whale Fishery," p. 57.

CHAPTER IV

FROM DOLDRUMS TO STU'N'S'LS

THE Revolutionary War brought independence to the colonies, but disaster to the whaling industry. The watchfulness of the British naval vessels necessitated a cessation of whaling activities which was well-nigh complete. All continental whaling ports were forced to forego any serious attempts to capture oil and bone throughout the duration of the war; and their vessels rotted at the wharves or were converted into privateers.

Nantucket alone, in spite of crushing losses and an overwhelming mortality rate, kept her harpoons free from rust. Although the odds were heavily weighted against success, whaler after whaler put out from port in brave hopes of finding whales and of evading the British. In truth, there was nothing else to do. For whaling had become the dominant means of livelihood for the entire island, and under the stress of a wartime blockade no adequate substitute could be devised. An attempted continuance of whaling seemed the only alternative to passive starvation.

Both the exposed and indefensible position of the island, and the Quaker beliefs of its inhabitants, prompted Nantucket to maintain a position of neutrality rather than to cast her lot on the side of the colonies. Save for some of her warm-blooded sons who followed their personal inclinations by serving on American privateers, Nantucket was careful, too, in observing her profession of neutrality. But neither British nor Americans would recognize or respect such a status. Year after year the island's vessels were captured or sunk by the British and interfered with by the Americans.

Finally, in 1781, the English Admiral Digby was prevailed upon to grant special permits which would allow twenty-four whalers to make voyages without molestation. Not until 1783, however, when too late to be of much practical benefit, were similar permits for thirty-five vessels secured from the Continental Congress.

In the meantime the war had proved ruinous. In 1784 Nantucket found herself with only twenty-eight whalers fit for sea — and many of those had been repaired after the close of hostilities. No less than 134 of her vessels had been captured by the British, and 15 more had gone down at sea. Some 1200 of her seamen had been lost and captured; and in a total white population of 4,269, comprising about 800 families, there were 202 widows.

All captured whalemen were given a choice between serving on a man-of-war and on a whaler. Because of the unspeakable conditions then existing on men-of-war, the choice was universally in favor of the latter; and in consequence England was enabled to use her prisoners of war to man a whaling fleet of respectable proportions. This fleet operated mainly off the east coast of South America; and in view of the peculiar character of the crews and of the distance from a main base during wartime, the success attained must have been unusually gratifying to the captors. John Adams, in writing to the Council of Massachusetts in September, 1779, recommended that an American naval vessel be commissioned to destroy these enemy whalers and to recapture their crews. Unfortunately, however, his advice went unheeded.[1]

After the conclusion of peace every effort was made to repair the shattered fortunes of the industry. Massachusetts granted a bounty of five pounds sterling per tun on white spermaceti oil, sixty shillings per tun on brown or yellow sperm oil, and forty shillings per tun on whale oil, provided these products were landed in a Massachusetts port by vessels which were owned and manned by residents of the state. But the optimistic expectations aroused by this act were destined

[1] The best accounts of whaling during the Revolution are to be found in Macy, O., "History of Nantucket," and in Starbuck, A., "History of the American Whale Fishery."

to be short-lived. For the increased production stimulated by the bounty was met by a disappointingly small demand. During the war, when whaling products were not to be had, the Americans had become accustomed to tallow candles; and they were slow in changing their habits of consumption. At the same time the British market, profitable for a decade before the Revolution, was now effectually barred by an import duty of eighteen pounds sterling per tun on sperm oil. This double drop in demand, coming in conjunction with the enlarged catches, soon glutted the market and reduced prices to such an extent that whaling became unprofitable in spite of the bounty.

The outlook at Nantucket was so gloomy that a number of the inhabitants determined to emigrate. After a protracted series of negotiations, one group of families sailed for Dunkirk, France. There, under the leadership of William Rotch and the patronage of the French Government, they carried on a small whale fishery from 1786 until 1793, when they were stopped by the events of the French Revolution. During these same years another colony of the islanders removed to Nova Scotia, where they founded the village of Dartmouth. This settlement fell apart within a few years; but some of the colonists were thereupon induced to migrate to Milford Haven, in England. And under the British Jack they finally secured an opportunity to practice their calling with some success.

The stay-at-home Nantucketers, together with the whalemen on the mainland, began to experience a brief return of prosperity about 1790. Sperm candles were being used more widely in the world of wealth and fashion, and the government was buying large quantities of sperm oil for the growing chain of light-houses. The foreign demand also showed a gratifying increase. France, in particular, began to consume whaling products in such amounts that her trade filled the gap caused by the loss of the pre-war British market. Nantucket lost her air of apathy and of ruin, and gradually came to regain some semblance of hopeful well-being.[2]

[2] An interesting account of Nantucket life about 1790 may be found in Crevecoeur, J. H. St. John de, "Letters From an American Farmer." Letters

At the same time the new and rich whaling grounds of the Pacific were being explored. The first sperm whaler to enter the Pacific was the British ship *Amelia,* owned by Charles Enderby of London, but manned largely by ex-Nantucket whalemen. Sailing in 1788, this vessel returned in 1790 with a full cargo and a glowing account of the newly-discovered grounds. During the following year no less than six American whalers rounded Cape Horn and found countless targets for their harpoons and lances. In the van of this small fleet was the ship *Beaver,* of Nantucket, Paul Worth, Master, which sailed in August, 1791, and returned on February 3, 1793. (In the later days of the industry, when Pacific voyages required from thirty-five to fifty months in which to produce a "full ship," such a cruise would have been considered little short of miraculous). And once around Cape Horn, the intrepid whalemen extended the scope of their explorations so rapidly that by 1800 the cachalot was being pursued along the whole western coast of South America.

Very soon after the *Beaver* dropped anchor, however, the fishery began to experience new difficulties. The promising French market was destroyed by the outbreak of the French Revolution; and once again prices slumped dangerously because of the over-supply of oil brought into port by an enlarged and far-cruising fleet. In such an industry as whaling, with its long voyages and peculiar equipment, it was impossible to adjust the scope of the operations to political and economic changes with any speed or flexibility. Enlargements or decreases in the size of the fleet were usually one season, and sometimes two, behind their dominant causes; and consequently, as in this instance, oil continued to flow into port long after the need for it had passed.

This time, however, the political and military chaos in Europe threatened a rapid and efficient trimming away of the surplus American whaling tonnage. British and French naval craft and privateers rendered the seas precarious for all merchant vessels; and neither nation was then in a mood to grant special favors to the United States. French privateers

4 to 8, inclusive, are devoted entirely to the interests and activities of the island,

were particularly active during the period of threatened hostilities between France and America in 1798–1799, when they seized a number of Nantucket whaling vessels for which no proper indemnity was ever paid. The British, not to be outdone in domineering conduct upon the high seas, extended their peculiarly insolent policy of search and impressment; and when opportunity offered they impressed whalemen as well as merchant seamen. With the turn of the century came some slackening in such depredations. But men-of-war, privateers, and full-fledged pirates continued to menace the seas for another decade and more.

In spite of losses and annoyances, however, the industry displayed signs of vitality and moderate prosperity. Sperm oil and spermaceti candles had routed the tallow candles of the Revolution throughout American polite society; and as a result the domestic demand was large and growing. Nantucket, which always manifested a strong liking for the sperm fishery, became even more than formerly the whaling center of the United States. By 1807 the island had a fleet of 120 vessels, some of which were making voyages of ten to thirty months. Eleven were whaling off the coast of Brazil, eleven more were near the Cape of Good Hope, eighteen were in the Pacific, and one had ventured as far away as the coast of New Holland (Australia). Other ports were also active, though on a smaller scale. New explorations were carrying the whalers from both island and mainland ever farther and farther afield. Before 1805 whales had been pursued in the high latitudes of the South Pacific and South Indian Oceans, in the vicinity of the Molucca Islands, and off the coast of antipodal New Zealand.

Then came the Embargo, which again eclipsed the fortunes of the whalemen. Returning vessels, instead of being discharged, overhauled, provisioned, and sent back to sea within a few weeks, remained restless at anchor or tied up to their wharves for indefinite periods. And the work of destruction begun by the Embargo was completed by the War of 1812. Many of the American whalers in the Pacific were captured either by the British or by Chilean and Peruvian pirates, who had become greatly emboldened by the wartime conditions.

Some of these vessels were recaptured in 1813 and 1814 by the American naval expedition under Admiral Porter; but most of them were burned, sunk, or transformed into British transports or pirate craft. Other whaleships, including many of those which sought to make port soon after the outbreak of hostilities, were captured or destroyed in the Atlantic.

And again, as during the Revolution, Nantucket bore the brunt of the losses. In fact, to the older inhabitants of the island this second war must have seemed but a dark and evil memory of the first. There was the same fatalistic perseverance in sending out vessels even while it was known that most of them would never return; the same suspense and agony of waiting and of ascending day after day to the boat-walk, spy-glass in hand; the same lack of provisions and of new materials; and the same terrifying increase in the number of widows and of fatherless children. Other ports could and did suspend operations for the duration of the war. But Nantucket had no other means of livelihood: she was forced to rely upon the small percentage of her vessels which did return in order to stave off complete destitution. And the result of this inevitable but ruinous policy, as before, was the loss of a crushing percentage of her entire fleet. In 1812 the inhabitants of the island owned and operated 116 whaling vessels: in 1815 only 23 strained and battered craft remained.

But the War of 1812 was to be the last major reverse suffered by the fishery for a half-century. In 1815 American whaling finally emerged from the long series of calamities which had beset it since 1775, and entered a decade of preliminary reconstruction and general clearing of decks. This stage of rehabilitation, in turn, gradually merged into the long period stretching from about 1825 to 1860, rightfully regarded as the golden era of New World whaling.

The first decade of peace was utilized to the full in repairing the ravages of war and in laying the foundations for future expansion. New and fertile whaling grounds were opened in rapid succession; and each newly-discovered whaling haunt drew to itself, as to a magnet, hundreds of new harpoons. In 1818 the ship *Globe*, of Nantucket, George W. Gardner,

Master, discovered the great Off-Shore Grounds, including hundreds of square miles in mid-Pacific. Two years later another Nantucket vessel, the ship *Maro*, followed hard upon the heels of the English *Syren* in opening up the prolific Japan Ground. And in 1823 the *Swan*, likewise flying the house-flag of the great English firm of Enderby, found excellent hunting on the Seychelles Islands Grounds, stretching from Madagascar to the Persian Gulf.

But these more distant grounds gave up their rich harvests only after exacting a heavy price in the form of longer and longer voyages. And such voyages demanded larger and stauncher vessels, with more width of beam and greater cargo capacity. Not only was it essential to carry provisions and equipment for many consecutive months: it was equally necessary to have space in which to stow away the captured oil and bone. Consequently ships and barks of 200 tons to 500 tons, carrying three or four whaleboats and crews of thirty men or more, rapidly supplanted the smaller schooners and brigs. Only in the short-voyage Atlantic fishery were the latter able to hold their own. The typical American whaler now rounded out the list of characteristics which were to distinguish her for the next half-century — the boats swung conspicuously from heavy wooden davits, the square lines (merchant seamen insisted derisively that whaling vessels were built by the mile and sawed off in the desired lengths), the large crew, the men at the mast-heads, the brick try-works on deck, the smoke-blackened sails, and the equipment for cutting-in and trying-out.

On these foundations the superstructure of expansion was quickly erected. By 1821 Nantucket had more than regained her pre-war position; and by 1825 every whaling port along the coast of New England, Long Island, and Martha's Vineyard was in the full flush of strenuous activity. The demand for whaling products was again increasing, and this time at an accelerated rate. The United States was entering upon a period of unexampled prosperity which enabled a growing percentage of its citizens to use the superior but relatively high-priced spermaceti candles. After the removal of Napoleon to St. Helena, the European markets recovered their purchasing power and once more absorbed American oil and bone in profitable quantities. Sperm oil and whale oil were employed more and more for both lighting and lubricating purposes. And the growing use of whalebone in the making of stays, corsets, whips, and umbrellas acted as a further spur to the right whale branch of the industry. With a heavy demand assured by these developments, a long period of peace and the marked Yankee aptitude for the business completed the requirements for the full fruition of New England whaling.

And fruition came in the form of the golden era of 1830 to 1860. Statistical evidence of the growth and prosperity of this period exists in abundance. Thus in 1833 the American fleet consisted of 392 vessels, aggregating 130,000 tons and carrying 10,000 seamen.[3] By January 1, 1844, the number of vessels had risen to 644, the tonnage to 200,484, and the number of men to 17,594. Two hundred and forty-two vessels, with an average crew of 28 men each and an average value of $55,000, including their catches, were in sperm whaling; 329 vessels, with a similar average of 28 men each but an average value of only $40,000, with their catches, were in right whaling; and the remaining 73 craft, with average crews of 22 men each and with vessels and catches valued at $18,000 each, were in Atlantic sperm whaling.[4] Two years later, with 233,189 tons distributed over 735 vessels, American whaling reached its apex. But this high mark of 1846 was almost repeated in 1847, when 722 whalers, displacing 230,218 tons, were claimed by 34 different ports. New Bedford, with 254 vessels, was so far in the lead that her only serious rivals were Nantucket, with 75 whalers; New London, with 70; and Sag Harbor, with 62.[5]

This scale of operations continued, without any essential change, until the outbreak of the Civil War. The record activity of 1846 and 1847 was never again reached, it is true;

[3] These figures were given by Williams, J. R., in an article published in the *North American Review* for January, 1834. See Vol. XXXVIII, pp. 103–108.

[4] Figures compiled by Grinnell, Joseph, and published in 1844 in a 16-page pamphlet entitled, "Speech on the Tariff, With Statistical Tables of the Whale Fishery."

[5] *Whalemen's Shipping List* for January, 1847, and January, 1848.

but for only two years during the period 1847–1860 did the annual statements of the *Whalemen's Shipping List* show less than 600 vessels and 190,000 tons. And even in 1850, one of the two lowest years, there were between 18,000 and 20,000 whalemen actually at sea.[6] New Bedford kept up her process of steady expansion, both absolutely and in proportion to the other ports, until in 1857 she controlled practically one-half of the entire American industry — 329 of the 655 vessels, 10,000 of the 20,000 seamen, and $12,000,000 of the $20,000,000 to $25,000,000 invested capital.[7]

Throughout these decades of ripening activity the economy of the individual vessel underwent surprisingly little change. The increased production of the fishery as a whole was secured through a multiplication of vessels of the same type, rather than through any essential improvements in rig, hull, size, type of equipment, or methods of capture. Just before the Civil War the average whaler was perhaps nearer 400 tons than 300 tons, and her crew had increased from about 26 men to about 30 men. There had been some tendency, too, to favor the bark rig at the expense of the full square-rigged ship. But from the standpoint of technology, methods pursued, and equipment employed, there had been no important change. In general, whaling craft and whaling gear remained standardized; and, with the true conservatism of the sea, whaling methods and whaling customs were neither lightly cast aside nor readily improved. The whaler of 1860 captured her oil and bone and brought her cargo into port in much the same manner as had her predecessor of 1830. In fact, in many instances she was the selfsame vessel. For the life of a whaler, heavily constructed out of irreproachable timber, might well be a long one.

Only two exceptions to this prevailing uniformity of method and equipment attracted the professional attention of whalemen. One was the recrudescence, during the forties and fifties, of the boat-whaling which had flourished in the early

[6] See the issues of the *Whalemen's Shipping List* for January of the appropriate years.

[7] Figures published in 1889 by Pease, Z. W., and Hough, G. A., in a booklet entitled, "New Bedford, Massachusetts: Its History, Industry, Institutions, and Attractions."

THE SPERMACETI WHALE OF THE SOUTHERN OCEAN
From "The Natural History of the Whale," Edinburgh, 1837

"CUTTING-IN" A WHALE
From an engraving by Rouarque (about 1850)

eighteenth century. Here and there along the coasts of Maine and of Massachusetts (and in a few other isolated spots such as Beaufort, North Carolina) small boats sought to capture the whales which came close in to shore. Boat-whaling, however, attained its greatest development in California. In 1851, at Monterey, a certain Captain Davenport began to hunt the whales which entered the coastal lagoons for breeding purposes. His activities attracted enough imitators to make this California lagoon-whaling a recognized branch of the industry — although the total output was always relatively unimportant.[8]

The second and more significant exception consisted of the bowhead whaling which came to be carried on primarily in the high latitudes of the North Pacific. Credit for the opening of these new grounds was usually given to the ship *Superior,* of Sag Harbor, Captain Roys, which penetrated the Arctic waters north of Behring Straits in 1848. At about the same time, however, the pursuit of the new game was begun in the Sea of Okhotsk and off the coast of Kamchatka. The Kodiak Ground, too, had been discovered in 1835 by the ship *Ganges,* of Nantucket, B. T. Folger, Master; but this area, off the Northwest Coast of North America, was frequented by the common right whale rather than by the bowhead. The latter proved to be more desirable for three reasons: it was more sluggish in its movements, and therefore easier to approach and to capture; its thicker coat of blubber afforded more oil per animal of given size; and it supplied longer and heavier whalebone.

Unfortunately, however, the bowhead was a lover of frigid waters; and the frozen regions which had to be explored in hunting it required the use of stauncher vessels, of heavier equipment, and of a technique capable of dealing not only with whales, but with ice-fields and bitter cold as well. Similar conditions had been encountered, it is true, in the earlier North Atlantic fisheries at Spitzbergen and in Davis Straits.

[8] The best accounts of this later boat-whaling are to be found in Clark, A. H., writing in "Fisheries and Fishery Industries of the United States," VII, pp. 40–63; and in Scammon, C. M., "Marine Mammals of the Northwestern Coast of North America," pp. 247–276.

But a bowhead in the North Pacific was by no means identical with a right whale in the North Atlantic. Consequently the later Arctic whaling had to develop certain technological changes which, though far from revolutionary, were not insignificant.[9]

There were also, of course, minor improvements. Such was the device of the toggle harpoon, which locked the iron in the whale's flesh, and so increased the chances of remaining fast to an animal which had been struck.[10] Such, too, was the gradual perfection of the strength of the whaleline, aristocrat of the world of ropes. And such, again, was the defeat of scurvy through a more liberal and intelligent use of antiscorbutics.

But in general it remained true that throughout its decades of greatest activity the industry was unaffected by any major or revolutionary changes. The crews of 1830 would have found little that was unfamiliar in the vessels and equipment of 1860. Some tendencies had been accentuated and others had disappeared; certain individual habits had entered the traditions of the fishery; here and there was a new method or an improved implement. But for descriptive and analytical purposes the period 1830 to 1860 formed a unit: it was made of whole cloth. The significant changes which did come about were quantitative rather than qualitative: the developing differences were not those of kind, but of degree. Figures were of the essence of the fishery during these decades.

Consequently the statistical summaries of whaling activity which were published annually (after 1843) by the *Whalemen's Shipping List and Merchant's Transcript*, the authoritative weekly organ of the industry, were of peculiar

[9] A good account of the beginnings of this North Pacific whaling may be found in *Hunt's Merchants' Magazine and Commercial Review* for August, 1852, Vol. XXVII, pp. 227 f. Further descriptive material, presenting this branch of the industry in some detail, is given by Clark, A. H., writing in "Fisheries and Fishery Industries of the United States," VII, pp. 73–84; by Starbuck, A., "History of the American Whale Fishery," pp. 96 ff.; and by Jenkins, J. T., "History of the Whale Fisheries," pp. 234 ff.

[10] Other experiments with the harpoon, however, came to naught. The average harpooner, justly proud of his own hand and eye, had little patience with such new-fangled ideas as shooting harpoons from a gun, exploding bombs containing everything from gunpowder to prussic acid within a whale's vitals, or paralyzing a victim with an electric shock.

significance. Each year this publication assiduously gathered figures concerning the total number of whaling vessels (including ships, barks, brigs, schooners, and sloops), the total tonnage, the amounts of sperm oil, whale oil, and whalebone brought into American ports (listed as imports), and the average annual price of each of these three products. As soon as possible after each thirty-first day of December these figures were then assembled and published in one of the regular January issues; and when placed in the proper sequence they spoke eloquently of whaling progress. The compilations which depicted the fishery at its period of highest development were these: [11]

American Whaling Vessels, Tonnage, Imports, and Prices, 1838–1860

Years	Number of Vessels	Total Tonnage	Imports of Sperm Oil (barrels)	Imports of Whale Oil (barrels)	Imports of Whalebone (pounds)	Aver. Price Sperm Oil (gal.)	Aver. Price Whale Oil (gal.)	Aver. Price Whale-Bone (lb.)
1838	—	—	132,356	226,552	—	$ .83	$ .32	$ .19½
1839	—	—	142,336	229,783	—	1.03	.34½	.18¼
1840	—	—	157,791	207,908	—	1.00	.30½	.19
1841	—	—	159,304	207,348	2,000,000 (estimated)	.94	.31¾	.19⅔
1842	—	—	165,637	161,041	1,600,000 (estimated)	.73	.33¾	.23
1843	675	199,192	166,985	206,727	2,000,000 (estimated)	.63	.34½	.35¾
1844	644	200,484	139,594	262,047	2,532,445	.90½	.36½	.40
1845	695	218,655	157,917	272,730	3,167,142	.88	.32⅞	.33⅝
1846	735	233,189	95,217	207,493	2,276,939	.87⅞	.33¾	.34
1847	722	230,218	120,753	313,150	3,341,680	1.00½	.36	.30⅞
1848	659	210,541	107,976	280,656	2,003,000	1.00½	.33	.25⅖
1849	614	196,110	100,944	248,492	2,281,100	1.08 9/10	.39 9/10	.31 8/10
1850	543	171,484	92,892	200,608	2,869,200	1.20 7/10	.49 1/10	.34 4/10
1851	553	171,971	99,591	328,483	3,916,500	1.27¼	.45 5/16	.34½
1852	620	193,990	78,872	84,211	1,259,900	1.23¾	.68⅙	.50¾
1853	661	206,286	103,077	260,114	5,652,300	1.24¾	.68⅙	.50¾
1854	668	208,029	76,696	319,837	3,445,200	1.48¾	.58⅝	.39⅕
1855	638	199,842	72,649	184,015	2,707,800	1.77 2/10	.71 3/10	.45¼
1856	635	199,141	80,941	197,890	2,592,700	1.62	.79½	.58
1857	655	204,209	78,440	230,941	2,058,850	1.28⅓	.73¼	.96¾
1858	654	203,148	81,941	182,223	1,540,600	1.21	.54	.90
1859	625	195,115	91,408	190,411	1,923,850	1.36¼	.48½	—
1860	571	177,049	73,708	140,005	1,337,650	1.41½	.49½	.80½

[11] The complete figures of the *Whalemen's Shipping List* began with the year 1843, although reports for some items go back to 1838. Alexander Star-

Those were the full, gala days of the industry, — when rugged New Bedford, in particular, was at the peak of a picturesque, salt-sprayed renown which took her name out into the seven seas. This New England town, which became the greatest whaling port in the world despite population figures which were unimpressive even in an era of villages, learned to blend the strange and exotic with the native and homely in a manner both alluring and ludricrous.

New Bedford knew bearded and tattooed harpooners who sea-legged their way into brothels and grog-shops; timid and unsophisticated green hands from the farms of the interior; veteran tars who knew the price of a harlot in Zanzibar and the cost of ale in London; mutineers and masters who should have been slave-drivers; sperm oil from the Seychelles Islands and whalebone from Kamchatka; barnacles acquired in every one of the seven seas; scrimshaw work and Chinese tea, Oriental silk and souvenirs from the Fiji Islands; bonanza voyages and penniless hands; log-books telling of stove boats and account-books telling of exorbitant charges; rope-walks and sail-lofts, outfitters and ship-chandlers; Quaker and Cape Verde half-breed, Puritan and Kanaka; pure sperm oil which had been baled out of the head of a cachalot and black and stinking whale oil which had been four years at sea; stories of murder and of rape in the South Seas; yarns of cheap love in Paita and of frozen noses in the Sea of Okhotsk; Seamen's Bethel and dens of drunken vice; counting-houses with high stools and lays of 1/200 of the net proceeds; toggle harpoons and tubs filled with coiled whaleline; nauseating forecastles and the spacious lawns of the leading whaling merchants; "lobscouse" and "duff," ship's bread filled with worms and salt-horse which had thrice crossed the equator!

Cosmopolitan and provincial, of the great outlying world and of pinched New England, pious and abandoned, æsthetic and ugly, alluring and repulsive, colorful and drab, adventurous and cautious, courting danger and loving security — such

buck, however, gathered other statistical material which carried the figures for annual imports and for average prices back to 1804. This material was published in his "History of the American Whale Fishery," pp. 660 f.

was New Bedford during the forties and fifties, when her favorite industry was plunging ahead full sail, with stu'n's'ls set and "every stitch of canvas drawing." Like her, too, though in smaller degree, were the other whaling ports.

Every branch of the fishery was in full career: full sail was everywhere the order of the day. And it is this period of maximum activity which must serve to characterize both the best and the worst in American whaling. These mid-century decades drew into themselves the heart and essence of the whaleman's calling. All preceding whaling was but preparation and launching: all later hunting was only aftermath and derelict.

PART II

FULL SAIL

CHAPTER V

FORECASTLE AND CABIN

THE human material entering into the crews changed slowly but radically during the very middle of American whaling development. So gradual was this change that it is impossible to assign it to any given year. Yet so unmistakable was it, too, that the industry was divided into two roughly defined but clearly recognizable periods. For want of a more precise boundary line these two may be separated by the half-decade 1825–1830. Before that time the crews were provincial and homogeneous: after 1830 they were cosmopolitan and heterogeneous. The early foremast hands were made up largely of Yankees from the New England seaboard, with an admixture of Gay Head Indians and a small representative of negroes; while during the second period individuals from every race and from a score of nationalities rubbed shoulders in the crowded forecastles and steerages. Coincident with this shift from provincialism to cosmopolitanism went a marked deterioration in skill, experience, efficiency, and morale, as well as a striking increase in the total number of men engaged in the fishery.

The earlier period saw the whaling industry virtually monopolized by native American stock, with the Yankee of the New England seaboard as the dominant type. The small whaling villages were concentrated on Long Island, Martha's Vineyard, Nantucket, and the Massachusetts mainland; and each of these communities drew an adequate labor supply, for the most part, from its own tiny hinterland of a few square miles. Not until the middle of the eighteenth century was Nantucket, the most prosperous of the whaling ports, compelled to import a few extra hands from her neighbors on Cape Cod and Long Island. Consequently these early crews were pervaded by a strong and intimate community spirit. The captain was likely to know the business and social position of the family of every man in his crew, and to be related, by birth or marriage, to several of them.

Whaling, too, was the recognized pathway to the local fame, fortune, and social prestige afforded by these isolated ports. Vessels were manned by the "sturdy, intelligent, and comparatively virtuous yeomanry of New England," as well as by the sons of the wealthiest and most prominent families.[1] The latter began as cabin-boys and foremast hands in order to acquire a first-hand knowledge of the industry before filling a captain's berth and then retiring to the life of a whaling merchant ashore. Whaling was at the very heart of community life; and the man who had actually handled lance and harpoon in action was a clear business and social favorite. So strong was this feeling in Nantucket, in particular, that at one time the most eligible maidens looked with scorn upon would-be suitors who had not harpooned at least one whale; while the chock-pin, worn on the lapel of the coat as the mark of a boatsteerer, was the ambition of every boy on the island.[2] As a result of this combined social and financial pressure the youths of the whaling villages went about their tasks with a serious purpose, and soon became skilled navigators and courageous whalemen. Since desertion was practically unknown, and most of the men sailed for several successive voyages, the crews contained a relatively high percentage of experienced and efficient hands.

In addition to the dominant Yankee, the early whaling forecastles housed two other significant elements — the Indian and

[1] See Cheever, H. T., "The Whale and His Captors," p. 303.

[2] The chock-pin was a small wooden peg used in keeping the whaleline straight while it was being taken out through the bow of a boat by a harpooned whale. At certain times and places it was worn on shore by boatsteerers as a mark of rank.

the negro. The Indians of Nantucket, Martha's Vineyard, and Long Island were accustomed to pursue whales in canoes even before the advent of the white man; and when the colonists first undertook a crude and hesitant form of boat whaling from the shore, they employed not only the Indians themselves, but also some of their methods and implements.[3] As the industry grew and larger vessels made longer and longer voyages, the Indians continued to be shipped as hands. Even after the Revolution the cry of the Nattick Indians, "Awaite Pawana" ("Here is a whale"), continued to ring out over waters far removed from the shores which had witnessed the aboriginal conflicts between birch bark and flukes.

The weak-willed red men of Nantucket succumbed so rapidly to disease and dissipation, induced largely by an excessive use of "fire-water," that the last member of the tribe died in 1822. But the Chilmark Indians of Gay Head, Martha's Vineyard, and of Montauk Point, Long Island, were of sterner stuff. Commonly referred to as Gay Head Indians, these men constituted a valuable and picturesque element in hundreds of whaling crews. With few exceptions, contemporary writers spoke of their skill and daring with respect and admiration. Consequently they were great favorites for the position of boatsteerer, though they seldom rose to be mates.

Negro whalemen seem to have been comparatively rare, at least on Nantucket vessels, until after the Revolution. Then the growing demands of the industry led to their importation from the mainland in considerable numbers. Living in a small area on the southern outskirts of Nantucket known as New Guinea, they were much given to drunkenness and improvidence; but some of them soon became excellent seamen.

Contemporary accounts indicate that some of the whaling owners and masters were guilty of sharp practices in their treatment of both blacks and Indians. Certain Nantucket negroes were shamefully exploited through exorbitant charges which were made to exceed the amounts of their lays, or earnings. The resultant debt to the owner was then used, wherever it seemed advisable to prevent possible escape, as a pretext for imprisonment during the short periods spent

[3] See Spears, J. R., "The Story of the New England Whalers."

ashore between voyages. The Martha's Vineyard whaling merchants, too, deliberately kept the Indians in debt, and thus forced them to ship for many successive voyages.

During the first two decades of the nineteenth century the whaling crews contained a much higher percentage of both negroes and Indians than in later years. About 1807 crews of sixteen to twenty-one men often included seven to nine negroes. And in 1820 it was estimated that about one-eighth of all Nantucket whalemen were pure Indians, while from two-eighths to three-eighths were negroes and Indian half-breeds.

But in spite of this inclusion of alien elements during the earlier decades of the industry, the size of the crews did not increase to the same degree as the tonnage of the vessels. About 1750 the typical Nantucket whaler was a small sloop of fifty to eighty tons, manned by some thirteen men; while about 1825 there were ships and barks of three hundred tons which carried crews of twenty-five hands. At the earlier date there was one man for every five or six tons of the vessel in which he sailed; but by 1825 this proportion had fallen to one hand for every twelve to fourteen tons. Roughly speaking, the tonnage of the typical whaler was quadrupled during the same period in which the size of the crew was doubled. And since tonnage bore a definite relationship to invested capital, it followed also that the amount of capital devoted to the industry had increased about twice as rapidly as the size of the labor force.

During the heyday and decline of American whaling, covering the second great period which began about 1825–30, the outstanding change in the crews was again not one of size, but of composition. Provincialism and homogeneity gave way to cosmopolitanism and heterogeneity. The neighborly, like-minded groups of the earlier days disappeared; and in their places were whole worlds of humanity in microcosm. Seldom have so many widely assorted specimens of the human family been packed into such small spaces as the forecastles of the latter-day whaleships. The mere enumeration of the diverse elements constitutes a list of formidable length. There were Americans from every part of the United States, Portuguese and mulattoes from the Azores and Cape Verde Islands, Spaniards, Swedes, Norwegians, Danes, Dutchmen, Germans, Frenchmen, Englishmen, Scotchmen, Irishmen, Gay Head Indians, negroes from the United States, Africa, and the West Indies, Maoris from New Zealand, Kanakas from the Sandwich Islands,[4] natives from other South Sea Islands, and half-breeds who represented the crossing of many different stocks. The colorful, picturesque, and exotic character of the crews resulting from this conglomerate mixture of races and nationalities is presented nowhere so well as in Herman Melville's classic novel of whaling life, "Moby Dick."[5]

The social and economic groups represented in the forecastles displayed the same extreme diversification. There were adventurous youths from inland farms and workshops who had been attracted by the mendacious advertising of the shipping-masters, by love of the sea, or by the lure of foreign travel; spoiled sons of indulgent parents who had run away from home or who had been sent into whaling in an effort to correct their indolent and spendthrift habits; reckless and impatient persons who were temperamentally unfitted for the conventional restraints imposed in a civilized community; and roving adventurers who chose a whaling voyage as one means of gratifying a dominating wanderlust. There were also immoral and unprincipled wretches who wanted to get "off soundings" in an attempt to elude the Ten Commandments. These included confirmed drunkards, vagrant ne'er-do-wells, unapprehended criminals, escaped convicts, and dissipated and diseased human derelicts of every description. And finally there were alert, intelligent lads from prosperous whaling families who were being initiated into the traditions of the industry and who expected to carry on the family fortunes; capable and self-respecting seamen who felt a real pride in their calling

[4] This term Kanaka seems to have been applied originally to natives of the Sandwich Islands, now more commonly known as the Hawaiian Islands; but gradually it came to be used more loosely in referring to any South Sea Islander.

[5] Melville actually made a voyage as a foremast hand on a sperm whaler; and consequently this volume, while properly classed as fiction, contains many descriptions of whaling life which are amongst the best, as they certainly are the most readable, which we possess.

and who gazed with critical appreciation upon the graceful lines and spick-and-span equipment of a "smart" vessel; and an occasional young man who had shipped for reasons of health or "education."

Three of the most bizarre and outlandish of these many elements comprised the South Sea Islanders, the "beach-combers," and the Portuguese. Of the many picturesque natives of the South Seas, the Sandwich Islanders were both most numerous and most desirable. They were commonly described as docile, good-natured, trustworthy, loyal, and moderately intelligent hands, and were universally conceded to be expert swimmers and boatmen. In 1844 it was estimated that there were from five to six hundred of these Kanakas on American whalers. At that time they were not allowed to embark without a license from the governor of the island to which they belonged; and each captain who shipped a native was expected to return him to the island within three years, if still alive, or to forfeit a bond of two hundred dollars. Though specific evidence is lacking, it is to be suspected that these regulations were honored more in the breach than in the observance.

Another type of recruit sometimes offered by the South Seas was the "beach-comber," "shoaler," or "seasoner," — a degenerate white man who had deserted or been cast adrift and who settled down into a lazy, licentious, and profligate life with the natives. When driven by dire need or prompted by the sheer monotony of his life, he would sometimes seek a berth on a whaler. Though he was usually the least satisfactory of hands, emergency requirements often caused a whaling master to hire him for a season or for a cruise between certain specified ports.

The Portuguese were usually shipped at the Azores and Cape Verde Islands, where the outward-bound whalers stopped in order to round out their crew-lists. Gradually, however, the better and steadier hands drifted into New Bedford and settled in a section of the city known as Fayal, in memory of the main port of the Azores. These men readily adapted themselves to life aboard a whaler, and soon established a reputation as capable boatmen and lookouts, though but indifferent seamen. Large numbers of them became boatsteerers; and often they were shipped as mates.

Appraisals of their personal characteristics differed with the vantage points of the observers. With few exceptions, contemporary writers who had actually lived in a forecastle insisted that the Portuguese were generally dirty and bestial; while the officers tended to favor them because of their long-suffering docility under poor food and abuse and their willingness to ship for "long lays" (low wages) under loose agreements whose terms could be readily reinterpreted. Certainly many of them were veritable marvels of thrift and of parsimony. Their slop-chest accounts, showing the supplies drawn during the course of a voyage, were often negligible in amount, and far below the average of the other hands.

Some of the Portuguese hands eked out their earnings by washing clothes for their less thrifty shipmates. Since the whalemen were often bathed in perspiration and exposed to innumerable forms of dirt and grease, and since their clothing was periodically saturated with whale oils and salt water, the task of washing such garments was hardly agreeable. It was rendered all the more laborious, too, because the scrubbing had to be done in salt water; for fresh water was available for cleansing purposes only when rain water could be caught upon deck. It was not surprising, therefore, that at times certain members of a crew were willing to pay trifling sums to escape such tedious laundry work.

While on shipboard both Portuguese and South Sea Islanders often answered to curious names which further accentuated the rough-and-ready cosmopolitanism of the crews. Whenever foreign names presented too many difficulties of pronunciation, their owners were dubbed with any sobriquets which came to mind. This was particularly true of the Kanakas. Many untutored Pacific whalemen bore names drawn from the honored aristocracy of New England; and numerous account-books bear witness to the fact that John Quincy Adams was shipped at a lay of 1/200, or that Samuel Adams drew divers quantities of tobacco from the slop-chest. In other cases they were given single names such as Peter, Henry, Frank, James—

or mere nicknames such as Joe, Jim, Sam, or Jack; and often they were christened facetiously, and became known as Spun-yarn, Peter Owyhee, John Hawaii, Henry Second, Jim Kan-acker, Ned Buntline, Bill Bowhead, Jim Maui, or William Bowline. If the native names were short and easily pro-nounceable, however, they were frequently retained; and con-sequently the whaling account-books contain such euphonious names as Kama, Kahana, Manu, Kavna, Kaune, Kohopu, Kalua, Pahia, Nahina, and Nakauna.

The confusion of names, nationalities, and races in the whal-ing forecastles was rendered still more pronounced by two other factors. Some foreign whalemen voluntarily adopted English surnames, or Anglicized their own names, as part of a naturalization process. And many other hands, in order to avoid a variety of undesirable contingencies, assumed fictitious names upon going to sea. Charles R. Tucker, one of the lead-ing whaling merchants of New Bedford, wrote under date of 7 Mo. 24, 1837: " — it has of late become very fashionable for Sailors to assume some fictitious name by which they ship and are known by before they sail — and thus may de-prive their friends from tracing them." He might have added another fact which was even more significant, — that shipping on a whaler under an assumed name was one of the most effective means of evading pursuit and capture by en-emies and by officers of the law.

More intimate views of specific crews brought out striking high-lights in this picture of strange cosmopolitanism. Con-trasting types and clashing personalities were everywhere. The forecastle of one whaler, for instance, contained during a single voyage the following kaleidoscopic group: Bill Mann, "son of Ed. Mann, sail-maker, New York" — a blustering, hard-drinking tar who knew a little Shakespeare as a result of playing minor rôles on the stage; Jack Smith, a young Eng-lishman, a kleptomaniac and an inveterate liar, but generous and fearless; Barzilla MacF——, a typical "down-easter" of nineteen years, awkward, lean, unselfish, imperturbable, with a quaint sense of humor and a keen desire, easily explained, to be back on the farm; Tom Vernon, a Philadelphia youth of some education who had been a hardware clerk before he had quarreled with his employers; Charley Clifford, the most capable man on board, who was all that a real sailor should be; John Blair, a New York stone-cutter and the bully of the crew — coarse, brutish, and ruffianly, boasting of a long list of rascalities which had finally driven him to sea; four Portuguese, ignorant and bestial; two Irishmen; and a well-educated young Kentuckian who had just left a position as reporter in the United States Senate in order to "see the world."

The steerage of the same vessel housed the cooper, a tall, gaunt, lame Mormon who was constantly attempting to make converts through interminable religious arguments; a boatsteerer named Tabor, sçion of a well-known Yankee whaling family, but also a heavy drinker and the loosest of Methodists; and other less colorful individuals. The cabin was dominated by the mate, an infidel who deemed all forms of religion to be mere humbug; and by the captain, "a mean, cunning, stingy Yankee," unmarried, who kept apart from the others and spent his time in the cabin or in walking the quarterdeck. During these appearances he was commonly clad in slip-shod shoes, a pair of soiled duck pantaloons, a green roundabout, and a broad-brimmed Panama hat.[6]

Another crew-list reflected the same fantastic jumble of races, nationalities, and social strata. Eighteen foremast hands included two boys who had escaped from a House of Correction in Pennsylvania; two dissipated youths whose military school training had been a complete failure and whose wealthy parents had chosen the sea as a more rigorous school of reform; a young orphan who was without a single known relative; two "down-easters" from the Maine coast; an Irishman who had just reached the United States; an English printer who had been shanghaied in New York; a Dutchman from Pennsylvania; and eight Portuguese. The boatsteerers, with the exception of one huge Maori from New Zealand, were Portuguese from the Azores and Cape Verde Islands. The steward and carpenter hailed from New York, the cooper from Ireland, and the cabin-boy presumably from Holland.

[6] This picturesque crew was described in Browne, J. R., "Etchings of a Whaling Cruise," pp. 34 and 182–192. The original crew-list, dated 1842, is now in the Old Dartmouth Historical Society Museum at New Bedford.

The captain (quite an unusual circumstance) and one mate were Portuguese; while the three other mates were natives of New Bedford.[7]

On a third large vessel the complement of twenty-four hands was made up of one professional gambler, six factory hands from New England mill towns, one Boston school-boy, two lawyers' clerks, one runaway from his father's counting-house in New York, one American who had sailed in both the navy and the merchant service, one New York "butcher-boy," six boys from the farms of New England and New York, one canal-boat man, and four Portuguese.[8]

Though greatly outnumbered in this flood of foreign elements, the Yankee of the New England seaboard retained a clear supremacy in the more skilled and responsible phases of the industry. He was shrewd, thrifty, self-reliant, ingenious, persevering, hard-working, reasonably intelligent, and, if necessary, courageous and daring. But this whaling paragon appeared to advantage only when in action. If at sea he was the personification of trained and alert activity, on shore he often became a caricature of awkwardness and ungainliness. Thus a typical "smart" harpooner was described by a mid-century writer as a man of "sallow cheek, with hands tanned a deep enduring saffron color"; very round-shouldered as a result of countless hours at the oars; listless in movement, careless in speech, and uncertain of gaze; obviously ill at ease in unaccustomed "shore togs"; and radiating a general air of shabbiness fully borne out by his dress. This consisted of a pair of rough shoes, with the ends of the shoe-strings trailing on the ground as he walked; a pair of trousers with several inches of gray woolen socks exposed at the bottom and a portion of red flannel drawers at the top; a shirt entirely open in front except where it was held in place by a rusty black handkerchief, which in turn was fastened at the throat by a large ring made of whalebone inlaid with mother-of-pearl; a cutaway coat of summer-cloth; and a small glazed cap.

But perhaps the most striking attribute of the typical whaleman was his youthfulness. Old men were virtually unknown

[7] Beane, J. F., "From Forecastle to Cabin," pp. 19–24.
[8] Nordhoff, Charles, "Life on the Ocean," p. 46.

at sea; and even middle-aged men were rare except amongst the masters and mates. Voyage after voyage whaling vessels sailed with crews whose average ages were little in excess of twenty years. It was exceptional to find a man of thirty in a forecastle; while countless hands were still in their 'teens. In one instance only the captain and one seaman, out of a crew of twenty-nine, were over twenty-six years of age. The ship *Acushnet*, when it sailed from Fairhaven on December 30, 1840, with Herman Melville as one of the foremast hands, carried a relatively mature crew; for six men had reached the age of thirty, and only four were under twenty.

Nor is it surprising that the whalers were manned so largely by young men. The life of the whalemen was one which involved adventure, hardship, daring, danger — experiences which have always been the birthright of youth. And in whaling, in particular, these experiences were so dearly bought that only the recklessness and ignorance of youth would refuse to count the cost. Two or three years before the Civil War the American consul at Paita stated in an official report that many hundreds of young men left the United States annually on whaling vessels and never returned because of the vicissitudes and degradation which they encountered in the Pacific. Some served as tributes to Davy Jones; some left their bones to rot on a tropical island; others bartered their manhood and will-power for the licentious and dissipated seductions of the South Seas; and many fell victims to the brutalities, hardships, and dangers of the industry itself.

A logical corollary of such extreme youth was found in lack of experience. The percentage of green hands carried by many whalers was truly astounding. In one vessel which left New Bedford in 1832 only four of the fourteen men in the forecastle had ever been to sea before. On another voyage, which began three decades later, fifteen of the eighteen men before the mast were rated as green hands. Joseph Grinnell, Member of Congress from New Bedford, stated on the floor of the House that on January 1, 1844, there were 17,594 men in the whaling crews of the United States, and that one-half of this number ranked as green hands.[9] This figure of fifty per

[9] This speech, made on May 1, 1844, was reprinted in the form of a pamphlet

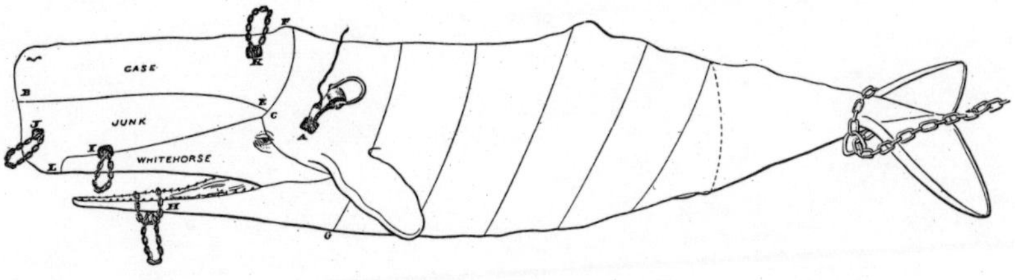

OUTLINE OF A SPERM WHALE, SHOWING THE MANNER OF CUTTING-IN.

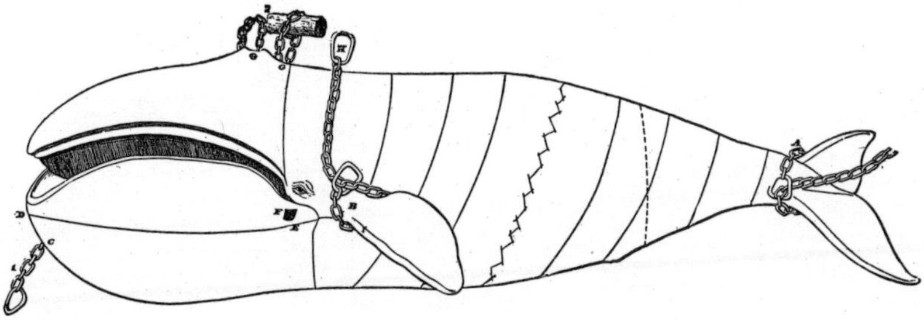

DIAGRAM SHOWING THE MANNER OF CUTTING-IN THE BOWHEAD AND RIGHT WHALE.

cent for the inexperienced men was also reached by Lieutenant Wilkes in an independent estimate made several years before.[10]

But inexperience was by no means the worst characteristic exhibited in the forecastles. Irresponsibility, vice, depravity, and criminality were also heavily represented. All too often the foremast hands came from the dregs of shore life. They were "made up to a great degree, and, of course, with some honorable exceptions, of the very refuse of humanity, gathered from every quarter, escaped from poor-houses and prisons, or gleaned from the receptacle of vagrancy and lazar-house corruption."[11] Virtually all other contemporary accounts of whaling life agree in placing similar emphasis upon the degenerate and deteriorated character of large sections of the crews.

This heavy dilution of the labor supply with inexperienced and degenerate elements brought about a notable decrease in both efficiency and morality. Closer supervision and more relentless driving were practised in an effort to secure the adequate performance of necessary tasks. In consequence the gulf between officers and men widened materially. Brutality and tyrannical abuse on the one hand were met by sullenness and growling discontent on the other. The average crew came to be composed of three main portions, viz., the green hands, the able and ordinary seamen, and the nondescript recruits picked up at the Azores and in the Pacific. And inefficiency and friction resulted both from the inexperience of the new men and from the resentment radiating from the more sophisticated hands to all the occupants of the forecastle.

Due to the necessity of stricter supervision of these less capable crews, the character of the officers escaped the lowering process to a marked degree. In their personal qualities, unfortunately, many masters and mates were harsh, brutal, profane, ignorant, and tyrannical. A few were intelligent gentlemen in the best sense of the term; while others were

entitled, "Speech on the Tariff, with Statistical Tables of the Whale Fishery." See pp. 9 and 13.

10 Wilkes, C., "Narrative of the U. S. Exploring Expedition," V, p. 485.

11 Cheever, H. T., "The Whale and His Captors," p. 304. This work was published in 1850.

men of good character, sound common sense, and homely wisdom, though much given to undue sternness and astounding parsimony on shipboard. But their fitness for the technical and material phases of their positions was beyond question. The average American whaling captain was a superb seaman, an unexcelled boatman, a consummate navigator, and a skilled and daring whaleman. And the ranks of the mates, in spite of the fact that they harbored many men of undeniably mediocre attainments, included also a large number of potential captains. The resulting enviable reputation of American officers was so well justified by a relatively small percentage of losses that the insurance premiums on American whalers were markedly lower than those paid by British whaling merchants.

A post as a whaling officer was anything but a sinecure. The successful master of a large whaler was necessarily a man of versatile attainments. During the course of an average voyage he was almost certain to act as a physician, surgeon, lawyer, diplomat, financial agent, entrepreneur, task-master, judge, peace-maker, sailor, whaleman, and navigator. The last function of navigator made perhaps the most severe and unrelenting demands upon his skill and judgment. With a vessel and outfit valued at thirty to sixty thousand dollars and a crew of twenty-five to thirty-five men entrusted to his care for a period of three to four years, he was expected to take vessel and crew into any or every known or unknown sea where whales might be found, encounter innumerable vicissitudes on sea and land, and in the end come sailing safely into the home port with a "full ship." And few indeed were the circumstances which were regarded as valid excuses for failing to do so!

Such a relatively high percentage of capable, though brutal and tyrannical, officers was due to the obvious advantage of trusting heavy responsibilities only to well-qualified persons. But in addition the deterioration of the foremast hands rendered it imperative. Inferior crews had to be counter-balanced by superior officers. Men who were inexperienced, reckless, lazy, depraved, or diseased had to be driven to work by means of stricter discipline, closer supervision, and more careful training. The ability and initiative of the masters and mates, therefore, was maintained because of the very fact that these same qualities were allowed to decline in the case of the foremast hands.

The deterioration of the crews was due partly to the operation of forces which were outside the industry, and partly to the acquiescence or only half-hearted resistance of the whaling merchants. The rapid expansion of the opportunities for exploiting natural resources, which came with the middle of the century, meant that both native American and immigrant found greater chances for advancement and independence on shore than at sea. A farm in Ohio or Illinois was more attractive to the normally ambitious individual than a whaleboat in the Pacific. And particularly so because whaling presented not only relative disadvantages, but also positive evils of the grossest and most unmistakable kind. Such were the wretched working and living conditions on board the average whaler; the brutalizing and cruel abuse often called discipline; the uncertain but in most cases pitiably low earnings; the countless privations, hardships, and dangers of the industry; the long absences from home, which made a normal family life impossible; and the deplorable system of shipping crews through agents and outfitters, which involved much misrepresentation, deceit, and fraud.

It was an unfortunate coincidence that the expansion of the industry demanded an enlarged labor supply during the very period when the best young blood of New England was simultaneously attracted by the potential wealth of the West and repelled by the barbaric, underpaid hardships of whaling. Only one result was possible. So many of the most promising young men heeded Horace Greeley's advice to "Go West" that the whaling forecastles had to be filled with the inferior human material already described.

Instead of attempting to compete for an adequate supply of labor by making the conditions of whaling life as attractive as possible, however, the whaling merchants sought to adjust the industry to the changing conditions by condoning the new elements. This may have been due to a recognition of the fact that whaling could not offer inducements to compare with the free land of the West; but more probably it was due to inertia, custom, and a canny realization that a more advantageous wage bargain could be driven with men who were alien, inexperienced, reckless, or dissipated.

In fact, the attitude assumed toward the green hand, at least, was one of definite encouragement. The shipping-agent preferred to deal with men ignorant of the actual conditions of the industry because they were more easily imposed upon, and also because they were more dependable in observing their contracts. And the owners were content to ship such hands because they could be hired for "longer lays" (lower rates of pay), and because capable and brutally aggressive mates could train them during the long months at sea. This training, though crude and unsystematic, was so vigorous and strenuous that when the whaling grounds were reached, after several months on passage, the erstwhile green hands were able to play an effective part in any of the required operations.

The thorough disillusionment of these neophytes, the chronic discontent of the sophisticated and unprincipled seamen, and the chafing resentment of all hands against brutality, poor food, and miserable living conditions, — all these factors combined to create a dominant unrest which resulted in an embarrassing number of desertions. This practice of desertion, in fact, added to changes caused by death and discharge, brought about a labor turnover so heavy that it constituted one of the chief problems of the industry.

A shifting personnel came to be a commonplace on board American whalers. Only in the rarest instances did a vessel return from a voyage with substantially the same crew with which she sailed. It was not unusual for a whaleship which never carried more than thirty-five men at any one time to come into port with several times that number of names in her account-books. At times the manning process for a single cruise required the hiring and discharge of the equivalent of three, four, or even five complete crews. During her fifth voyage the ship *James Maury* presented the unedifying spectacle of having two mates and three boatsteerers join a large number of foremast hands in deserting; while the fourth and fifth voyages of the same vessel saw the discharge of two first mates and several other officers.[12]

The accounts of numerous whalers reflect a kaleidoscopic succession of hands which was little short of chaotic. During the course of one voyage extending from 1866 to 1869 the ship *Adeline*, carrying a normal crew of thirty-one, shipped one hundred hands. Of this number, twenty-two deserted, forty-nine were discharged (including twenty "seasoners" who were hired only for a season), and one died. For another voyage, made between 1854 and 1859, the accounts of the *Isaac Howland* showed that nineteen hands deserted, eighteen were discharged, eight were put ashore because of an attempted mutiny, one refused duty, two "disappeared," thirty had no indicated settlements whatever, and only ten received cash balances.

But the master of the *Montreal* capped the climax by shipping no less than one hundred and fifty-eight hands for a single cruise extending from 1857 to 1862. There were seventy-nine discharges (including that of the first mate) and thirty desertions. In addition, two men were taken out of the ship by an American consul, one was put on board another vessel, five were "left" at foreign ports, seven died, and fifteen were not even deemed worthy of having their accounts closed. As a result of such hectic shifting, only five members of the original crew of thirty-nine were still on board when the *Montreal* returned to her home port.

Because such instances are admittedly exceptional, the average figures for larger numbers of voyages are even more significant. Thus thirty-six voyages made by thirteen vessels during the period 1839–1879 were responsible for three hundred and twenty-six desertions, three hundred and forty-four discharges, and thirty-seven deaths. These figures represent an approximate average of nine desertions, nine and one-half discharges, and one death per voyage — or an average of almost twenty men out of original crews which seldom con-

[12] Figures and statements applying to all vessels mentioned by name throughout the following pages have been taken directly from the original manuscript accounts of the whalers in question. Unless otherwise stated, these accounts have been deposited in the New Bedford Public Library.

tained more than thirty-three hands. Another group of fifteen voyages, made by eight whalers between the years 1843 and 1862, accounted for one hundred and forty-three desertions, one hundred and sixty-six discharges, and sixteen deaths. Stated in terms of percentages of the original crews of four hundred and eighty-nine men, practically three-tenths of the hands deserted, one-third were discharged, and three per cent died. In other words, the average whaler lost almost two-thirds of her original crew during the course of each voyage.[13]

These changes were disastrous alike to morale, efficiency, and the owners' pocketbooks. Among men who spent such long periods in inescapably close contact, the element of group psychology was all-important. Nowhere was the question of morale, of esprit de corps, more vitally connected with the smooth conduct of affairs than in the little worlds bounded by the decks of whaleships at sea: and nowhere was it more difficult to control. Even at best the nerves and tempers of men could not fail to be tried severely by unending confinement in inadequate living quarters, poor food, hard and dangerous work, brutal driving, vulgar and obscene companionship, and unutterable monotony. And when irregular and disrupting changes in personnel were added to the other factors affecting these small, isolated groups, it became exceedingly difficult to create and to maintain a normal attitude of mind. Consequently the morale of the average whaling crew was notoriously low. Suspicion and intolerance towards each other and bitter resentment and hatred of the officers were mingled with a keen sense of indignation against the impositions of far-away owners and landsharks.

In a few extreme cases these smouldering resentments broke out into murder and mutiny. But normally the feeling of discontent resulted in waste, sabotage, inefficiency, and a low standard of achievement. A spirit of enthusiastic coöperation and of trained and disciplined team-work, so essential to whal-

[13] The detailed figures from which these results were obtained, together with the list of vessels, voyages, and dates, are given in Appendix B. The voyages represent a sampling of manuscript account-books which chanced to contain the necessary figures. All of the manuscripts were found in the New Bedford Public Library.

ing activities, could hardly be maintained in groups which were frequently disorganized by subtractions and additions. Hands who had become proficient at certain strategic tasks were lost, and new men who had neither experience nor aptitude had to be fitted into their places. This was particularly disrupting in the case of the small and closely-linked whaleboat crews, where the loss of a single cool and expert oarsman often cut down materially the captures made by a certain boat. Foremast hands as a class were as ineffective as the savage driving of the mates would allow them to be; and in spite of constant supervision by men whose type has become a byword for parsimony, they contrived to waste not only countless opportunities for telling effort, but a certain amount of equipment and provisions as well. In general, they had neither a sense of loyalty toward officers or owners nor a feeling of responsibility for the success of any phase of a voyage.

But the heavy labor turnover brought another form of loss far more evident than the potential gains missed through efficiency and low morale. As a result of the system whereby men were provided in advance with outfits which ranged from seventy to one hundred dollars in price, each member of a crew was indebted to the owners for that amount when he sailed. If he remained on board until his undeclared but readibly calculable share in the total catch was equal to the price of his outfit plus any subsequent charges, the owners would be able to recover the amount of their advances. But if he left the vessel before such time, the difference between the sums advanced and the amount of his earnings constituted a net loss to the owners. Often such losses ran into hundreds of dollars for a single voyage. They were partially offset, it is true, by the net gains realized on the accounts of those deserters whose earnings exceeded their debits. But these gains were commonly smaller than the losses; for a man who had remained on board long enough to amass relatively large earnings was less likely to desert than those who had smaller stakes at issue.

No one realized the significance of desertion in rolling up losses more keenly than the men who watched the ledgers at home. Letters of instruction contained frequent admonitions to watch the men so closely that there would be no opportunity to leave the vessel during the early part of a voyage, when the catch was still small. Consequently many captains deliberately avoided those ports which offered the greatest temptations; or, if compelled to put in, they refused to grant shore leave and weighed anchor at the earliest possible moment. In certain South Sea harbors the natives were sometimes hired to watch seamen while ashore; and it was a common practice to offer rewards for the apprehension and return of deserters. Whalers were anchored well beyond easy swimming distance from the shore, and a careful watch was kept by the mate in charge of the deck. The question was even injected into the original shipping of a crew in the home port; for one of the reasons for preferring green hands lay in the fact that they were less resourceful in matters relating to desertion.

But when a vessel was well filled with oil and bone, so that the lays of those men who had been longest on board were appreciably greater than their obligations to the owners, the captain's vigilance often relaxed significantly. For desertion under such circumstances meant that the confiscation of the entire earnings would leave a tempting sum as net gain to owners or masters. In fact, charges were made that certain of the more grasping and unscrupulous officers not only looked with equanimity upon such cases of desertion, but actually took steps to bring them about. This might be done either by means of "working a man up" (subjecting him to such cruel and abusive treatment and such long periods of gruelling work that he was driven to leave the vessel in sheer desperation), or by employing some real or alleged offense as a pretext for sailing without him.

The part played by the quarter-deck in stimulating such late-voyage desertion was mentioned by so careful and informed an observer as Lieutenant Wilkes. Writing in the forties, he said: "Many Americans are found on the different islands, who have been turned ashore from whale-ships, or left because they have broken their liberty a single time, near the end of a voyage. Such treatment leaves too much ground to believe that they are purposely left, in order to increase the profits of the ship-master or owners."[14] This belief can only be strengthened, too, by occasional entries in certain crew-account books which are still extant; for here and there, occurring at rare intervals, are individual accounts which, instead of being balanced and closed, are terminated with the single word "left." The records of the ship *Montreal,* of New Bedford, afford a case in point. During the course of her third whaling voyage, 1857–1862, this vessel "left" no less than five members of her crew; and the evidence seems to imply clearly that in these instances the ship deserted the men, rather than vice versa!

This practice of desertion, if it involved losses for the owners and disciplinary problems for the officers, was indeed a desperate measure for the seamen. Its mere attempt constituted a violation of the legally enforceable contract which was embodied in the agreement to ship for an entire voyage, and rendered a violator liable to pursuit and capture by the police forces of the United States and of most friendly nations. The deserter was thus placed upon the same footing as a common outlaw. If recaptured, he was automatically returned to his vessel, where the mates took care to make his lot even worse than before. If, on the other hand, he was successful in making his escape, often after taking desperate chances and undergoing severe hardships, other dangers and difficulties surrounded him as if by magic.

As long as his vessel was in the harbor he was compelled to remain in hiding or to take to the open country. The former carried with it a constant fear of betrayal: the latter presented possibilities of fever, disease, lack of food and water, and dubious association with hostile savages. And even if he contrived to elude capture until after the departure of his shipmates, there was still the pressing problem of a future means of livelihood. With both his earnings and personal effects confiscated by the owners of the vessel which he had just left, this was not a matter to be taken lightly. In the out-of-the-way South Sea ports most often visited by whalemen there was only the choice between shipping on another whaler or

[14] Wilkes, Charles, "Narrative of the United States Exploring Expedition," V, p. 498.

sinking into the lethargic and degenerate life of a "beach-comber," with a native wife and native drink as the only solaces.

In spite of its desperate character, however, desertion was often the readiest means of escape. For other probabilities included only death and discharge. Death came, on the average, to one member of a crew during the course of each long whaling voyage; and it came in a variety of forms. As the result of illness or disease, due to exposure, hardship, and privation; after weeks of thirst and starvation in an open boat; in the midst of battle with wind, wave, or an "ugly" whale; or in the wake of a foul line, a stove boat, or a razor-edged cutting-spade, — in some guise death visited each crew, on the average, at least once.

Discharges, although wholly within human control, were by no means easy to secure. The difficulties of hiring new hands, together with the statutory requirement of advance wages with each discharge, were enough to insure that. A man suffering from chronic illness or permanent incapacity might expect to be honorably released. At rare intervals a consul might be induced to listen favorably to a recital of grievances and to order the discharge of certain seamen. And at times the financial or disciplinary strategy of a cruise might cause a master to dismiss certain hands, particularly in the case of the "seasoners" who were shipped for relatively short-run periods. But aside from these factors, the whaling hands found distressingly few reasons which appealed to the captains as constituting adequate grounds for discharge.

As month after month of privation, abuse, and monotony wore away, the most spirited and the most sensitive members of a crew gradually came to feel that nothing could be worse than life on a whaler. Youth, vitality, and the instinct of self-preservation prevented them from actively courting death as a method of escape; and consequently they wooed discharge or desertion with a desperate ardor. And since, upon the whole, the inflexible will of the captain was even more difficult to attack than the obstacles of their environment, desertion came more and more into favor. It is safe to say that the heyday of the whaling industry witnessed scarcely a single long voyage during which desertion was not regarded with longing by a goodly number of the foremast hands. So desperate and insistent was this desire that it was translated into action by three out of every ten men, on the average, who were shipped in a vessel's home port.

The whalemen who contemplated desertion with such passionate intensity were actuated both by a pull and by a push. The pull came from the real and fancied delights of life ashore, especially when cruising in the South Pacific. To men imprisoned for months on the drab and narrow decks of a slow-moving whaleship, the languor and luxuriance of the South Sea Islands made a powerful appeal. During the dog-watches the older hands recounted yarn after yarn which glorified the pleasures of those islands by a grotesque mingling of fact and fancy. With imaginations thus stimulated, many of the long night-watches were spent in discussing both licit and illicit satisfactions and the means of attaining them. And gradually yarns, discussions, and imagination merged into a burning desire to get ashore, no matter what the conditions or consequences.

But an even stronger motive lay in the push. Life on board the average whaler was hemmed in by physical limitations which were often revolting and at times well-nigh intolerable. The forecastle was small, ill-ventilated, dirty, and often filled with vermin. The food was cheap, coarse, miserably prepared, maddeningly monotonous, and at times spoiled and indigestible. On passage there were interminable watches filled with hard, dirty, and dangerous work; and on the whaling grounds stretches of furious and exhausting exertion alternated with periods of unutterable tedium and boredom. Rigid discipline was intensified into brutal driving, cruel abuse, and profane threats. The original outfit, provided at extortionate prices, included articles of such poor quality that they had to be replaced long before the end of the cruise; and the consequent resort to the slop-chest involved the purchase of more high-priced but low-quality goods.

These conditions, together with the resultant labor turnover and its waste and inefficiency, were due in part to what may be called the labor policy of the industry. Whaling, it is true, could never be made an easy or pleasurable occupation. Storms and wrecks, danger and monotony, hard work and cramped quarters could not be abolished by the mere fiat of man. But coincident with these fundamental factors were other evils more or less amenable to human control, and which, if honestly attacked and even slightly ameliorated, would have removed many of the most vital grievances of the crews.

One of the most flagrant series of abuses was connected with the hiring and outfitting of crews at the outset of a voyage. This important function was gradually surrendered by the whaling merchants to the outfitters, who were content to undertake the work of assembling and equipping the hands in exchange for the privilege of retaining the profits accruing from the sale of the outfits. In consequence the owners were relieved of a large amount of onerous and tedious detail; but they also secured men who went to sea with grievances and resentments which smouldered throughout the course of each cruise. For the outfitters, being concerned with profits rather than with the character and attitude of the crews, exploited the necessities and inexperience of the seamen in every possible way; and a ramifying system of commercialized vice not only fed upon dissipated appetites, but deliberately created opportunities for the organized robbery of whalemen's purses. Even the owners were often imposed upon by having worthless and incapable hands foisted upon them.

Financial impositions, thus begun by the outfitters, were continued by owners and agents. Earnings, reckoned on the basis of a fraction of the net proceeds of an entire voyage, were so low that it was not uncommon for a whaleman to find himself actually in debt to the owners at the conclusion of a long cruise. Nor were the relatively few bonanza voyages sufficient to carry average wages above a point at once tragic and ludicrous. And with low earnings went high charges. All cash advances were made at generous rates of interest, although no interest was allowed upon accumulated but unpaid earnings; and most other advances were accompanied by interest charges. Every plausible pretext for exacting payments from the foremast hands, including even the provision of a medicine-chest and the loading and unloading of the cargo and equipment, was seized upon avidly.

The same parsimonious policy was pursued in regulating other phases of shipboard life. The food was so coarse, standardized, and limited in quantity that during the middle decades of the century it cost the owners only fifteen to thirty cents per day to feed each member of the crew.[15] So little of the precious cargo space was given over to the forecastle, and so little effort was made to keep it decently clean, that the living quarters were not only cramped, but nauseating. A ship's library, which might have been provided easily and cheaply by simply dumping old books and magazines on board, was virtually unknown. A trained physician and surgeon, found as regularly as the first mate on British sperm whalers, was not even seriously considered for American vessels. In general, it was the policy of the whaling merchants to provide, in the cheapest possible manner, only those things which were unquestionably essential to the pursuit of whales and the preservation of life.

The smouldering resentment inevitably accompanying such conditions was fanned into flame by the day-to-day treatment received at the hands (and feet) of the officers. Blows, kicks, and curses were so common that on many vessels they came to be expected as a normal accompaniment of every order. Threats, vituperation, and the assignment of deadening tasks followed the slightest relaxation of discipline or the least indisposition on the part of a mate. Flogging, confinement, and more ingenious forms of formal punishment were all too frequent. The two-fisted, bludgeon-footed type of mate, who boasted long and loudly that he feared neither man nor devil, was regarded as the ideal disciplinarian.

But in this, as in the other phases of whaling labor policy, the methods pursued would seem to have overshot the mark. Discipline, and that of a stern and rigid sort, was undeniably essential in handling crews containing such a large percentage of unprincipled and shiftless hands. Gentle persuasion, a meek demeanor, and soft answers had no place amongst men

[15] The question of food costs, including the estimates from which these figures were derived, is considered in the chapter on Earnings and the Lay.

who responded only to sharp commands backed by unquestioned authority. But in practice firm and effective discipline, tempered by uniform justice and fair dealing, was too often replaced by blundering brutality and indiscriminate abuse. In consequence the means defeated the end; for instead of giving willing and eager effort, the crews responded to inescapable commands with a resentment which at times amounted to hatred.

And so it was with other matters as well. A reasonably careful supervision over the activities of outfitters and landsharks would have removed the most flagrant of the impositions which sent whaling hands to sea with burning indignation. A less grasping financial policy—including slop-chest prices based upon cost or nominal profits, lower interest rates, the abolition of minor charges such as that for a medicine-chest, and a slightly better rate of wages—would have removed serious causes of low crew morale without adding heavily to the owners' burdens. And the provision of somewhat better and more varied food, a slightly larger and much cleaner forecastle, a reasonable amount of reading matter, and adequate medical care would have prevented a vast amount of discontent and resentment. In all these respects the attitude of the whaling merchants would seem to have been penny-wise and pound-foolish.

But was it so in reality? Undoubtedly such changes, slight as they were, would have been admirable from the standpoint of humanitarianism; but their probable effect upon profits is not so clear. In fact, it is quite possible that the shrewd tradesmen of the whaling ports, whose emotions seldom interfered with their acumen, and who appreciated the value of a dollar as much as any class of persons in America, knew full well what they were about.

For American whaling, throughout the period of its dominance and decline, had all the essential characteristics of a sweated industry. The mere composition and character of the crews were sufficient to insure that. And then, as now, the only hope of securing profits out of a sweated industry lay in reducing the labor costs to a point which was even lower than that of the workers' efficiency. But inefficient workers and low labor costs could be combined only through the maintenance of very low wage rates and the provision of miserably inadequate working conditions. Both of these things the whaling owners brought about; and then, by adding the third factor of stringent and driving supervision, they were able to wring profits out of the industry in spite of the scarecrow crews and their inefficiency.

Whether the payment of better wages, the provision of decent working and living conditions, and the maintenance of a less brutalized discipline would have attracted hands whose superior efficiency would have offset the increased expenses, is problematical. In spite of the ill-assorted jumble of unpromising hands in the forecastles, the supervision was so close and the discipline so relentless that these crews took part in thoroughly significant accomplishments. Every sea was explored, whales captured and slain with reckless daring, boats and larger vessels handled with consummate skill, and "full ships" brought home in the face of countless perils. A certain reckless abandon, which was an outgrowth of the very degradation of the worst hands, may well have been a definite asset in the dangerous crises which accompanied such achievements.

And unless the effectiveness of better hands was definitely greater than that of sweated labor driven to the last ounce of endurance, it is doubtful whether the shipping of more capable crews would have warranted the greater expenses involved in attracting them. For in spite of extreme fluctuations, the average rate of profit for the industry as a whole was modest. And the possibility of recovering heavier outlays through higher prices would have required the control of complicated economic factors quite beyond the reach of the whaling owners.

But even if better hands had been sought, it is not at all certain that the whaling industry, because of its very nature, could have offered conditions to appeal to the higher types of shore labor after the opening of the West. Thus it would seem to remain a moot question whether the execrable conditions which prevailed in whaling should have weighed heavily upon the consciences of owners and merchants, or whether they were wholly traceable to economic factors which carried both seamen and employers in their wake.

The failure of the whalemen themselves to make any effective demands for the amelioration of their lot was due to utter lack of bargaining power. This, in turn, resulted from a long series of disabilities which included economic, social, racial, national, political, and legal factors. These drawbacks, together with the intense individualism of mid-century America, prevented even the thought of collective bargaining. And in individual bargaining the average whaleman was hopelessly handicapped by ignorance, inexperience, poverty, dissipation, and that peculiar spirit of reckless abandon which so often characterized the seaman ashore.

The character, composition, and environment of the whaling crews provided an inevitable guarantee that they would suffer from many obvious economic and social handicaps. Nor could these be remedied by a show of political strength. Voting power was reduced to a minimum by long absences from home, wide geographical dispersion, lack of residence requirements, racial and national obstacles to the acquisition of citizenship, and ignorance and indifference to political parties and issues. And even if the manpower of the industry could have been organized into a whaling bloc, the number of votes mustered would have formed only an unimportant element in American politics.

As a result of such combined political and economic weakness, both the legislative consideration and the legal status given to whalemen were inadequate. The first legislative attempt to attack the evils of sea life in a comprehensive, thoroughgoing manner came only in the form of the Shipping Commissioners' Act of June 7, 1872. This inclusive act virtually codified all important previous legislation applying to seamen in the merchant services, created the office of Shipping Commissioner, and added a number of other much-needed regulations. Had it not come too late in the century to be of any real service in the rapidly-declining whaling industry, and had it not been emasculated two years later by exempting from its provisions "any case where the seamen are by custom or agreement entitled to participate in the profits or result of a cruise or voyage," its sixty-eight sections might have rendered whaling life somewhat less intolerable. As it was, the effect upon whaling was negligible.

Throughout the years before 1872 the two original statutes of April 30 and of July 20, 1790, remained the fundamental pronouncements on matters relating to American seamen. Although often enlarged and amended, they were not superseded. The body of laws then laid down included the definition and punishment of crimes at sea; the requirement of a written agreement, known as the shipping articles, between a master and each member of his crew; provisions for the apprehension and treatment of deserters; the protection of wage payments; the conditions necessary for a survey of seaworthiness; and the insistence upon the inclusion of minimum amounts of fresh water and provisions in the cargoes of all seagoing vessels.

In 1825, and again in 1835, laws were passed which sought to protect seamen from their own officers. Under certain circumstances fines and imprisonment were provided for assault, beating, wounding, the infliction of "any cruel and unusual punishment," and forcing ashore or leaving a seaman in a foreign country. In 1850 the culmination of a long period of agitation was marked by writing into the statutes a single sentence: "Flogging on board vessels of commerce is hereby abolished." The prohibition of sheath-knives, sixteen years later, protected officers as well as seamen; for the sheath-knife, an ugly if romanticized instrument responsible for countless crimes of violence, had become the traditional weapon of the foremast hand. These provisions, together with an extended series of regulations for the consular protection of sailors, constituted the most significant attempts to safeguard the personal rights of seamen for three-quarters of a century.

But even the laws which were upon the statute books, inadequate at best, were commonly weakened by loose interpretation and poor enforcement. Ambiguous phraseology, ignorance of the law by those wronged, and the prejudices, preconceptions, and indifference of judges, juries, and consuls, all combined to rob the protective legislation of its full effect. A striking illustration of self-defeating legal verbiage is afforded by a section of the act of March 3, 1835, which sought to prevent the maltreatment of seamen by masters and mates. The statute provided that: "Every master or other officer of any American vessel on the high seas, or on any other waters within the admiralty and maritime jurisdiction of the United States, who, from malice, hatred, or revenge, and without justifiable cause, beats, wounds, or imprisons any of the crew of such vessel, or withholds from them suitable food and nourishment, or inflicts upon them any cruel and unusual punishment, shall be punished by a fine of not more than one thousand dollars, or by imprisonment not more than five years, or by both."

The difficulties involved in prosecuting an officer under this clause are obvious. It was necessary to prove not only the commission of an offense, but that such action had been accompanied by "malice, hatred, or revenge"—all subjective attributes widely removed from the type of evidence required in a court of law. What, too, was "suitable food and nourishment?" What forms of punishment were both "cruel and unusual?" And when was mistreatment "without justifiable cause?" These ambiguities, coupled with the fact that the witnesses were either fellow-officers with an unconcealed bias or seamen who were easily intimidated and confused, meant that convictions were rarely secured. The widespread knowledge of this truth, together with the passage of weeks and months before a court of competent jurisdiction could be reached, resulted in a general disregard of the enactment which virtually gave the officers a free hand.

The ambiguity of many statutes was all the more unfortunate because the atmosphere of the average courtroom was decidedly uncongenial to the seaman. Judges, juries, and lawyers were alike suspicious of a sailor's knowledge and veracity. Owners and masters not only possessed far greater political and economic power than the men in their crews, but were also enabled to retain more capable legal advisers and to enjoy greater prestige in the eyes of the court. Whether from ignorance, inexperience, dissipation, or lack of assurance, the average sailor was unable to give testimony in a clear, concise manner, was easily confused by opposing counsel, and consequently failed to present his case in a manner which satisfied the technical rules of evidence.

The seaman who was a party to a lawsuit also suffered from two widely-held tenets of public opinion. One was the careless and unthinking belief that sailors as a class were social outcasts and moral pariahs who were well beyond the bounds of common decency and ordinary consideration. The other was the widespread feeling that a too-strict enforcement of the laws designed to protect seamen would result in a serious impairment of discipline, dangerous not only to lives, but also to vessels and cargoes. It was held to be the function of the law to uphold and to strengthen the officers' authority, and not to weaken it by inquiring too closely into the nature of the disciplinary measures which might be deemed advisable. Reported instances of cruelty and of brutality were either discounted as gross exaggerations or were defended as being unavoidable in dealing with such degenerate and desperate crews. Making a voyage was vaguely regarded as a sort of continuous battle of muscle and of wits between forecastle and cabin, in which the forces of law, order, and respectability were needed on the side of the officers, while the foremast hands might safely be allowed to shift for themselves.

The resulting miscarriage of justice was admirably exemplified in the case of Marion vs. Moody, which was tried in the Court of Common Pleas at Boston in June, 1854. A seaman brought an action against the master of his vessel for a flogging which was the culmination of a chain of circumstances originating in the refusal of the entire crew to obey an order. There was no doubt concerning the commission of the offense; and for once the wording of the law was clear and explicit. Only four years before, on September 28, 1850, a statute had been passed which declared sweepingly, without qualification or allowance for extenuating circumstances, that "Flogging on board vessels of commerce is hereby abolished." And yet the judge, after making the obvious and irrelevant ruling that while the act did abolish flogging, it did not prohibit the use of force in any form as a means of coercion, instructed the jury that there were two possible findings. If the action under review were regarded as a case of flogging for the sake of punishment, then the master was to be found guilty as charged; but if it were held to be a form of chastisement intended to induce a return to duty, and one which seemed an appropriate means towards such an end, as well as one which had been adopted in good faith and reasonable judgment, then the captain was to be exonerated. In other words, he pointed out to the jury that by calling flogging a form of chastisement, the application of the law could be evaded. And this in spite of the fact that tying a man to the rigging and striking him on the back with a rope, the offense charged and proved in this case, constituted the very definition of flogging! The jury, taking a hint from such a devious and questionable interpretation of the law, returned a verdict under which a fine of one cent was to be paid to the plaintiff.

Curiously enough, this legal indifference to the sufferings of seamen, together with the accompanying suspicion of their personal rights, ran parallel to a strongly paternalistic watch over the formal terms of their agreements. The seaman was commonly regarded as a so-called ward of admiralty, whose numerous legal and political disabilities entitled him to the special protection of the courts in the making and observing of his contractual relationships. Consequently the average judge sternly insisted that owners and masters must carry out their share of the bargain embodied in the shipping articles, particularly with regard to rank, length of voyage, and the payment of wages duly earned. Thus was presented the interesting spectacle of having identical courts refuse to protect foremast hands against the most harrowing maltreatment, and yet exhibit meticulous concern over the receipt of a few dollars in wages.

But if the courts were often remiss in the interpretation of protective legislation, the consuls were guilty of even greater sins of omission. This was particularly reprehensible because these officials had been appointed by law as the accredited guardians of American seamen in foreign countries. As early as April 14, 1792, they were especially empowered to receive protests or declarations concerning treatment and conditions on shipboard; and on February 28, 1803, they were ordered to collect three months' advance wages on behalf of sailors discharged under certain stipulated circumstances, as well as to employ part of the resulting fund in providing for the subsistence and return passage of destitute seamen. On July 20, 1840, it was made the duty of all consular officers to assist in reclaiming deserters. But if it appeared that desertion had been caused by cruel and abusive treatment, they were authorized to order the discharge of the offender and to demand three months' extra pay. And again on August 18, 1856, the protective and supervisory functions of the consuls were emphasized and clarified in a series of provisions which also sought to limit the influence of personal bias and of private interests. These pronouncements, together with minor ones of the same tenor which were made during the pre-Civil War days of whaling prosperity, left no doubt that the consul's office was meant to be a haven of refuge for American sailors when abroad.

There was no doubt, either, concerning the fact that it was regarded in this light by the supposed beneficiaries. Foremast hands in varying degrees of distress appealed for assistance in such numbers that the consuls in the larger ports came to look upon this phase of their duties as an onerous burden. The Consular Letters are filled with complaints upon this score. Two excerpts from the Liverpool correspondence are typical of many other passages. In a letter written on March 7, 1832, to Edward Livingston, Secretary of State, the consul reported that he was "Called upon constantly to mediate between often unreasonable Captains and unruly men — obliged to protect the interests of both against harpies of Landlords and rapacious Jews, having my office constantly overrun with destitute Seamen, whose stories I must listen to and (no easy task) distinguish between those really deserving and the hundreds of impostors that importune me from morning 'till night — for each shilling doled out compelled to take triplicate receipts." And again on October 15, 1833, the same official insisted that "Scarcely a ship comes to the port, that there does not exist some difference between the Master and the Seamen, to the stories and complaints of both the Consul is compelled to listen, to mediate and settle amicably if possible and if not to enforce the Law. Sailors are constantly getting into trouble on shore and require his interference, and petty suits are instituted against the Masters for the purpose of extortion to which he is expected to give his personal attention." [16]

In spite of the plain intent of the law and the importunities of the seamen, however, the average consul provided small reason to be admired or respected in the forecastle. If he did not look upon his unwelcome wards with antagonism or open contempt, he was at least likely to be indifferent to their interests. Not only did he come from a social class whose interests and ideas were virtually the same as those of merchant, master, and shipowner, but his own immediate livelihood was largely dependent upon the favor and custom of the same persons. From the seamen, on the other hand, he had nothing to expect and little to fear, since they had neither money nor connections. Consequently he found it difficult if not impossible to maintain the impartiality which his rôle as an arbiter between masters and men demanded.

This unfortunate situation was due largely to the scheme of consular remuneration in force throughout the greater part of the century. Instead of a regular salary, the emoluments of the office consisted of the fees which were collected for the performance of stipulated official services. Typical fee-yielding functions included such routine tasks as the certification of ships' papers, the administration of oaths, the issuance of certificates of discharge and of desertion, the shipping of new hands, and the collection of advance wages. Statements of consuls and entries in ships' accounts agree in showing that the individual fees ranged from twenty-five cents to four dollars, with two dollars as an average amount. In a small handful of world-famous ports such sums, when accumulated throughout the course of a year, amounted to a respectable total. But in other places, where the arrival of an American vessel was a rare event, the annual totals were more akin to pin-money than to a source of livelihood.

[16] These excerpts were taken from the manuscript Consular Letters, which are bound and preserved in the Library of the Department of State.

Consequently most consuls were obliged to supplement their official earnings in some other manner. This was commonly done by engaging in one or more of the regular channels of international trade or by undertaking to fulfill many of the functions of a ship-chandler. If he chose the former course his attitude inevitably became that of the merchant rather than of the seaman; while if he adopted the latter alternative the merchant-customer relationship between himself and the captains of visiting vessels caused him to refrain from displeasing his customers by unfavorable decisions. The masters, on the other hand, realized full well that the purchase of needed supplies from the consul was likely to induce an official predisposition to look with favor upon their side of any dispute.

The notorious consular tendency to advance the interests of masters and owners in preference to those of seamen was particularly marked in the case of whaling vessels. Although due in part to national and racial prejudices aroused by the polyglot nature of the crews, this fact was due primarily to the environment of the consuls who commonly dealt with whalemen. For reasons which were inherent in the nature of the industry, whalers touched customarily at the smaller, more remote ports of the world. Because the fees to be secured in such places were limited and the opportunities for trading correspondingly restricted, the resident consuls found themselves largely dependent for their custom upon the whaling captains who were their most frequent visitors. As a result the good-will of these masters was cultivated even more assiduously, and decisions adverse to the crews rendered even more frequently, than in the larger ports where a more varied merchant marine prevented such an intimate intermingling of financial and personal relationships.

The shortcomings of such underpaid consuls were plain enough to unbiased contemporary observers; and these men realized, too, that the foremast hands were not the only losers under the system. At times the interests of the industry as a whole suffered appreciably through the lack of dignity and prestige attaching to these hireling, eager-for-personal-business officials. Said one writer, after intimate observation:

"The difficulties to which the whaling fleet is exposed are often aggravated by the position of our consuls; for if engaged in trade, as they almost always are, they lose that influence and standing with the authorities which they otherwise would have, whether civilized or savage, as well as with their own countrymen.

"The whole system is wrong: those appointed to such situations should not be suffered to engage in trade, but should receive a salary adequate to their support." [17]

Instances of consular prejudice and injustice were many. One episode, unusual only in that the decision of the consul was later reversed by higher authority, occurred at the Hawaiian Islands. While cruising in the Pacific the captain of a whaler had attempted to flog one of the foremast hands as a punishment for fighting. The remaining members of the crew, insisting correctly that flogging was forbidden by law, rescued the man and refused duty. After an armed truce which continued for several days, the master finally agreed to make for Hawaii and to refer the matter to the consul for settlement. Upon arrival, that official completely ignored the request for a trial, blustered and browbeat the men, and terminated the matter by sending an entire boat's crew to prison for six months. After some six weeks, however, the affair chanced to come to the attention of the commanding officer of the sloop-of-war *Peacock,* who held that the men had been improperly confined and ordered their immediate release. The captain of the whaler, who had placed three men in irons, was held liable for false imprisonment; and subsequently the seamen were enabled to recover the earnings due them by means of an action in the federal courts.[18]

Another case which was more truly typical, in that it failed to result in a vindication of the rights of the foremast hands, was that of a whaler which put into the port of Zanzibar in 1838. The men had been flogged, beaten, abused, and starved by a drunken captain for many months; and such exorbitant prices had been charged for slop-chest articles that

[17] Wilkes, Charles, "Narrative of United States Exploring Expedition," V, p. 500.

[18] Davis, W. M., "Nimrod of the Sea," pp. 220 *et passim* and p. 323.

their debts to the ship exceeded their lays in amount. In these circumstances nine hands refused duty and complained to the consul. But after a farcical trial the plaintiffs were sent to the prison fort, and the master was left free to continue his abuses with even greater assurance than before. The combination of sickening heat, filth, and effluvia caused the prisoners to contract the deadly coast fever; and several died in consequence. But when, some time later, the same captain returned to Zanzibar, the surviving invalids were again placed on board the vessel, there to undergo a continuation of the same outrageous treatment which had caused them to make the original appeal to the consul.[19]

Such illustrations of consular discrimination, which might be multiplied indefinitely, were but imperfectly offset by the far smaller number of decisions which went in favor of the foremast hands. Now and again, however, grave charges made by seamen against their officers were sustained. The story of the whaling bark *William Gifford* affords a case in point. One day in May, 1872, this vessel, with the captain confined to his quarters and the crew in charge, put into the port of Papiati, Society Islands. During the course of the resulting consular inquiry it developed that the master had abused and ill-treated his men with such long-continued refinements of cruelty that at length he was deposed and kept under guard while the crew navigated the bark into the nearest port. Cruel and inhuman treatment of whaling hands was hardly a new story for consular ears, however; and it must have surprised the captain beyond reason when this particular consul found that the members of the crew had acted in self-defense, freed them from all legal responsibility, including a possible charge of mutiny, and sent the vessel to San Francisco under another master.[20]

[19] See Browne, J. R., "Etchings of a Whaling Cruise," pp. 501 ff. and 373. Further material regarding the relations between consuls and seamen may be found in Wilkes, C., "Narrative of the U. S. Exploring Expedition," V, pp. 490 *et passim;* Olmstead, F. A., "Incidents of a Whaling Voyage," pp. 239 and 278; Davis, W. M., "Nimrod of the Sea," pp. 220–256; and Jewell, J. G., "Among Our Sailors." The last-named volume, which is especially rich in concrete cases, devotes a large amount of space to this particular subject.

[20] See Jewell, J. G., "Among Our Sailors." p. 137.

CHAPTER VI

THE WHALEMAN ASHORE

THE evolution of the American whaleman left a record of ever-decreasing periods of time spent on shore. From the two-day absences of seventeenth century boat-whaling to the fifty-month voyages of post-Civil War sperm whalers, a "full ship" demanded steadily increasing stretches of unremitting vigilance and devotion. More and more the whaleman's life was consecrated to the sea. He worked and lived on the water: he made occasional visits, sometimes brief, sometimes extended, to the land. This amphibian existence inevitably colored and conditioned his whole outlook; and it is impossible to understand him or his manner of life without making generous allowance for the physical and psychological changes wrought by the none-too-subtle alchemy of the sea.

A whaling voyage was a venture not to be entered upon lightly, nor terminated speedily. A report made to the Secretary of the Navy in 1828 showed that during the years 1815–1824, inclusive, the average length of 178 cruises had been twenty-nine months.[1] And figures compiled by the *Whalemen's Shipping List*, the authentic organ of the industry, made it clear that from 1842 to 1857, inclusive, the average absence of 865 sperm whalers was no less than forty-two months per voyage.[2] Three and one-half years of steady cruising in order to fill a single vessel with oil! And with the passing

[1] This "Report on Islands Discovered by Whalers in the Pacific," made by Reynolds, J. N., was transmitted on September 24, 1828, and printed in 1835 as House Executive Document III, No. 105, 23rd Congress, 2nd Session.

[2] In January of each year, beginning in 1843, this publication printed invaluable statistical summaries of the industry's progress which had been collected throughout the preceding year.

years and growing scarcity of game, voyages of four to four and one-half years became increasingly common.

Perhaps no one had a greater right to speak with authority and feeling on this subject of prolonged absences than the women of the whaling ports. Month after month and even year after year the whaleman's wife was, in all outward respects, the whaleman's widow. One woman, married for eleven years, had had her husband at home on just 360 scattered days. During a seafaring career of forty-one years Captain Benjamin Worth, of Nantucket, helped to capture 20,000 barrels of oil, visited more than forty islands in the Atlantic and Pacific, and sailed some 1,191,000 miles at an average rate of four miles per hour — but found time for only seven years at home. Another Nantucket captain, George W. Gardner, saw even less of his family and friends. During thirty-seven years he traveled about 1,000,000 miles and captured 23,000 barrels of oil, but spent only four years and eight months, in scattered fragments of time, under his own roof.[3] The women of the whaling families found in the sea a rival both jealous and exacting.

Of course there were shorter voyages by other types of whalers. During the mid-century decades a two-season right whaler was gone from twenty-four to thirty-two months; and the one-season right whalers and small Atlantic brigs and schooners managed to return at the end of eight to fourteen months. But these smaller craft never enjoyed the prestige attached to the larger blubber-hunters which roamed all the seas of the world for three to four consecutive years. The contempt of the Pacific whaleman for the short-voyage Atlantic "plum-pud'ners" was well illustrated by the exaggerated yarn of the veteran master who found himself compelled to make a short cruise. He insisted upon regarding the whole voyage as a humorous interlude, and finally embarked without troubling to bid farewell even to his wife, because, as he remarked casually, he would "only be gone for a year anyway."[4]

[3] See Macy, Obed, "History of Nantucket." pp. 220ff.

[4] See Macy, W. F., and Hussey, R. B., "The Nantucket Scrap Basket," for a collection of quaint and colorful whaling yarns.

Now and then, too, a Pacific sperm whaler would shorten her voyage by returning without a full cargo. But this was a drastic measure, to be resorted to only *in extremis,* and seldom considered justified except in cases of shipwreck, mutiny, or the loss of many men and of several boats. Yankee thrift and perseverance, as well as professional pride so strong that it became virtually an unwritten law, demanded that a vessel remain out until her casks were filled. And filled they were, though it required four years or more! The whaling captain who dared to return to his home port with neither a "full ship" nor an impregnable excuse was a brave skipper indeed. For the Spartan mother who told her son to return with his shield or upon it would have found many spiritual followers in the ranks of whaling owners and officers, who expected a master to come back with a "full ship" or not to come back at all.

But whether a vessel had been gone two years or four years, her arrival was an occasion of great excitement and much rejoicing. Lookout stations, making use of a semaphore telegraph system, often enabled the return to be announced several hours before the whaler herself dropped anchor in the harbor. In Nantucket, as soon as the vessel's name was ascertained, the small boys of the town vied with each other for the privilege of being the first to inform the captain's wife, as well as for the ownership of the coveted silver dollar which was the traditional reward for bringing such long-awaited news. This speedy messenger service, together with the unfurling of the national flag, soon placed the tidings on every tongue.

In the smaller ports the entire community turned out to welcome the returned wanderers. The townsfolk were eager with news of what had transpired since the last batch of letters went out. They had marriages, births, and deaths to announce, and they had many choice bits of town gossip to retail. And they bartered their homely tidings for the colorful and exotic yarns of the whalemen. There were tales of chance meetings in strange seas and stranger ports, of daring feats of the chase, of miraculous escapes, of wild and picturesque adventures, of heavy weather, of tragic sequels to encounters with "ugly" whales, and of shipmates who would never return.

Usually, too, there were letters from other vessels which had crossed their homeward path.

These events, however old in time, were fresh and vivid to the listeners. For mid-century whaling news traveled with a slowness unimagined in an age which uses the telegraph, cable, and radio. Letters were received only at long and irregular intervals, and then, often, as the result of chance or coincidence. The general rule was to send a letter by every vessel sailing from nearby ports for the general part of the world in which friends or relatives were believed to be cruising. Letter-bags were kept in the leading stores or counting-rooms, and the contents despatched whenever opportunity offered. The men at sea, in turn, forwarded messages by every vessel which seemed likely to arrive at home before they themselves could do so. Outgoing mail was transferred from one whaler to any other which might be more likely to meet the addressees; and homeward-bound letters were passed from captain to captain until at length they came into the hands of one whose vessel was bursting with oil, who had thrown overboard his brick try-works, and who was bending all sail for Buzzard's Bay or for Long Island Sound.

The quaintest and crudest of post-offices, consisting of a large covered shell, was established on one of the Galapagos Islands. Here passing vessels left mail to be taken aboard by other craft bound for still more distant parts of the world, and took away packets of letters which they might be able to deliver, or, with luck, found messages addressed to members of their own crews. The volume of business transacted by this uninhabited mail transfer spot was great enough to make it a favorite point of call for American whalers.

But in spite of (or perhaps because of) such an informal and necessarily irregular scheme of mail delivery, a large percentage of all letters were hopelessly delayed or went astray entirely. Not until October did the first news of the previous summer's Arctic whaling reach the Atlantic seaboard, and then only because of a few early arrivals at the "Islands" (Hawaii). In November further information became available; but it was not until December or January that definite reports from all vessels could be expected. In cases of accident or of unusual delays the owners might remain in ignorance of the fate of some of their whalers until February or March. And in sperm whaling, with its longer and more irregular voyages, there were even stronger probabilities of delay and of complete loss. One faithful wife, for instance, wrote more than a hundred letters to her husband during the course of a three-years' voyage. But when they compared notes upon his return, it was found that he had actually received but six of the hundred messages. The others had been lost in shipwrecks or had traveled thousands of miles in vessels which never chanced to cross his path.

In the larger ports, however, there was a very different type of homecoming. There the self-appointed reception committees were made up of runners, outfitters, grog-dispensers, boarding-house keepers, prostitutes, and other species of landsharks; and the details of the welcome were in full accord with the character of the committee members. But whether he was received by a community of friends and relatives in a small town or by a group of landsharks in a larger one, the stay of the whaleman in his home port was likely to consist of a few bewildered weeks sandwiched in between two extended voyages.

To the whaleman the shore meant a holiday, a vacation, or a debauch. In port, and only there, was he able to find freedom from the restrictions of discipline, release from dullness and monotony, respite from hardship and danger, and deliverance from vile food and abominable living quarters. The stored-up yearnings and repressed desires of months gushed forth in unrestrained exuberance, and the whole psychological stage was set for an emotional explosion which was inevitable unless the person in question was a confirmed stoic. Coming to land presented the only opportunity for satisfying desires, gratifying appetites, and indulging in a mad round of activities which could be lived over in retrospect during many weary months at sea. Discretion and caution were thrown to the winds, and utter abandon became the order of the day. The very recklessness of much shore behavior, ranging from strenuous and extravagant pleasures to hectic and degenerate dissipation, showed that the participants were in an abnormally uncritical and irresponsible frame of mind.

Such an attitude, based on certain elementary principles of psychology, went far towards explaining many phases of the average whaleman's situation, conduct, and treatment while on land; and it was particularly illuminating in clarifying the ramifications of the system through which the entire shore life of the foremast hand was controlled. For in a different state of mind it is unlikely that this individual would have allowed himself to become completely enmeshed in the toils of that system, in spite of his penniless economic dependence and consequent lack of bargaining power. Nor can it be objected that the inclusion of the green hand in the same toils invalidates such a conclusion; for his ignorance and credulity rendered him a mere pawn in the landsharks' game as fully as the abandoned irresponsibility of his more experienced comrade the foremast hand.

The organization through which men were secured for the whaling crews and through which they were catered to whenever they returned to port consisted of three major elements, or groups of persons: viz., the shipping-agents, the boarding-house keepers, and the combined outfitters and infitters. These groups maintained connections, in turn, with an array of allies, satellites, and hirelings which included the keepers of grog-shops and brothels, pimps, prostitutes, runners, and nondescript hangers-on. Frequently one person or firm combined two or more of these activities, and performed the functions, for instance, of both an outfitter and a shipping-agent. Principals and subordinates alike were known collectively to the seamen by the unlovely but suggestive term, "landsharks." And so descriptive was this epithet that the record of the average whaleman's life in his home port can be written largely in terms of his exploitation by the various functionaries so denominated. The first to be encountered was usually the shipping-agent.

Throughout the second and third quarters of the nineteenth century a steady, organized effort was made to attract young and inexperienced men from the farms, workshops, and offices into the whaling forecastles. The printed advertising involved in this effort consisted largely of cheap, untruthful posters which set forth in glowing terms the joys, comforts, and benefits to be derived from a whaling voyage. Many notices not unlike the following were tacked up both in the larger cities and in the country villages.[5]

LANDSMEN WANTED

One thousand stout young men, Americans, wanted for the fleet of whaleships, now fitting out for the North and South Pacific Fisheries. Extra chances given to Coopers, Carpenters, and Blacksmiths. None but industrious young men with good recommendations, taken. Such will have superior chances for advancement. Outfits, to the amount of Seventy-Five Dollars furnished to each individual before proceeding to sea. Persons desirous to avail themselves of the present splendid opportunity of seeing the world and at the same time acquiring a profitable business, will do well to make early application to the undersigned.

"The undersigned" was in most instances a shipping-agent in some Atlantic or Great Lakes port who conducted his affairs in a poorly furnished room divided into unequal portions by a low railing. The larger space served as a reception room for prospective hands, while the smaller area, equipped with a desk and stool, was honored in being termed an office. Here the candidates for whaling berths were collected, preparatory to sending them on to a central shipping-master in New Bedford, Sag Harbor, or New London. It was a funnel-like system in which men from all parts of the country flowed into the forecastles, with the whaling ports at the receptive end.

These shipping-agents were marvels of suave deceit and shameless misrepresentation, if contemporary accounts of their methods may be credited. False promises and fluent mendacity were resorted to whenever it seemed advisable; and blandishments and exaggerated inducements formed part of their regular stock in trade. The benefits and delights of whaling, when not created out of whole cloth, were pictured with every embellishment; while the darker aspects of the fishery were scrupulously ignored.

Obviously, these campaigns of misrepresentation could be

[5] See Nordhoff, Charles, "Life on the Ocean," p. 11.

successful only with persons who were both credulous and inexperienced. No man who had ever been to sea could have been deceived for a moment by such manifest falsehoods. Consequently the shipping-agents never spoke to an old tar about a whaling voyage. Seamen who had been in the navy or in the merchant service sometimes attempted to bait an agent by pretending to be simple and gullible yokels. But the experienced eye of the shipper invariably detected their ruse and passed on in search of less sophisticated forecastle-fodder.

The ideal whaling candidate, from the standpoint of the procurer of labor, was a green hand who knew nothing of the sea and its ways. Such hands were favored not only because they were more easily impressed and persuaded, but also because they involved less financial risk. The agents received a fee for every man sent to the whaling ports; but if a given hand did not actually go on a voyage, part or all of the fee was forfeited. Consequently all prospective whalemen were carefully watched until they were finally established in the forecastle of an outward-bound vessel. And green hands, because of inexperience and timidity, were less likely to take French leave than the more sophisticated men.

The fees received by the shippers varied with general business conditions and with the needs of the industry. When men were plentiful, as during a period of unemployment, they went down; and in times of labor scarcity they rose in an effort to induce the agents to adopt stronger methods of persuasion. In the panic year 1837 the payment for men sent from New York to Sag Harbor was only three dollars; whereas in 1880 the price of hands delivered in New Bedford was sixteen dollars per head. The representative figure changed from about ten dollars per head for the two decades preceding the Civil War to a point between ten and twenty dollars per head for the war and post-war years.

If an aspirant proved to be particularly unsophisticated and really anxious to engage in whaling, he might be provided with a ticket and allowed to make the journey to the port alone. But such instances were exceptional. Whenever convenient, new men were sent on by boat, so that they could be turned over to the regular officers of the vessel for surveillance. Or they were marshalled into groups and placed in charge of one of the agent's henchmen, who was responsible for their arrival and who guarded them jealously. Jacob A. Hazen, sent from Philadelphia to New York in company with some twenty "drunken, desperate-looking characters," described one such journey, made in 1837, in "Five Years Before the Mast." The human units of this labor consignment were closely watched all the way, and upon their arrival were marched through the streets of New York to the shipping office in column of two's. The agent was determined to prevent the loss of a single fee.

Upon reaching the journey's end, the whaling neophytes were turned over to the resident shipping-master or outfitter. This functionary sent them at once to a sailors' boarding-house, where they might secure board and lodging while awaiting their sailing dates. And he was also responsible for finding outward-bound vessels which were in need of men, as well as for arranging for his protégés to be on board when the vessels weighed anchor.

This last task was by no means an easy one, and often required the exercise of ingenuity, tact, cunning, or force. Men who had been cajoled into undertaking a voyage were likely to exhibit, when sober, a strong desire to slip away. Enthusiastic green hands who had been over-persuaded by the agents' alluring promises were harshly disillusioned by the atmosphere of the boarding-houses. Unsophisticated youths were pursued and rescued by irate or tearful parents. And the general air of psychological deflation about the activities of the waterfront brought such strong last-minute misgivings that many others sought to escape the final entrance into a forecastle.

Still other prospects, though they themselves failed to initiate attempts to escape, were "stolen" or enticed away by rival shipping firms. In Sag Harbor this practice became so common and so inconvenient that the shippers were compelled to place restrictions upon their own activities. In 1837, for instance, they were operating under an agreement which specified that any signer who shipped hands belonging to another agent should be compelled to pay a fine of $100 for each of-

View of New Bedford from Fair Haven in 1853

From an engraving by W. Wellstood, after a drawing by J. W. Hill

fense. Such agreements, however, were neither universal nor easily enforced. Ordinarily it was necessary for shipping-masters to keep their recruits in hand by means of cajolery, exaggerated promises, liquor, prostitutes, intimidation, or forcible detention. The many men who required little watching were offset by some who were so troublesome that even a short jail sentence was welcomed by their guardians.

But in spite of the watchfulness of the shipping-masters there were many hands who began but failed to complete the journey which led into a whaling forecastle. These failures rolled up financial losses which were divided between the shippers and their agents in such a way that the latter bore the heavier share of the burden. Two notations, jotted down in a memorandum-book of a shipping firm, provided typical illustrations of this unequal division of risk. On the first page is the information that "J. E. Williams Letter on File of May 11th, 1863, Says will furnish men for us for Passage and $12.00 if they go and $4.00 if they don't go, and nothing if they are not what we can use." And later on is this entry, under date of June 15, 1865: "After this date we agree to pay for all men from N. T. $15.75 if they go to sea and their ½ Board if they don't and nothing more." [6]

This same memorandum-book affords an insight, too, into the relations existing between the shippers and their prospective hands. It contains the condensed record of some 2,000 men who were handled by one New Bedford firm between 1854 and 1872,—a number sufficient to offer an excellent cross-section view of this phase of the industry. A detailed analysis of this long list shows that 558 hands, or about twenty-eight per cent of the total, were not placed. Of these, 248 deserted, or, in whaling parlance, "run"; 153 were turned off because no vessels could be found for them; and 157 were not shipped for other reasons. These "other reasons," in the

[6] This manuscript booklet, worn with use, was found in the New Bedford Public Library, catalogued under the title, "Mss. Memorandum Book of a New Bedford Firm, Giving List of Men Shipped on Whaling Voyages 1854–1872." Unfortunately the manuscript contains no trace of the name of its owner. But from internal evidence, such as the names of vessels, boarding-houses, and other agents, it is certain that the firm which used it operated from New Bedford and that it placed its men on a score of well-known whalers during the years indicated.

ENTRIES TAKEN FROM A MEMORANDUM-BOOK

Date	Name	Station	Who From
6 Mo. 20, 1857	James McCloud	......	?
9 Mo. 16, 1857	George Williams	Cook	....
9 Mo. 21, 1857	John Berrlin	Green Hand	....
May 10, 1862	Chas. H. Wilson	"	D. L. Pearl
May 23, 1862	Henry Smith	"	J. E. Williams
May 23, 1862	Thomas Moore	"	" " "
May 8, 1862	Charles Connor	"	" " "
7 Mo. 20, 1862	Wm. H. Conklin	"	D. L. Pearl
8 Mo. 10, 1862	Harlan P. Boyd	"	" " "
8 Mo. 8, 1862	Chas. H. Philips	Boat Steerer	James
8 Mo. 10, 1862	Charles Wheeler	Green Hand	D. L. Pearl
July 2, 1863	Frank Loroz	Green Hand	H. James
April 18, 1865	John W. Barton	"	" "
April 18, 1865	Wm. H. Whitford	"	Hy. James
May 3, 1865	Julius Kraft	"	" "
May 17, 1865	Thomas Crowell	Cook	Harris
Oct. 2, 1866	Thos. Hercules	Seaman	S. S. Butters
6 Mo. 17, 1869	Wm. Lambert	Green Hand	Morrison
4 Mo. 7, 1870	Edward Butler, Jr.	"	J. Morrison
11 Mo. 2, 1870	Edw. Seepers	"	Morrison

OF A NEW BEDFORD FIRM OF SHIPPING AGENTS

Board Where	What Ship	Time Come	Remarks
Long	Responsible to 15th, and then Long runs the risk.		
J. C. Carter	*Kensington*		$10 if he goes
Mrs. Dodge	Did not come back.	Gone home, 9/22.	
Manix. 87	*Marcella*		Break. Went jail
Davis	*Awashonks*		Tea $12
Davis ½ of 5 days $1.45	Run 5 Mo.		Tea Break 28th aft.
E. H. D. ½ is $1.16	Turned off 5 Mo. 12th		Tea $12 Poor thing
Miller ½ $3.00	Gone in Army. Run. was Rejected $3		
Miller's paid his board.	Paid his board	Father came took him.	
Dodge's $3.62	Davis ½ is 5.04		Went in army. Green stole him.
Miller ½ is $4.00	*Gazelle* *California*	Was taken out of *Gazelle* and stowed on board *Calif.* 15	
M. Home	U. S. Navy	M.	Would not go whaling 12
Wilcox			Did not go. Sick. Pox.
Wilcox ½ Day @ $3.00	Turned off. Stealing		Went Jail.
Butts	Run May 6		James to pay back passage
Dodge ½ is .86		Turned off May 19 Old Runaway	
None	T, off		Sucker
Mack		Sold him to Mack for $5	
Wm. Slocum	T. off Apl. 13		Pd. fare in N. Y. No. Brok. only Fare to charge.
Slocum	*Niger*		Gave coat to pay Fare in N. Y.

laconic phrases of the record itself, were as follows: "refused to ship; went back; father took him home; went to hospital; no chance; transferred to other shippers; did not arrive; went to jail; joined navy; went in army." The accompanying list of representative entries will convey some concept of the atmosphere which pervaded these whaling labor agencies.[7]

While the shippers were thus juggling his accounts and arranging for his sailing date, the new hand had ample opportunity to become familiar with the life of the sailors' boarding-houses. The keepers of these establishments, comprising the second major division of landsharks, were usually either whalemen's widows or shrewd and unscrupulous men who soon became adepts in playing upon the weaknesses and necessities of the seamen. A few houses, frequented by mates, boatsteerers, and the more thrifty foremast hands, were able to boast of good food, cleanliness, and a certain air of gentility; but the great majority were sadly lacking in all of these attributes. In the latter places the green hands came into contact with veteran tars who were known to their intimates by such euphonious names as Blue John, Long-Legged Bill, Big-Foot Jack, Chaw-o'-Tobacco Jim, Handsome Tom, and Bully Clincher; and he also met housemaids, known as Mag, Moll, Bet, or Peg, to whom formal introductions were entirely superfluous.

Throughout the middle of the century the price of accommodations in most of the boarding-houses ranged from two to four dollars per week; and in view of the cuisine and the condition of the quarters these sums would seem to have afforded ample compensation to the owners. Even in the Mariners' Home at New Bedford there was one lone tin wash-basin in a rusty iron sink and a single towel on a roller for the use of as many as fifty men thrice daily. A typical day's menu at the same place included fat-soaked sausage and fried potatoes for breakfast; boiled beef, cabbage, and soggy potatoes for dinner; and a hash made of the scraps of the two preceding meals for supper. Repasts which were amply sufficient in quantity, but

[7] While the entries here reproduced were chosen from those cases in which men were not successfully shipped, they are representative in the sense that each item could be matched by many others belonging to the same category. Some of the more extreme occurrences, in which the "Remarks" were couched in terms hardly appropriate for the printed page, have been omitted.

which lacked much in quality and in excellence of preparation! Seldom, if ever, did the boarding-house keepers serve fresh meat, in spite of the fact that salt beef and salt pork were the inevitable monotonous, scurvy-producing standards of shipboard food throughout endless months at sea.

While still on land, too, the green hands were given their "first lesson in Lunars." This initiation into whaling life required that all of their fellow-boarders be taken to the nearest saloon for a cigar and a drink. Everyone had either a "close-reef" (some strong drink such as whiskey or brandy) or a "sea-breeze" (some milder concoction such as beer). Afterwards there were songs, yarns, and toasts, in which the new men were admonished to

Be cheery, my lads, may your hearts never fail
While the bold harpooner is striking the whale!

The last and strongest link in the chain of persons interested in securing men for the whaling forecastles was the outfitter. This individual may be characterized briefly as the enterpreneur of the labor supply phase of the industry. He was the intermediary between the agents of the vessels and the shipping-agents, boarding-house keepers, catering parasites, and men themselves. Usually, too, he was the organizer, guiding spirit, and main beneficiary of the system of commercialized exploitation which fed upon the whalemen's earnings. And in many instances he acted as shipping-master as well. In this capacity he was called upon to advance considerable sums by paying the shipping-agents and boarding-house keepers for their services before he was able to collect his own profits.

But his main business was that of a merchant. His store or shop, crammed with articles likely to be desired by whalemen, formed the base of supplies for the outfits which were sold to all hands just before embarking. And the prices he charged for his goods yielded a rate of profit which was more than sufficient to repay him for his many activities, the undeniably heavy risks which he assumed, and the sting of the uncomplimentary epithet, "landshark," which was generally applied to him.[8]

[8] So widely was this term employed that at one time the New Bedford Out-

The detailed administration of the affairs of a whaler was usually entrusted to a firm which was also part owner of the vessels which it operated. When such an agent wished to ship a crew it was customary to place an order with one or more outfitters for the number and type of men wanted. The green hands were then secured through the network of shipping-agents, while the experienced seamen were commonly found in the boarding-houses, brothels, or grog-shops. After recruiting a crew from these sources, and after watching the individual members vigilantly for several days or weeks, the outfitter hailed the vessel's sailing date with a sigh of relief. At some time during the last twenty-four hours before weighing anchor the future shipmates were corralled, provided with their outfits, and dumped into the forecastle.

When the whaler was well out to sea the outfitter made out a bill against each man whom he had furnished and presented it to the agent. The overshadowing item in this bill was always the price of the outfit; but there might be other charges as well. If the various items were satisfactory, the agent gave the outfitter his note, payable from one to six months after date, for the amount in full. The agent, in turn, charged the amount of this note, with interest, against the seaman's account; and, together with numerous other entries, it was subtracted from the lay at the end of the voyage, several years later. By this circuitous and complicated method each man shipped, even though he had not a single penny, was made to pay for the outlay involved in placing him in a whaler's forecastle.

The outfitters' bills varied in amount according to the needs and desires of the individual hands. Seldom, however, were they less than $60 or more than $100. The one figure most often mentioned by contemporary writers was $75; and an analysis of hundreds of accounts showed that a majority of the men were charged with amounts ranging from $70 to $90. The ship *Maria*, starting out on a whaling voyage in April, 1836, carried twenty-six men for whose outfits the owner, Charles W. Morgan, had signed six-months' notes amounting

fitters' Association was compelled to resort to fines in order to prevent its members from referring to themselves as landsharks in their own meetings!

to $1832.72 — an average of $70.49 per man.[9] Four months later the bark *Minerva* sailed out of New Bedford with eighteen men whose outfits were valued at $1289.08, or an average of $71.62 per head.[10] Boatsteerers and mates were likely to be somewhat more extravagant in their tastes. The former often spent from $100 to $200 on an outfit; while the latter occasionally exceeded even $200 by a generous margin. But such figures were exceptional, and were confined to the dandies of the business or to those who had lost everything through shipwreck or other trick of fate.

What items were included in the outfits sold to the ordinary foremast hands? The following list of articles, given to one man for an even $100, was representative: a cheap white pine sea-chest, or "donkey," containing even cheaper, shoddy clothing; one dozen needles; a hank of linen thread; two spools of sewing cotton; a shaving outfit; a knife and fork; two combs; eleven extra buttons; a tin spoon and a tin plate; a quart dipper; one sheath knife and belt; two bars of yellow soap and a box of oil soap; one pair of the cheapest blankets; and a striped bed-tick and a pillow, both scantily filled with hay.[11] Unfortunately this writer did not itemize his articles of clothing with the same meticulous care. But many other accounts indicate that they included, in most instances, one or two jackets, a Guernsey frock, an oil suit, and a motley assortment of heavy and light trousers, shirts, and underwear, together with several pairs of shoes and socks.

The number, quality, and price of these articles varied with the ignorance, need, or gullibility of the purchaser and with the shrewdness, rapacity, or dishonesty of the seller. In most cases, however, the outfits were exceedingly scanty (in view of the length of the voyage for which they were intended), shamelessly inferior in quality, and extortionate in price. At times the prices were only twenty to thirty per cent above the going

[9] Figures taken from an original manuscript account-book of Charles W. Morgan, labeled "Bill Payable, 1834–37," which is now in the New Bedford Public Library.

[10] From a manuscript Day Book of Charles R. Tucker, covering the period from 4th Month 1, 1836, to 6th Month 30, 1837. This volume is also in the New Bedford Public Library.

[11] Beane, J. F., "From Forecastle to Cabin," p. 14. This was in the sixties.

rates charged for similar goods in ordinary stores; but in countless instances the discrepancy rose to one hundred per cent and more. In general, the coarsest and cheapest materials were provided for amounts which would have been more than sufficient, elsewhere, to purchase goods of excellent quality and workmanship.

So lucrative was the outfitting business that strenuous competition developed between rival firms. When the bark *Minerva* weighed anchor in 1836 she had on board eighteen hands who had been furnished by seven different New Bedford outfitters. Such cutthroat rivalry led to the gradual establishment of certain customary rules of the game. Thus in the post-war years, at least, it was understood that if an outfitter furnished a captain or first mate for a given vessel, he was entitled to provide, in addition, four foremast hands. If he secured a second mate, he was privileged to fill three bunks in the forecastle; if a third mate, two bunks; and if a fourth mate, boatsteerer, steward, or cook, one bunk.[12]

Most of the contemporary writers who actually made voyages as foremast hands were unsparing in their criticism of the average outfitter.[13] They charged that he was heartless, unscrupulous, and avaricious; that he was guilty of flagrant misrepresentation; that he demanded extortionate prices for inadequate outfits; and that he was sometimes guilty of deliberate dishonesty. "This system, from its novel and somewhat singular operation," said one author, "is like the vine, which entwines itself around the huge and gigantic oak, and thus it grows and expands according to the height and dimensions of its support."[14] Perhaps the most generous contemporary judgment of the outfitters was that they were honest enough when dealing with intelligent, experienced men who knew what they wanted and who were able to pay for their purchases. But, it was added, the green hands were "so glaringly ignorant and so vastly gullible," and many of the experienced men were so recklessly improvident and so drunkenly be-

[12] Brown, J. T., writing in "Fisheries and Fishery Industries of the United States," VII, p. 225.

[13] See almost any of the narratives and accounts of whaling life mentioned in the bibliography.

[14] See Holmes, Lewis, "The Arctic Whaleman," p. 276.

fuddled, that the temptation to cheat them thoroughly was irresistible. And since these two groups, taken together, comprised the overwhelming majority of all whaling crews, the outfitters found but a small field in which to practice honesty and fair dealing!

Unfortunately, serious abuses would seem to have been inherent in the very nature of the situation. The outfitter's occupation involved heavy risks, arduous and aggravating toil, constant vigilance, and many unpleasant associations. For these activities his only reward came from the profits obtained through the sale of outfits. His customers, on the other hand, were either inexperienced, unfortunate, or dissipated and reckless; and in practically all cases they were penniless at the time when they required his goods. These considerations, coupled with the fact that a relatively expensive outfit was a *sine qua non* for a whaling voyage, meant that the foremast hands were left without a shred of bargaining power. And their economic helplessness constituted a temptation to resort to sharp practices which proved quite too strong for the shrewd, penny-loving Yankee merchants of the whaling ports. In fact, in many instances the temptation would seem to have been welcomed and cherished, rather than resisted.

But contributions to the outfitters' profits came not only from consumers, but also from producers. For much of the clothing sold to the crews was made in the whaling ports under a plan known to economic historians as the "putting-out system." Cloth was purchased by the outfitters and given out to women of the town in order to be made up into garments. These women did the work in their homes, and were paid by the piece at rates strangely prophetic of the later sweated trades in the clothing industry. Thus during the years 1855 to 1870, inclusive, the piece-rates paid by a New Bedford firm were 12½¢ to 15¢ for trowsers (denim, plaid, or duck); 15¢ to 20¢ for "thick trowsers"; 8¢ for drawers; 9¢ or 10¢ for undershirts; 12½¢ for "denim frocks;" 10½¢ to 12½¢ for shirts (blue, kersey, cotton, or mixed); and 35¢ to 58¢ for reefing jackets and monkey jackets.[15] One contemporary writer re-

[15] These figures were found in the accounts of Cook and Snow, a New Bedford outfitting firm, for the years given above. The manuscript "Sewing-

marked, apropos of such figures, that "the price for making is the minimum, or starving price; and therefore the garments are made accordingly. There are two losers by this arrangement, and one winner. The maker and the buyer are the losers, while the profit passes into the hands of the seller." [16]

Many other articles which found their way into the outfits were also made in the town or surrounding countryside. In fact, a goodly percentage of the shore labor of a whaling port was engaged in catering to the needs of the whaleman under the general direction and to the financial advantage of the outfitters. These shrewd tradesmen succeeded in building up a system in which they bought the coarsest raw materials, had them made into poorly-fashioned articles by cheap labor, and then sold the resulting low-cost goods to economically helpless customers at exorbitant prices.

The self-same tradesmen, too, were infitters as well as outfitters. Not only were they the last to speed the whaleman upon his outward journey, but they were also the first to greet him upon his return. For both in beginning and in ending a voyage the crews stood in need of, or could be made to feel that they wanted, certain goods and services which the combined outfitters-infitters were anxious to supply.

So eager were they, in fact, that there was the keenest competition in boarding incoming vessels. Now and again the rivalry became so strenuous and so bitter that regulatory agreements were virtually forced into being. In New Bedford during the sixties, for instance, the Outfitters' Association sought to control the methods employed in boarding vessels which arrived during the night. All members were to put out from certain specified points along the shore, and then only after a signal had been given by a central watchman. Another rule, carrying with it a penalty of one hundred dollars for each violation, provided that all members must use the same boat, that no intoxicants were to be carried, and that the boarding must be done "quietly and peacefully." Such agreements, however,

Book" in which the accounts were kept is now in the New Bedford Public Library.

16 Holmes, Lewis, "The Arctic Whaleman," p. 284. This author describes the abuses and operation of the outfitting system in some detail.

were more honored in the breach than in the observance.[17]

During the daytime the infitters, in common with all others interested in whaling, made use of the semaphore telegraph system which announced the incoming vessels. By means of observation stations, telescopes, and a set of signal flags it was possible to broadcast the name of a returning whaler several hours before her arrival in the harbor. The New Bedford Harbor Signal Book, for instance, described an elaborate scheme of signal flags which was in use at that place. In addition to the regular house flags of the various whaling firms, there were flags for each number from one to six, inclusive, and for the differing rigs which were in use. Each merchant firm, as well as every vessel which it operated, was given a number made up of four digits, e. g., 5234, 1642, etc. Armed with this Signal Book, then, an outer observation station would display the appropriate set of flags, giving rig and number, as soon as an incoming whaler was sighted in the offing. This signal was read, with the aid of a telescope, by an inner observation station, and was similarly relayed into the town itself, where the news was heralded by means of flag poles.[18]

This advance information afforded ample time for the infitters to arrange receptions which made small pretence of concealing boisterously competitive business under the guise of effusive cordiality. Long before a returning whaler dropped anchor her crew had been greeted by competing runners who were fairly bursting with expansive comradeship (although some captains refused to allow them to come aboard until the vessel was actually in her berth). Each runner assiduously assisted the men in packing their chests, retailed the port gossip, drew upon an extensive repertoire of unprintable stories, and

[17] See the Old Dartmouth Historical Sketches, No. 44, pp. 23–26, for an interesting account of the manner in which the competition between seventeen New Bedford firms was regulated between 1859 and 1873.

[18] A copy of the New Bedford Harbor Signal Book for 1848, part of the Daniel B. Fearing Collection, is now in the Treasure Room of Widener Library, Harvard University. In the foreword the proprietor of this "Telegraph Establishment" complained that he was forced to conduct the business at a loss because of inadequate income. Individual users of the system subscribed only one dollar per year, while agents paid fifty cents per year per vessel. Similar booklets for other years are in the stacks of the New Bedford Library.

attempted to attach to himself as many followers as possible. Upon reaching the wharf, he steered his customer-conquests past the runners of various boarding-houses and brothels to his own firm's establishment. There they were given facilities for a bath and a thorough cleansing, after which they were furnished with an outfit of "long togs," or shore clothes. Then they were shorn of the superfluous hair which had accumulated on heads and faces during long months at sea. And after a further bit of attention in the form of an advance of spending-money, they were at length ready to sally forth in search of the delights of life on shore.

All this, of course, the infitter did with the understanding that he would receive generous compensation when the men obtained their lays, always paid within a few days after landing. But what of those men who had no net earnings awaiting them, because they had returned from a two to four years' voyage actually in debt to the owners? Such hands, upon landing, found it possible to secure attention only upon one condition, namely, that they agree to undertake another voyage within the immediate future. With this understanding, the infitter, suddenly turned outfitter again, would make advances upon the terms regularly granted to outgoing seamen. And the men in question, after a short period of hectic revelry or furious dissipation, would find themselves once more at sea, bound on a cruise of three to four years' duration.

But perhaps these hands, after all, were not less fortunate than their comrades who spent several weeks ashore in dissipating the small savings of the previous voyage. For the average seaman was emphatically not in the hands of his friends while in port with money in his possession. The keepers of the boarding-houses, brothels, and grog-shops, assisted by their parasites and agents, and aided and abetted by the outfitters, conspired to relieve him of his earnings at the earliest possible moment. And with the aid of the spirit of pecuniary recklessness which characterized the mariner ashore, and of the chronic state of drunken irresponsibility which was all too common, this was a task easily and speedily accomplished.

The multitudinous and voracious landsharks formed the nucleus of an organization which existed for the sake of robbing and defrauding the sailor at every turn. Both amazing and revolting were the experiences of many men caught in the toils of this system. "The pages of a Smollet, himself a sailor, and drawing from the life, might furnish us with truths so startling and revolting, that the reader of the present day regards them as exaggerations."[19] From the moment they set foot on shore until their money was gone, the returned whaling hands were subjected to multifold temptations and encouraged to indulge in every form of vice and dissipation. Divorcing the seaman from his money became a fine art, practiced with striking success by experts who were parts of a systematized and commercialized scheme of exploitation. Even the detailed methods were largely standardized, in spite of the fact that they ran the whole gamut from mere encouragement of unwise generosity through shameless overcharging and cheating to downright theft. There was ample justification for the warning issued by one author, "Seamen, beware! There are shoals, quicksands, and death-pointed rocks upon the land as well as upon the ocean."[20]

Unfortunately, many of the returned voyagers were only too eager to throw themselves headlong into this mad career of drunken sensuality. After many weary months of danger, hardship, monotony, poor food, and cramped quarters, subject at best to an unvarying discipline and at worst to cruel and inhuman treatment, they were in a mood for hectic revelry and complete forgetfulness. And what more easily available means of attaining these ends than the harlot and the bottle — especially when both were paraded and thrust forward by an organized system of exploitation which employed vice and drunkenness as material aids in its programme? To many men hard liquor seemed the prime solvent for the harsh and ugly facts of their environment, as well as the only means of regaining even temporarily that feeling of self-respect and of personal importance so difficult to retain in a whaler's forecastle. For "with a shilling's worth he becomes a man; with

[19] Harris, John, "Zebulon; or, The Moral Claims of Seamen Stated and Enforced," p. 20 (first American edition, Boston, 1837). This work was written originally about British conditions; but the American edition introduced much material drawn from American ports as well.

[20] Holmes, Lewis, "The Arctic Whaleman," p. 282.

twice that quantity the old boyhood is with him again; and with thrice that . . . he feels, for the time, 'every inch a man.'"[21] And if that feeling was fleeting, counterfeit, and creative of an aftermath which reduced life to the plane of a beast, nevertheless it was sought again and again, with and without the encouragement of the landsharks. For spiritual as well as physical escape from the evils of shipboard life was imperative; and in seeking such escape the average seaman followed the path of least resistance, particularly when this path was smoothed and graded for him by his exploiters.

The more worthy and respectable institutions of the community did practically nothing to meet this crying need of the sailor for wholesome recreation and decent care while ashore. The one organization which seemed to have a clear conception of the demands of the situation was the American Seamen's Friend Society, which was formed in 1828 "to improve the social and moral condition of seamen, by uniting the efforts of the wise and the good in their behalf, by promoting in every port boarding-houses of good character, Savings Banks, Register Offices, Libraries, Museums, Reading Rooms and Schools, with the ministrations of the gospel and other religious blessings."[22] The Society also published the *Sailor's Magazine*, which appeared monthly throughout the greater part of the century.

But unfortunately the aims set forth in this admirable programme far outran the actual accomplishments of the Society. Of libraries, reading rooms, savings banks, and decent amusement places there was no hint in the whaling ports, though such institutions sometimes gained a precarious footing in the large maritime centers. New York, Boston, and other prominent mercantile ports also boasted of a few national hospitals, supported in part by a levy of a few cents per month on the wages of merchant seamen. But again the whaling ports were neglected, so that the average whaleman who became ill while

[21] See Davis, W. M., "Nimrod of the Sea," p. 92, for a discussion of the relation between drinking and the seaman's insistent desire to escape from the ugliness and sordidness of reality.

[22] See Harris, John, "Zebulon: or, The Moral Claims of Seamen Stated and Enforced," for an extended discussion of the moral and ethical phases of a sailor's life while ashore.

ashore had to pay his own bill or be content with the heartless neglect of the boarding-houses.

Religious and temperance societies made certain abortive efforts, largely negative in purpose, to reach the sailor through a form of condescending philanthropy. Marine Bible Societies distributed Bibles and religious tracts amongst the crews of outgoing vessels; and an occasional minister sought to interest himself in the religious welfare of the foremast hands. In general, however, the orthodox churches, ministers and members alike, regarded the sailor as a moral pariah, and remained comfortably aloof from the forecastles and the waterfront. In some instances they even repulsed the hesitant inclination of some lone seaman to attend their services. The only ecclesiastical doors definitely and invitingly open to the whaleman were those of the Seamen's Bethel, half-church, half-mission, where there was a strong tang of salt water in the discourses of the preacher and in the whole atmosphere of services and building. These Seamen's Bethels, in conjunction with the neighboring Mariners' Homes, or supposedly model boarding-houses, formed the centers of the better type of waterfront life. And together they comprised the only religious and philanthropic activities which reached the whaling crews in any vital manner.

Here and there an isolated voice was raised in behalf of better treatment and conditions. Thus a certain Mr. J. Girdwood, of New Bedford, wrote in 1857: "When the extent and value of the interests involved are thought of, it seems surprising that efforts are not made to improve the character and condition of the sailor. Millions of property are entrusted to his care. Thousands of precious lives are in his hands for weeks and months; yet many sailors are the refuse of jails, penitentiaries, and state prisons. The sentiment often prevails that the worse man makes the better sailor. Hence we may easily account for many shipwrecks, vessels cast away, sunk, and burned. This is becoming too expensive." [23] And Lieutenant Wilkes, writing of the morals of seamen during the forties, said: "The field for improvement is wide, and those

[23] These sentences form part of the introduction to the volume by Holmes, Lewis, "The Arctic Whaleman."

who first labor in it must reap a most satisfactory harvest. To none does it more appertain to take the first step, and push earnestly onwards, than the owners of our mercantile marine, and of our whaling fleet in particular."[24] But such occasional expressions were like voices crying in the wilderness; and appeals to efficiency and to humanitarianism alike fell upon deaf ears.

Not all whalemen, of course, belonged to the class of dissipated, social outcasts whose weaknesses and sins were encouraged and preyed upon by a system of vicious exploitation. There were always those who had happy and respectable homes, and who spent their time ashore in a normal manner. In the early days of the industry, up to 1825, this type of man was in the clear majority. But in the succeeding decades he gave way more and more to the homeless, irresponsible, weak-willed hand whose brief, intermittent periods on land were so sadly misspent; and long before the Civil War the relative proportions were reversed. At any time after 1840 the discovery in a single forecastle of two or three sober, thrifty, self-respecting men who were happily married would have been unusual enough to excite surprised comment. The ranks of the captains and mates contained a goodly percentage of such men, since the best elements in the fishery naturally gravitated towards the posts of responsibility. Nor was it uncommon to find them amongst the coopers and ship-keepers, who were often the oldest and most reliable members of a crew. But it would have required a long search through many vessels to find a corporal's guard of such mariners amongst the foremast hands.

But bad as conditions were in the seaman's home port, they were even worse in the small, little-frequented places at which the whalers commonly touched during the course of their long and tedious wanderings. In his "Nimrod of the Sea" W. M. Davis asserted that on the west coasts of both North and South America there was not a single port in which there were any doors to welcome the whaleman other than those of gaming houses, grog-shops, and brothels. The coast of Australia dur-

[24] Wilkes, Charles, "Narrative of the United States Exploring Expedition," V, p. 502.

ing the middle years of the century was notorious for its vice and degenerate forms of dissipation. The Orient welcomed participation in its sophisticated and voluptuous sensuality, but kept its higher life, characterized by age-old wisdom and a subtle appreciation of beauty, scrupulously concealed from the contaminating influence of the foreigner. And in many ports of the East Indies and of Southeast Africa the danger from the deadly coast fevers was further enhanced by drink, prostitution, degrading indulgence, and a bestial lack of the first rudiments of sanitation. Zanzibar was a prime illustration. At that place the combination of the terrific heat, the exhalations from the marshes, and the unsanitary filth was sufficient to cause an unacclimated white man, unschooled in the necessary precautions, to contract the fever in a single night. The conditions in a certain fort, sometimes used for the imprisonment of whalemen, were so unspeakable that confinement there for more than a few days was virtually equivalent to a death sentence.[25]

The only places reached by the whalers which offered ready opportunities for natural, wholesome recreation were the numerous islands scattered about the South Pacific. But even here, with cruel irony, the whalemen brought their habits of vice and intemperance with them, and soon taught the natives, usually innocent and friendly though often promiscuous, unashamed, and weak-willed, to outdo their instructors in sensual excesses. The South Seas became the scene of drunken revels and bestial orgies of the most revolting and degrading character. Venereal disease, theretofore virtually unknown amongst the simple islanders, spread widely and took a heavy toll of misery and of life.

Going ashore for licentious purposes was a commonplace; and where the natives showed signs of resentment parties often went armed for the purpose. When the missionaries came and attempted, with but little success, to interfere with these practices, they were often threatened and abused for their pains. At Lahaina, one of the Sandwich Islands, armed parties of

[25] See Browne, J. R., "Etchings of a Whaling Cruise," pp. 371–378, for a description of life and conditions at Zanzibar. The author spent several months on the island after leaving a whaler there.

whalemen were reported to have gone ashore several times for the avowed purpose of driving away the missionaries and burning their homes. Fortunately their plans were frustrated by the natives, who protected their benefactors.[26] Many a subsequent massacre of whites was directly traceable to the hatred and resentment engendered amongst the islanders by these unprovoked assaults upon their women. Not infrequently the aggressions of the whalemen changed well-disposed natives into bitterly vindictive enemies who were incapable of discriminating between individuals in their hatred of the white race as a whole. Taken all in all, the atrocious behavior of the whaling crews in the South Seas constituted one of the most shameful chapters in the long story of Anglo-Saxon expansion.[27]

While their companions on regular shore leave were conducting themselves in this infamous manner, those whalemen who failed to complete a voyage, but left their vessels permanently in a foreign port, were in positions even less enviable. If they were deserters there was the probability of pursuit, capture, and punishment, or of the cowering terror and slinking hardships of a hunted animal. If they had been left behind by their captains (a practice by no means unknown in whaling), there was the task of obtaining a berth on another vessel, and in the meantime the necessity of finding precarious food, clothing, and shelter.[28] Even if they had been honorably discharged by mutual consent, they were confronted with the problem of future employment and livelihood. And in remote, isolated spots which were known only to natives, whalemen, and "beach-combers," and where it was often necessary to wait for weeks even for the dubious privilege of ship-

[26] Harris, John, in his work "Zebulon: or, The Moral Claims of Seamen Stated and Enforced," p. 69, gives some discussion of the relations existing between natives, missionaries, and whalemen.

[27] See Wilkes, Lieutenant Charles, U.S.N., "Narrative of the U. S. Exploring Expedition During the Years, 1838 . . . 42," for a forceful and authoritative discussion of the relations between whalemen and the natives of the South Seas.

[28] Because of the right to confiscate the effects and the lay of a deserter, certain unscrupulous masters were sometimes tempted to force an occasional hand into the position of a deserter in order to secure the lay. See Chapter V, on Forecastle and Cabin, for a more detailed discussion of this point.

ping on another whaler, this was a problem by no means easy of solution.

The worst plight of all, however, was that of the ill or incapacitated. Owing to the various fees and charges which were required in discharging a seaman in a foreign port, it was often less expensive for the owners to keep a man on board, even though completely incapacitated for some time, than to let him go. This was particularly true of the early stages of a whaling voyage, when the hands were heavily indebted to the owners because their slowly-growing lays were still appreciably less than the sums which had been advanced for their outfits. In many cases, consequently, men who were ill were able to secure their discharges only after the most unpleasant controversies with their captains.[29] Whaling officers experienced great difficulty (and not without some reason) in distinguishing clearly between illness and malingering!

But when and if a seaman was discharged, the American consul became responsible for providing medical assistance, accommodations, and essential supplies, as well as for securing a return passage to some port in the United States. Some consuls attempted to perform these troublesome and often aggravating duties as well as circumstances permitted. Many others, however, were so strongly suspicious of the claims of the ordinary seamen, and so notoriously apathetic in coming to their assistance, that they failed miserably in providing either redress of grievances or adequate care for the incapacitated hands who came under their observation.[30]

The greatest of all foreign whaling centers, however, was Honolulu. This picturesque Pacific rendezvous of the whalemen reached the height of its renown after 1850, when the Arctic whaling was in full swing. Twice each year the spacious harbor would be crowded for several consecutive weeks

[29] See Browne, J. R., "Etchings of a Whaling Cruise," pp. 102–104, for material bearing on this point. The author cited one instance in which his captain stubbornly refused to grant a discharge to a sick shipmate until at length, in desperation, the man offered his gold watch in return for the coveted release. Thereupon he was finally put ashore at Fayal, after the master had paid a four dollar fee to the port doctor, ten dollars in boat charges, and thirty-six dollars advance wages to the consul.

[30] See Chapter V, on Forecastle and Cabin, for further material pertaining to the consular treatment of seamen.

with scores of whalers fitting out and recruiting for their next season — in March preparing for a summer in Behring's Straits or the Sea of Okhotsk, and in November and December getting ready for several months of sperm whaling in tropical and sub-tropical waters. This close proximity of so many vessels in one of the world's fairest harbors, still relatively untouched by the materialism of an industrial age, afforded an unexampled opportunity for boisterous debauchery and voluptuous dissipation on the part of the reckless and pleasure-starved crews. Amidst the languorous, tropical surroundings and amongst the comely, ingenuous natives the sensual abandon of the sailor ashore reached its full fruition.

But not, if contemporary accounts may be credited, without some opposition from the forces of law and order. In attempting to cope with such large numbers of hard-fisted, reckless, and intoxicated seamen, the police force of the port found an unenviable task which was none too well accomplished. But, partly because of this very laxity of enforcement, a copy of the regulations supposedly in force at Honolulu during the forties presents an illuminating suggestion of the social and moral conditions which prevailed in the port. The offenses and penalties applying to seamen are reproduced in full.[31]

Hanging, as a murderer, for knowingly and maliciously violating those laws whereby a contagious disease is communicated on shore.

$60.00 fine on any captain who leaves on shore any men without written leave from the Government.

$10.00 for coming ashore with a knife, sword-cane, or any other dangerous weapon.

$2.00 for every seaman seized ashore after 9:30 P. M., at firing of second gun from fort.

$1.00 to $5.00 for hallooing or making a noise in the streets at night.

$6.00 for striking another in a quarrel.

$5.00 for racing or swift riding in the streets or frequented roads.

[31] These regulations were printed in the "Honolulu Friend" in September, 1844, when they were included in an article on Hawaiian conditions written by a certain Mr. R. C. Wyllie. They are reproduced in Browne, J. R., "Etchings of a Whaling Cruise," p. 546. The King of Hawaii encouraged the whalemen to visit the islands, since they purchased large quantities of foodstuffs from the natives. But evidently his police force was instructed to draw the line between encouragement and license!

$6.00 for desecrating Sabbath the first time; $2.00 for second time; and fine doubled for every repetition.
$6.00 for catching deserter near harbor; $12, if 10 miles off.
$6.00 for drunkenness.
$10.00 for lewd, seductive, and lascivious conduct.
$5.00 for fornication.
$30.00 for adultery.
$50.00 for rape.

The religious agencies, too, sought to combat vice and immorality with rather more than ordinary energy and insight. W. M. Davis, for instance, in his "Nimrod of the Sea," spoke with real enthusiasm of the work of a Reverend Mr. Deil, who succeeded in boarding incoming vessels ahead of the land-sharks (no mean accomplishment), offered reading and writing facilities, and befriended the men in a vital manner. But, in any large sense, the religious forces of the port were no more successful than the police in curbing the boisterous spirits and reckless self-indulgence of the seaman on shore. And so, in spite of the activities of the Reverend Mr. Deil and of the King of Hawaii's police force, Honolulu remained for many years a whaleman's paradise!

CHAPTER VII

AT SEA: ON PASSAGE

THE decks of a whaler weighing anchor for an extended cruise presented a scene of confusion and disorder. Many of the experienced members of the crew had come aboard only a few hours before in a drunken stupor or had been literally dumped into the forecastle, where they lay grog-soaked and unmoving. Some of these hands still resisted all efforts to rouse them out upon deck, while others were able to respond but feebly and clumsily to the familiar orders and imprecations of the mates. The green hands, though displaying a somewhat higher percentage of sobriety, were even more helpless because of their ignorance of everything pertaining to the intricate world of spar and rigging. Faced with these twin handicaps of inebriety and inexperience, the officers were forced to undertake the routine operations of getting the vessel under way with the small handful of men who were serviceable, at times assisting in the execution of their own orders. Meanwhile, they kept up a succession of admonitions, curses, and blows, calculated to elicit as much activity as possible.

After several hours of such pandemonium, when the vessel stood boldly out to sea, the green hands experienced their first touch of seasickness. With this familiar malady added to their ignorance of the ship they were worse than useless. A few captains recognized this fact frankly and allowed the new men to spend the first day or two in getting their sea-legs. But in most instances no such consideration was shown. Clumsy and painful though their efforts might be, they were forced to go through the form of taking part in every task demanded of the crew. If they pleaded illness or fear, a curse or a blow from one of the mates served as an irresistible argument.

And certainly no incentive less violent and menacing would have succeeded. For few situations were less likely to encourage voluntary activity. The nauseated and enfeebled condition of the sufferers, the coarse and poorly prepared food, the ugly, cramped, and dreary living quarters, the terror of contemplating the first trip into the dizzy heights of the rigging of a rolling vessel — all these elements combined to produce an abject state in which any exertion seemed repellent and impossible. The annals of whaling, like those of the merchant marine, are filled with admittedly inadequate attempts to depict the utter misery and hopeless dejection of the first forty-eight hours on shipboard.

The task of the officers, too, was no light or pleasant one. Starting out with a crew whose members, with few exceptions, were intoxicated, seasick, or inexperienced, they faced the necessity of creating and preserving a spirit of discipline, of familiarizing the men with manifold duties, and of building up a state of collective efficiency. Inebriety and seasickness soon wore away; but the transformation of inexperience and callous indifference into effective labor was a long, slow process accompanied by the ruthless and often brutal enforcement of a stern policy.

The first official steps in this process were taken during the dog-watches of the first day out, when the whole crew was called aft to be divided into watches. The two watches were headed usually by the first and second mates; and these officers made alternate choices until all the names on the crew list had been drawn. If the boat-crews were to be organized at the same time, the captain and each of the mates, in order of rank, chose the hands who were to man their respective whaleboats. Each of the three or four boats carried by the average whaling vessel was manned by a crew of six, including the officer in command. Theoretically these original assignments to watches and boat-crews were to remain unchanged throughout the entire voyage; but in actual practice transfers and changes were often necessitated by desertion, illness, accident, and death.

When these preliminaries had been arranged, the captain made a brief speech in which he attempted to strike the key-note of the voyage, especially in the relationship between officers and men. The subject-matter of these talks was rather stereotyped, though the manner of presentation varied with the eccentricities of the individual captains. There was often an explosive effort to impress or to intimidate the crew through sophisticated bravado. The following example, together with a colorful description of the attendant circumstances, was set down by J. Ross Browne, who had been a stenographic reporter in the United States Senate before starting out upon his whaling voyage.[1]

"The captain deliberately stalked the quarter-deck, exulting in the 'pomp and circumstance' of his high and responsible position. Every step he took bespoke the internal workings of a man swelling with authority. . . . With his hands stuck in his breeches pockets, he then approached the break of the quarter-deck, and, straddling out his legs to guard against lee-lurches, asked if all hands were present. One of the officers replied in the affirmative.

"The scene was at once grotesque and impressive. Fourteen men, comprising the whole crew, were huddled together in the waist, at the starboard gangway. Of these four were Portuguese, two Irish, and eight Americans; and certainly a more uncouth-looking set, including my friend and myself, never met in one group. The Portuguese wore sennet hats with sugar-loaf crowns, striped bed-ticking pantaloons patched with duck, blue shirts, and knives and belts. They were all barefooted, and their hands and faces smeared with tar. On their chins they wore black, matted beards, which had apparently never been combed. The color of their skin was a dark, greenish-brown, if the reader can imagine such a color, and was calculated to create the impression that they never made use of soap and water. The variety of dress in which the rest of the crew were habited was fully as striking as that of the Portuguese. Some wore Scotch caps, duck trowsers, red shirts, and big horse-leather boots; others, tarpaulin hats, Guernsey frocks, tight-fitting cloth pantaloons, and red neckerchiefs. Several were bareheaded and barefooted, having lost their hats and shoes in the late gale. All the green hands, which included most of the Americans and the two Irishmen, were still cadaverous and ghastly after their sea-sickness, and not more than two had yet entirely 'squared accounts with old Nep.' Altogether we were the most extraordinary looking set of half-sailor nondescripts possible

[1] Browne, J. R., "Etchings of a Whaling Cruise," pp. 33 ff.

to conceive. Thus situated, and thus equipped for sea life, we stood gaping at the captain in silent admiration.

"The mates and boat-steerers, consisting of the chief mate, an Englishman, the second mate, an American, two Portuguese boat-steerers, and an American of the same grade, stood near the main-mast, looking on with the air of men who were used to such things, and took no particular interest in them.

"The captain, after considerable deliberation, and a great show of contempt toward everybody within range of his visual rays, then addressed us in a sharp nasal voice, fixing his eyes upon each man alternately.

" 'I suppose you all know what you came a whaling for. If you don't, I'll tell you. You came to make a voyage, and I intend you shall make one. You didn't come to play; no, you came for oil; you came to work.' (While making these remarks the captain stood upon the quarter-deck, which he paced during frequent breaks in his line of thought. The crew formed a closely-grouped knot of men in the waist.) 'You must do as the officers tell you, and work when there's work to be done. We didn't ship you to be idle here. No, no, that ain't what we shipped you for, by a grand sight. If you think it is, you'll find yourselves mistaken. You will that—*some,* I guess. . . . I allow no fighting aboard this ship. Come aft to me when you have any quarrels and I'll settle 'em. I'll do the quarreling for you—I will. . . . I'll have no swearing, neither. I don't want to hear nobody swear. It's a bad practice—an infernal bad one. It breeds ill will, and don't do no kind o' good. If I catch anyone at it, damme, I'll flog him, that's all. . . . When it's your watch below, you can stay below or for'ed, just as you please . . . I won't have no skulking. If I see sogers here, I'll soger 'em with a rope's end . . . You shall have good grub to eat, and plenty of it. I'll give you vittles if you work; if you don't work, you may starve. Don't grumble about your grub neither. You'd better not, I reckon. If you don't get enough, come aft and apply to me. . . . Now the sooner you get a cargo of oil, the sooner you'll get home. . . . Do your duty, and act well your part toward me, and I'll treat you well; but if you show any obstinacy or cut up any extras, I'll be d—d if it won't be the worse for you! Look out! I ain't a man that's going to be trifled with. No, I ain't—not myself, I ain't. The officers will all treat you well, and I intend you shall do as they order you. If you don't, *I'll* see about it. That's all. Go for'ed, where you belong!' "

This speech, quoted thus fully, not only betrays the type of man who made it, but also portrays in an illuminating manner the paternalistic, serf-like relationship existing between master and crew. There were, of course, some captains who dealt with their men in a quiet, dignified manner which inspired genuine respect and esteem. But the master of a vessel at sea wielded such untrammeled authority that he became inevitably a small-scale despot; and in controlling ships, as in governing nations, the men who could exercise supreme authority without abusing it were in the minority.

The work of the first few weeks combined routine, preparation, and training. All the regular tasks connected with the working of the vessel had to be carried on unfailingly. There were yards to be squared, sails to be reefed, furled, or shaken out, look-outs to be posted at the mast-heads, the wheel to be manned, decks to be scrubbed,[2] and a multitude of other duties to be performed in endless repetition. There was a constant breaking out of some articles and a reciprocal stowing away of others. Many objects hastily taken aboard just before sailing had to be more securely stowed, while all the paraphernalia actually used in whaling had to be made ready for instant action. To this end the harpoons, lances, spades, and knives were ground to a razor edge, the whale line was coiled very carefully in the tubs, the oars, paddles, and sails were overhauled, and the boats were conditioned for immediate service.

On days when the weather was fine and the sea was calm the boats would be lowered and the green hands instructed in the maneuvers and tactics involved in the pursuit and capture of whales. In rough weather the older seamen were engaged in "sailorizing"—a collective term designating those activities requiring dexterity and expert knowledge, such as mending the rigging, patching old sails, or making use of intricate knots. The new hands, meantime, were set to picking oakum and making spun-yarn — two phrases applied to the double process of picking old rope to pieces and re-twisting the strands into smaller and shorter lengths. The real sailor was of necessity a Jack-of-all-trades, ready to turn his hand at a moment's notice to the most widely varied tasks. For a sailing vessel

[2] The work of scrubbing decks was particularly laborious on whalers because the water was drawn up in huge canvas buckets by men stationed at the side. In most merchant vessels the far easier method of rigging a head-pump was employed.

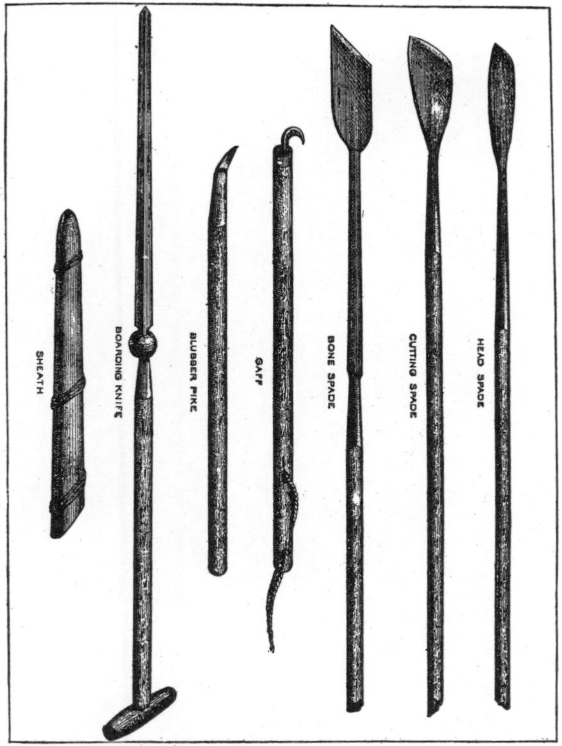

IMPLEMENTS USED IN CUTTING IN A WHALE.

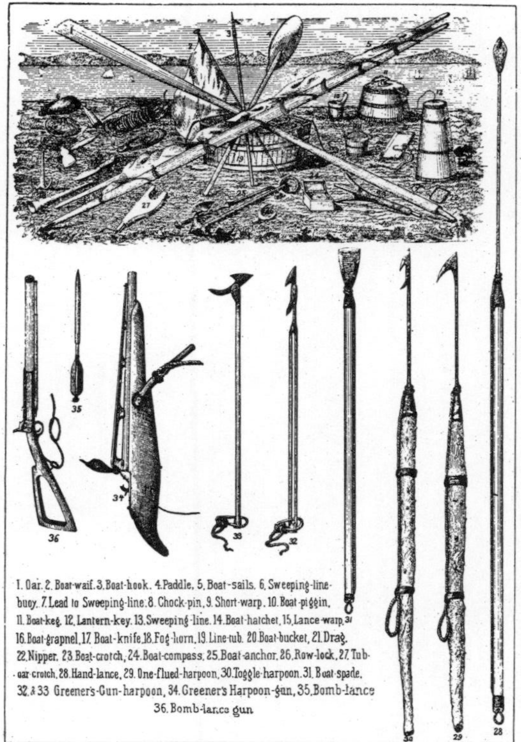

C. M. Scammon, del. Lith. Britton & Rey, S.F.

IMPLEMENTS BELONGING TO A WHALE BOAT

exacted such constant attention and required such interminable adjustments and repairs that a sailor's work, like a housewife's, was never done.

During this early part of a voyage the green hands became acquainted also with the ranks and duties of all the persons on shipboard. The captain was given full responsibility and supreme authority over every detail of the cruise and every member of the crew. Although answerable for his conduct to courts and owners ashore, he was the most unbounded of autocrats while at sea. Sometimes he headed a boat in the pursuit of whales: at other times he remained aboard to direct the strategy of the chase more effectively.

The duties of the three or four mates (according to the size of the vessel) were manifold. When lowering for whales each mate commanded a boat, directed the pursuit, and attacked the prey with his lance; and he also performed the tasks requiring the greatest skill and experience in the long, arduous process of converting the carcasses into oil and whalebone. On shipboard these officers were responsible for the maintenance of discipline and efficiency, exercised general supervision over the work of the crew, performed the most skilled and delicate tasks themselves, and stood watches which placed them successively in charge of the deck. The first mate was usually the busiest man on board. For in addition to being foremost in action every time whales were sighted, he was directly responsible for the proper conduct of all routine affairs, as well as for carrying out the captain's orders, keeping the log-book, and directing the navigation of the vessel.

After the mates, the most important group of men on board consisted of the boatsteerers, or harpooners—one for each boat. These were skilled and experienced whalemen who were given the posts of greatest danger and responsibility in performing the actual labor of a cruise. Their alternative names grew out of their two most important functions — those of flinging the harpoon and thereupon going aft to steer the whaleboat while the mate came forward to wield the lance. Together with the steward, and sometimes the cooper or shipkeeper, they lived in separate quarters amidships, known as the steerage. In location as well as in rank, therefore, they were half-way between the officers, whose quarters were aft in the cabin, and the foremast hands, who were housed forward in the forecastle. Roughly, their position was analogous to that of petty officers in the navy or of non-commissioned officers in the army.

The cooper, as his name implies, devoted himself mainly to the task of making and keeping in repair the casks in which the oil was stowed. In an industry in which leakage was so costly, his function was extremely important. The ship-keeper took charge of the vessel and manœuvred it, with the aid of a few other hands, while the boats were down in pursuit of whales. Many whalers, however, did not carry a special ship-keeper. In such cases either the captain remained aboard or some regular member of the crew, usually the cooper, directed the movements of the vessel while the boats were away. The steward acted as the personal servant of the captain, waited on the officers' table, and was in charge of the officers' and boatsteerers' provisions. The carpenter, blacksmith, and cook performed the duties which their names suggest. Most whalers also carried one or two boys who were, after a fashion, whalemen's apprentices.

There remained only the main body of foremast hands, differentiated ordinarily into seamen and green hands. At first the former were assigned to the work requiring previous experience, and the latter were entrusted only with the simpler routine jobs. But as the voyage wore on and the green hands improved through repetition and observation the functions of the two were gradually merged. When pursuing whales the foremast hands, four in a boat and with their backs turned to the game, did the rowing; in transforming the carcasses into oil and whalebone they again performed the more laborious tasks; while the duties which fell to them in the ordinary routine of shipboard life were legion.

The relationship between officers and men was far from wholesome. Impressed with the necessity of maintaining discipline through their own efforts, the captain and mates were commonly domineering and autocratic, often bullying and unreasonable, and upon occasion brutal and cruel. The combination of unhampered authority and serious responsibility, together with the prevalent belief (sometimes justified) that discipline could be maintained only by cowing and terrorizing the foremast hands, led to an attitude which even in the better officers was harsh and condescending, and in the coarser ones was vindictive and brutal. Men who were respectable citizens ashore not infrequently became niggardly and cruel tyrants at sea. Except in extreme cases of undoubted criminality, the sailor had no adequate means of redress for any form of mistreatment, and could protect himself only at the risk of incurring a charge of insubordination or even mutiny. The knowledge of this fact by both parties paved the way for numerous flagrant abuses of authority.

One of the commonest and least defensible of these abuses was that of "hazing" or "working up" certain individuals to whom one of the mates had taken an especial dislike. This consisted of assigning such men to all the dirty, disagreeable, and dangerous tasks, keeping them at work for inordinately long hours, and annoying and harassing them in every conceivable manner. Sometimes there were deliberate attempts to provoke a man who was being "hazed" to break through the bounds of discipline and attack the officer responsible for his plight. When this occurred the usual result was a severe beating or flogging for the attacker. "Hazing" was also employed more impersonally for the purpose of disciplining a particularly refractory or criminal hand; but for the most part it was used by the mates as a means of settling their personal grudges.

Physical brutality in some form or other was well-nigh universal in the handling of a crew. It was a rare whaler indeed which was entirely free from it; and on many vessels it was a daily occurrence. The three standard forms of punishment were beating, flogging, and confinement in the "run." Beatings were usually administered on the spur of the moment, when one of the officers was in a fit of anger or impatience. The hapless hand who had incurred displeasure, justifiably or not, would be struck with whatever first came to hand. Sometimes it was a belaying-pin; sometimes a piece of rope, long or short, heavy or light; sometimes a heavy fist or a heavier sea-boot. Some of the men escaped with only a stinging bruise; others were incapacitated for days or weeks by painful and serious injuries; a smaller number were disfigured or disabled for life. But in whatever form or degree of severity, impromptu or premeditated, administered as a reproof for a definite fault or called forth as the result of an officer's peevishness, corporal punishment was the commonly recognized method of enforcing discipline and of forestalling potential insubordination. The annals of whaling, as of the merchant marine, are replete with accounts of brutality, ranging from minor reprimands emphasized by blows to the most severe cases of cruelty, resulting in an appreciable number of instances in death.[3]

Flogging and confinement in the "run" were more formal, standardized modes of punishment. Theoretically they were reserved for the more flagrant infractions of discipline; but too often they were employed whenever it suited the master's fancy. In administering a flogging the body was bared down to the waist. The wrists were tied together and lashed to a ratline with the hands and arms extended full length above the head at such a height that the feet barely touched the deck. This was known as "seizing up." After having been secured in this manner, the man was given as many blows on the bare back with a piece of heavy rope as the captain might direct: for the severity of the punishment was such that the exact number of blows was usually fixed beforehand. When wielded by a strong, angry officer, the impact of the rope on the bare flesh was excruciatingly painful, and every blow drew blood or raised a distinct welt. Finally, with his back lacerated and the torn flesh matted with blood, the man would be cut down and sent forward to care for his wounds as best he might.

Not infrequently a flogging would be followed by confinement in the "run," a small hole under the deck which was utterly without light or ventilation and in which even a very short man could not stand upright. Invariably such confinement, extending over a period of several days or several weeks,

[3] See practically any of the contemporary accounts of whaling voyages given in the Bibliography, as well as the Consular Letters in the State Department Library, Washington, D. C.

was accompanied by a diet of bread and water or worse. Both flogging and the use of the "run" as a prison were widely prevalent practices during the first half of the nineteenth century; but later they fell more and more into disfavor, and public opinion gradually forced their discontinuance.

Added to these orthodox forms of chastisement were many novel ones developed by ingenious captains. In one instance, when two men were reported for fighting, they were compelled to engage in a real fistic encounter, with the understanding that the one who was beaten would receive a flogging in addition. Because he chose to believe that the men were striking too lightly, the captain finally ordered them to strip to the waist and gave each eighteen heavy blows with a whip of tarred cords. The master of another vessel, in dealing with a similar offense, tied the left wrists of the combatants together and forced them to strike each other on the back with pieces of rope held in their right hands. This was continued for thirty minutes; and thereafter one of the men was sent to sit astride the flying jib-boom and the other to occupy a similar position on the spanker boom (the extreme forward and after points of a vessel, which were both uncomfortable and precarious seats). At times these novel punishments were pointedly facetious, as when a cook was ordered to consume single-handed the whole of a poorly-prepared mess of beans and was allowed no other food whatever during the several days required to complete the feast. But for the most part they were grim, bitter, and cruel affairs.

However much the punishment might vary in conception or execution, there was one element which was seldom absent. This was a generous use of profanity and vituperation. It would seem to have been the firm belief of many officers that a choice assortment of the vilest oaths and the foulest epithets was as necessary in handling a crew as a ready fist. Though the habit of addressing the foremast hands in profane terms was not literally universal, it was so common that a sailor simply expected it. Most reprimands and many routine orders were accompanied by minor and commonplace oaths; while a beating or a flogging called forth a storm of studied curses and imprecations. Although it was supposed that such explosive profanity was a material aid in cowing the men into a state of submissive discipline, the very frequency of its use went far towards defeating its purpose; for the seamen became so accustomed to the endless repetition of abuse that the net impression left upon them was far less than was intended.

Chief amongst the offenses giving rise to punishment were fighting, insubordination, slowness or stupidity in obeying orders, refusal of duty, insulting or attacking an officer, attempted mutiny, and all kinds and degrees of criminality, from petty theft to murder. Such matters obviously merited punitive treatment, though they seldom deserved its ferocious manner. But often the most trivial fault, or even a wholly imaginary one, was used as the pretext for a severe beating. This was especially likely to occur when one of the officers chanced to be peevish or partially intoxicated. In such circumstances many persons seemed to derive a positive satisfaction from the process of inflicting suffering upon others; and with unchecked authority in their hands they often gave their impulses free rein.

In other instances the difficulty was caused by misunderstanding or by stubborn stupidity. Thus at the beginning of one voyage a man who had been sick in the forecastle was ordered to take the wheel. While at this post the captain directed him to put the wheel to leeward. The man did so; but having never been to sea before he was ignorant of the universal rule requiring the helmsman to repeat all orders, and therefore said nothing. Instead of merely instructing the green hand in the proper method of response, the master chose to consider it a case of stubborn neglect, and worked up a scene in which the victim was beaten severely. At another time the same captain became enraged because the Portuguese boy at the helm, who had been very ill, was not strong enough to keep the vessel's head steady to the wind in the heavy sea which was running, and because he did not know enough English either to understand the commands and threats or to reply to them. Again the commanding officer insisted that this was a manifestation of stubbornness and of insolence, and struck the boy repeatedly on the face and head with his fists and with a heavy rope.

Fortunately such cruel and abusive treatment was not universal. There were many officers, of course, who were both whalemen and gentlemen, and who were able to hold the respect and confidence of their crews and to maintain a high standard of discipline and efficiency without resorting to brutal and blustering tactics. Such officers conducted the affairs of their vessels in a spirit of firm but just paternalism. They made some attempt, at least, to carry out the advice of a well-known whaling owner, who said in a letter of instructions to one of his captains: "Spare not precept and admonition, and bring us home men, better and happier than when they left us."[4] It was one of the great tragedies of whaling, however, that in an overwhelming number of instances the exact opposite occurred. For as a rule men returned from a voyage not better and happier, but penniless, broken and diseased in body, and corrupted in mind and spirit.

The living quarters on board a whaler were arranged to effect a rigid separation of officers and men. The former occupied the cabin, situated in the extreme after part of the vessel; while the latter lived forward, in a space known as the forecastle. Between the two, but somewhat nearer the stern than the bow, was the steerage, reserved for the boatsteerers, the cooper, and the steward. There was thus a regular gradation of rank from one end of a whaler to the other.

Conditions in the cabin were usually adequate, if not wholly commendable. The captain occupied a large stateroom on the starboard side, with a bed so swung that the rocking of the vessel was counteracted. His quarters were large enough to be shared with his wife if she chose to accompany him. A few hardy and courageous wives did make one or more voyages with their husbands; but in general the whaling masters sailed alone. The mates had smaller staterooms, with ordinary bunks, just forward of the captain's quarters.

[4] Manuscript Letter-Book of Charles W. Morgan, 4 Mo. 1830 to 9 Mo. 1833, now in the New Bedford Public Library. This sentence is taken from a letter of instructions written by Mr. Morgan to George B. Worth, who was about to undertake a voyage as master of the ship *Magnolia*. The letter is dated 1 Mo. 1, 1831. The importance of the Quaker element in the whaling industry, especially in New Bedford, is interestingly reflected in the fact that most whaling manuscripts and documents followed the Quaker custom of referring to the months by number instead of by name.

Still farther forward, and completely separated from the officers' staterooms, was the steerage, an irregular compartment ordinarily containing eight plainly constructed bunks. It was small, poorly ventilated and lighted, and allowed no privacy; but with care and favoring conditions it might be made passably comfortable.

In the forecastle, however, conditions were universally inadequate and often squalid and filthy. The average forecastle was a very low compartment, just under the main deck in the extreme forward part of the vessel, which followed the curve of the bows back some sixteen to twenty-five feet and enclosed the lower portion of the foremast, thus diminishing still further the small deck space. The bunks, crudely constructed of rough planking, were ranged along the sides of the compartment in a double tier. The only ventilation and light came from the hole cut in the deck above for the purpose of giving access to the ladder which was the sole means of entrance and egress. This hole was thus entrance, exit, ventilator, and skylight. In cold or stormy weather, when it had to be kept closed, there was no ventilation or daylight whatever. Such quarters commonly housed from twelve to twenty men (a number at once tragic and ridiculous.)

The resultant conditions were such as could be described only by those who had known them at first-hand. One writer declared: "It would be difficult to give any idea of our forecastle. In wet weather, when most of the hands were below, cursing, smoking, singing, and spinning yarns, it was a perfect Bedlam. Think of three or four Portuguese, a couple of Irishmen, and five or six rough Americans, in a hole about sixteen feet wide, and as many, perhaps, from the bulk-heads to the fore-peak; so low that a full-grown person could not stand upright in it, and so wedged up with rubbish as to leave scarcely room for a foothold. It contained twelve small berths; and with fourteen chests in the little area around the ladder, seldom admitted of being cleaned. In warm weather it was insufferably close.

"In this loathsome den, the Portuguese were in their element, revelling in filth, beating harsh discord on an old viola, jabbering in their native tongue, smoking, cursing, and black-

THE FORECASTLE OF THE WHALER *Charles W. Morgan*, OF NEW BEDFORD

Photograph showing the bunks for the foremast hands

guarding. Their chief recreation, however, was quarreling, at which they were incessantly engaged. Nor was it confined to week-days, for not the slightest attention was paid to the Sabbath. The most horrible profanity was indulged in, and to an excess that was truly revolting." [5]

Elsewhere in his narrative the same author said: "The forecastle was black and slimy with filth, very small, and as hot as an oven. It was filled with a compound of foul air, smoke, sea-chests, soap-kegs, greasy pans, tainted meat, Portuguese ruffians, and sea-sick Americans. The Portuguese were smoking, laughing, chattering, and cursing the green hands who were sick. With groans on one side, and yells, oaths, laughter, and smoke on the other, it . . . did not impress W—— and myself as a very pleasant home for the next year or two." [6]

Another description of forecastle life contained this sentence: "Here, with no possibility of classification and separate quarters, with few or no books, or opportunity to use them if they were possessed, with the constant din of roystering disorder, superabundant profanity, and teeming lasciviousness of conversation and songs . . . three-fourths of their forty months' absence are passed." [7] In the same passage the author stated that he had visited all of the principal prisons in more than half of the United States, and insisted that at the time he wrote (1850) the cells in these prisons were preferable to the average forecastle with respect to roominess, light, ventilation, and possibility of privacy. This picture of forecastle life would not be complete without the statement that many of the older vessels were infested with vermin — a source of endless torment which became particularly acute in the sickening heat of the tropics.

As an offset to such an array of scathing criticisms, at least one instance may be cited in which a forecastle was described as being decently clean and roomy. It boasted of a table, a lamp, and a library of about two hundred volumes. And (*mirabile dictu*) it was scrubbed out every morning! [8]

[5] Browne, J. R., "Etchings of a Whaling Cruse," p. 43.
[6] *Ibid.*, p. 24.
[7] Cheever, H. W., "The Whale and His Captors," pp. 305–306.
[8] Olmsted, F. A., "Incidents of a Whaling Voyage," p. 52.

The amount of time which the men spent in these crowded, nauseating forecastles depended upon the arrangement of the hours of duty. In this regard the whalers, while on passage to and from the whaling grounds, commonly followed the time-honored custom of the sea known as "watch and watch." Under this plan the crew was divided into two watches which were on duty during alternate four-hour periods throughout each twenty-four hours. The sequence was changed daily by substituting two shorter intervals of two hours each, known as the dog-watches, for the period from 4 P.M. to 8 P.M. Thus during any given twenty-four hours, beginning with the first dog-watch, each of the two watches would be on deck, either actually at work or subject to call, during the following periods:

Starboard Watch

4 P. M. to 6 P. M.	= 2 hours
8 P. M. to 12 midnight	= 4 "
4 A. M. to 8 A. M.	= 4 "
12 noon to 4 P. M.	= 4 "
Total	= 14 "

Larboard Watch

6 P. M. to 8 P. M.	= 2 hours
12 midnight to 4 A. M.	= 4 "
8 A. M. to 12 noon	= 4 "
Total	= 10 "

On the following day, owing to the interposition of the dog-watches, the two schedules would be reversed; and this daily alternation would continue indefinitely.

Certain members of the crew, however, were not obliged to stand "watch and watch." The cooper, carpenter, blacksmith, cook, and steward ordinarily worked at their respective tasks during the day and then slept throughout the night, though they might be called upon to assist in handling the ship whenever it seemed advisable. The captain, of course, came and went as he pleased. But in practice most masters felt their responsibilities so keenly that they spent a large proportion of their time on deck, and were always within easy call.

The nature of the duties demanding attention during these various groups of hours differed widely. The watches on deck from 8 A.M. to 12 Noon and from 12 Noon to 4 P.M. were kept steadily at work in the performance of the myriad services which a vessel at sea demanded so insistently. Three men were always at the mast-heads as look-outs, and one man at the helm. These hands were relieved every two hours. The first dog-watch, from 4 P.M. to 6 P.M., was devoted to eating the evening meal, shortening sail, recalling the look-outs, and putting the vessel into condition for the night. The second dog-watch, from 6 P.M. to 8 P.M., was the invariable time for recreation and relaxation. In fine weather everyone was on deck, reading, singing, spinning yarns, dancing, or what not.

During the two night-watches, from 8 P.M. to 4 A.M., the helm was manned and a look-out posted on the knightheads. The remaining members of the watch, though forced to remain on deck, were required to work only in case the weather necessitated changes in the sails and yards. In stormy weather this sometimes meant heavy and dangerous labor. But under normal conditions the officer in charge of the deck was not too inquisitive as to how the men were spending the time, provided they remained within easy call, and kept the wheel and look-out properly manned. Consequently they soon learned to sleep while lying on deck. The morning watch, from 4 A.M., to 8 A.M., was spent in scrubbing the decks, spreading sail, posting the look-outs, eating breakfast, and making preparations for the day.

Though "watch and watch" was the prevailing system on passage, there were occasions when the crew was required to work all day and to stand regular watches at night in addition. This occurred when a "hard" master instituted such a régime either to break the spirits of his men or to secure the greatest possible amount of work from them; and it might be required by emergency conditions such as a long period of severe weather or the necessity of making vital preparations or repairs. Needless to say, such an arrangement, if prolonged, sapped the energy of even the strongest men; for it permitted an average of only five hours per night for sleep.

But no matter what the conditions, there was no escaping a prompt response when the watch was called. Partly for the sake of discipline, partly because the men on deck were anxious to be relieved, there was always a stern insistence that the foremast hands "tumble out" instantaneously.

A Sunday on passage was commonly a welcome day of rest. One phase of the regular week-day routine — posting look-outs at the mast-heads — was carried on as usual, unless the entire vessel was fairly bursting with oil and bone; but otherwise only work incident to the actual handling of the ship was performed. Those who had been foresighted enough to save a little fresh water from their week's allowance enjoyed the luxury of a shave and donned fresh clothing. Thereafter each man, according to his tastes, made the most of the slender recreational resources of a whaler.

But holidays, other than Sundays, received scant attention. Christmas, Thanksgiving, and the Fourth of July were the only holidays enjoying any wide observance in New England during the days of whaling prosperity; and even these were ignored at sea, except for adding some luxury, such as cheese or mince-pie, to the day's menu.

One of the greatest evils of whaling life was the food. In many instances it was not only unpalatable, but positively nauseating. At best the meals were insufferably monotonous and indifferently prepared. The average cook aboard a whaler was so far from expert that in many instances the coarse dishes which he served would have been more digestible, if not more appetizing, in their raw state. Scarcity and poor quality of the food formed the basis of more complaints than any other phase of the whaling industry; and, significantly enough, such complaints seem to have occurred most frequently when the master of a vessel was also part owner.[9] It was only after whaling had entered into its later and decadent stages that laws regulating the quantity and quality of food served to seamen attained any real importance.

The commissary department of a whaler functioned through three sections or divisions — one catering to the officers, a second to the boatsteerers, and a third to the foremast hands. And usually there were significant differences in the meals pro-

[9] See Wilkes, Charles (Lieutenant), "Narrative of the United States Exploring Expedition," Vol. V, Chap. 12.

vided for these three groups. Thus for breakfast the officers had salt beef or salt pork (commonly referred to as "salt horse" or "salt junk"), hard or soft bread, coffee, sugar, butter, and sometimes potato hash; the boatsteerers had the same, but without butter and with molasses substituted for sugar; and the foremast hands were given salt beef or salt pork, hard bread, coffee, and molasses. Molasses was always used by the crew for all sweetening purposes, even in tea and coffee, because it was cheaper than sugar. Dinner, the lightest of the three meals, brought hard bread and salt pork or salt beef to all hands except the officers, who usually added some form of heavy dessert. For supper in the forecastle there was the same "salt horse" or "salt junk," hard tack, and tea or coffee; while in the cabin the salt meat was re-enforced by warm soft bread, butter, sugar, tea, and sometimes pie or hash. The boatsteerers shared the fare of the cabin, but without sugar and butter.

On many vessels certain standardized menus followed each other in such regular sequence that the various days of the week were synonymous with certain articles of food. On one voyage Monday brought beans; Tuesday, rice; Wednesday, potatoes; Thursday, "duff"; Friday, rice; Saturday, codfish; and Sunday, "duff" again. Another weekly schedule was as follows: Monday and Thursday, pork, beans, and corn; Tuesday and Saturday, codfish and potatoes; Wednesday, mush and beef; Friday, rice and beef; Sunday, beef and "duff."

"Duff" was a dish widely known at sea. It was made by boiling a quantity of flour, lard, and yeast in equal parts of fresh and salt water until the mixture was quite hard. Sometimes salt pork fat, chopped fine, was added. A crew which did not receive "duff" and molasses at least once each week was deeply aggrieved. Among the few other delicacies in a whaleman's fare were "lobscouse" (made of hard bread and salt meat, well sprinkled with pepper, chopped into pieces and boiled in water) and "sea-pies" or "dough-boys" (a kind of flour dumpling containing the flesh and bones of porpoises).

At first glance such a drab series of menus does not seem to agree with the long and rather imposing lists of foodstuffs shipped by certain whalers. But a more detailed examination discloses the fact that the more delectable and expensive articles were taken in such small quantities as to be hopelessly inadequate for any purpose except an occasional repast in the cabin. In fact they were used only upon such rare occasions as a visit by the captain of another vessel or an entertainment given in honor of some South Sea chieftain or port official. Never did they find their way into the forecastle; and even the mates were fortunate if they secured the merest taste of them. When a whaler with a crew of thirty men, for instance, started out upon a three to four years' voyage with six pounds of ginger, two dozen cans of lobsters, four pounds of sage, or even one barrel of cucumber pickles and four dozen cans of green corn, it is not to be supposed that any of these articles were ever tasted by a foremast hand.

The staple diet was occasionally supplemented, however, by other articles both more perishable and more enjoyable. Whenever convenient the whalers made port for the purpose of taking on supplies of fresh fruits, vegetables, and meats. When near a favorable coast a few hours' fishing furnished much-needed relief from salt meat; while at sea the flesh of porpoises which had been captured with the harpoon was often eaten with a real relish. A few of the hardened whalemen even enjoyed an occasional whale steak, in spite of its toughness and ultra-gamey flavor. Live animals, particularly chickens and pigs, were frequently carried on deck in specially-constructed pens, whence one or more would be taken from time to time to grace the officers' table. Usually the fresh provisions, like the staple commodities, were regarded as part of the ship's stores, to be given out whenever the captain might deem it advisable; but at times the crew was put upon a regular allowance, or a certain stock would be fully divided, and each man would be permitted to consume his share as he saw fit.

But at best such fresh provisions could not prevent the enforced consumption of other articles which were all but inedible. Long before a vessel was ready to turn homeward, many of the foodstuffs on board were rotten. This was true particularly of the meat, bread, butter, and water. In certain cases crews were compelled to eat meat which was undeniably spoiled. Amazing statements were made regarding the ravages of worms in the bread. One writer declared that it was a common practice to soak the bread in hot tea or coffee, sweetened with molasses, in order to scald out the worms, which were then skimmed from the top of the cup before drinking. And even such treatment yielded bread which was "water-bewitched, coffee-begrudged, and molasses-wasted."[10] Another described an attempt to check the increase of the worms by knocking the bungs out of the bread-casks and pouring half-a-pint of rum into each. The condition of butter which had spent weeks and months in the torrid zone without refrigeration may be imagined. The fresh water, kept in casks for long periods under the same conditions, developed a brackish taste and a disgusting odor. Even "duff," the traditional delicacy for whaling crews, was not always satisfying; for "often this substitute for pudding" was "so hard that it" was "not only indigestible, but difficult to masticate, and more fit to be used as shot for storming forts or towns, than to be eaten."[11]

In addition to these normal and more or less unavoidable circumstances there were innumerable cases in which accidents and stress of weather necessitated a short allowance of even worse food. In all too many instances, too, the niggardliness of owners or masters resulted in the serving of matter which was all but poisonous in its state of advanced decay.

The etiquette observed at meal-time varied directly with the rank of the diners. The captain and mates ate at a table in the cabin, and enjoyed the luxuries of heavy chinaware and a butler disguised as a steward. They were punctilious in seating themselves in order of rank, and in arising in inverse order. As a result the fourth mate was often compelled to eat with prodigious rapidity, in order to satisfy his appetite before the others became impatient. When the officers had finished, the boatsteerers were admitted to the cabin, but only after such articles as sugar and butter had been safely re-

[10] See "Scraps From the Log Book of George Lightcraft, Who Was More Than Twenty Years a Sailor; an Account of The Whale Fishery, With Many Thrilling Incidents in the Life of the Author," pp. 24 and 80.

[11] *Ibid.*

moved; and these trenchermen fell to with scant regard for ceremony.

In the course of his chapter on "The Cabin-Table," in "Moby Dick," Herman Melville gave an inimitable description of the dining-room etiquette of officers and boatsteerers. The dramatis personæ included Ahab, the captain; Mr. Starbuck, the first mate; Mr. Stubb, the second mate; and Mr. Flask, the third mate.

It is noon; and Dough-Boy, the steward, thrusting his pale loaf-of-bread face from the cabin-scuttle, announces dinner to his lord and master; who . . . presently, catching hold of the mizzen shrouds, swings himself to the deck, and in an even, unexhilarated voice, saying, "Dinner, Mr. Starbuck," and disappears into the cabin.

When the last echo of his sultan's step has died away, and Starbuck, the first Emir, has every reason to suppose that he is seated, then Starbuck rouses from his quietude, takes a few turns along the planks, and, after a grave peep into the binnacle, says, with some touch of pleasantness, "Dinner, Mr. Stubb," and descends the scuttle. The second Emir lounges about the rigging awhile, and then slightly shaking the main brace, to see whether it be all right with that important rope, he likewise takes up the old burden, and with a rapid "Dinner, Mr. Flask," follows after his predecessors. . . .

Thus Flask was the last person down at the dinner, and Flask is the first man up. Consider! For hereby Flask's dinner was badly jammed in point of time. Starbuck and Stubb both had the start of him; and yet they also have the privilege of lounging in the rear. If Stubb even, who is but a peg higher than Flask, happens to have but a small appetite, and soon shows symptoms of concluding his repast, then Flask must bestir himself, he will not get more than three mouthfuls that day; for it is against holy usage for Stubb to precede Flask to the deck. . . .

Now, Ahab and his three mates formed what may be called the first table in the *Pequod's* cabin. After their departure, taking place in inverted order to their arrival, the canvas cloth was cleared, or rather was restored to some hurried order by the pallid steward. And then the three harpooners were bidden to the feast, they being its residuary legatees. They made a sort of temporary servants' hall in the high and mighty cabin.

In strange contrast to the hardly tolerable constraint and nameless invisible domineerings of the captain's table, was the entire care-free licence and ease, the almost frantic democracy of those inferior fellows the harpooners. While their masters, the mates, seemed afraid of the sound of the hinges of their own jaws, the harpooners chewed their food with such a relish that there was a report to it. They dined like lords; they filled their bellies like Indian ships all day loading with spices.

But in the forecastle etiquette and ceremony were entirely eliminated. Breakfast was commonly served at seven o'clock, dinner at twelve, and supper at five. At these hours the cook dumped the salt meat into one wooden kid, the potatoes or vegetables (if any) into another, poured the tea or coffee into a large bucket, and vociferously demanded that the food be taken forward by several of the foremast hands. This done, it was deposited on deck or taken into the forecastle, according to the state of the weather. Here the contents of the kids and the bucket were fallen upon by the hands, each of whom was equipped with his ever-present sheath-knife, a tin plate, quart cup, fork, and spoon.

In most cases an inherent sense of rough justice insured a reasonably fair division of the food. Frequently, however, the rule was, "First come, first served"; and on this basis there was always a mad scramble for the kids, accompanied by a disgraceful amount of elbowing and jostling. In order to avoid such scenes some one member of the crew was sometimes chosen to divide the meal into equal parts. Having obtained his share in one of these ways, each man consumed it in any convenient spot on deck or in the forecastle. Thereafter the tin utensils were replaced in the netting above his bunk, where they were often cleaned by the cockroaches before the next meal.

After consuming coarse and often nauseating foodstuffs in this barbarous manner, the sailor found solace in tobacco. The typical veteran whaleman kept a quid of tobacco in his mouth throughout most of the day; and in the evening and at other moments when not on duty he was seldom without his pipe. Many men consumed between one hundred and two hundred pounds of tobacco during the course of a three years' voyage.[12] It was stated that 515,844 pounds of tobacco were used in one year by the 17,594 men of the American whaling fleet at sea

[12] Olmsted, F. A., "Incidents of a Whaling Voyage," p. 83; and Jarman, R., "Journal of a Voyage to the South Seas," p. 89. Many slop-chest accounts, too, serve to substantiate these figures.

on January 1, 1844—an annual average per man of almost thirty pounds.[13] Tobacco was the standard means of alleviating the pangs of an unsatisfied or outraged stomach and the insufferable monotony and discomfort of long months at sea.

A more violent means of reaction lay in the use of strong drink. In the earlier days of whaling, whiskey and rum constituted a regular and supposedly indispensable part of every outfit; and throughout each voyage the men were given a definite allowance, at intervals and in quantities specified by the master. But during the course of the decade 1830–1840 the temperance movement made some headway on American whalers, in spite of active opposition and the deadening influence of tradition. When a certain Captain West refused to carry whiskey on a voyage from New Bedford in 1831 he was considered so crack-brained that he was able to obtain a crew only after the greatest perseverance. Major Williams, the pioneer temperance owner of New London, met with similar opposition in his early attempts to eliminate liquor from his outfits. Both owners and masters, however, soon learned that the absence of strong drink on shipboard was a material aid in preserving discipline and efficiency, as well as a means of economy. After 1840, consequently, there was a steady increase in the number of whalers sailing without intoxicants; and in a few years more this number became a clear majority.

This did not mean, of course, that there was no strong drink whatever on board. The captain and mates usually had a certain quantity for medical and hospitable purposes, as well as for personal consumption. Many foremast hands, too, smuggled small amounts into the forecastle when they came aboard. But it did mean that liquor was no longer doled out to the men, and that after the first few days away from port, when their meagre supplies had been exhausted, they were forced to remain sober.[14]

Facilities for the care of the sick and injured were extremely

[13] Grinnell, Joseph, "Speech on the Tariff, With Statistical Tables on the Whale Fishery." This pamphlet is a revised reprint of a speech delivered in the National House of Representatives, May 1, 1844.

[14] For a discussion of drink and temperance on the whalers see Brown, J. T., in "Fisheries and Fishery Industries of the U. S." (Goode, G. B., Editor), VII, p. 228; Olmsted, F. A., "Incidents of a Whaling Voyage," p. 12; and Whitecar, W. B., "Four Years Aboard the Whaleship," p. 123.

crude. The captain was always *ex officio* physician and surgeon, although he was utterly untrained for such offices and was forced to follow some naïve rule of thumb or his common sense in ministering to his patients. As physician he was guided by the contents of a standardized medicine-chest. This included a limited quantity of the more common drugs and medicines, each of which was given a separate number. Accompanying each chest was a book of directions describing the symptoms of those diseases most common at sea and prescribing a dose of a certain numbered medicine as a cure for each disease so listed. The duties of a physician then became a "simple" matter of reading the book and following directions. For instance, if a man complained of a certain set of symptoms, the captain turned to the guide and found that the appropriate treatment consisted of a dose of No. 7 or of No. 4, which might be calomel or what not. This was promptly administered in the amount specified, and the case considered closed. But at times even this process was felt to be too complicated, and a generous dose of glauber salts, calomel, or castor oil was prescribed for any and every complaint, regardless of the symptoms.

As surgeon the master was even more untrained in his methods and crude in his equipment. With a few exceptions, so rare as to be negligible, no American whaler carried adequate surgical instruments. Nor was there anyone on board who had had the slightest professional training in surgery. This was particularly surprising because accidents of all degrees of seriousness occurred frequently, and because the British whaling industry set a familiar example by carrying an accredited surgeon on every large vessel. But whereas the captain as physician was either ludicrous or grotesque, as a surgeon the very inadequacy of his training and equipment often forced him to exhibit downright heroism. Dozens of instances might be cited in which terrifying operations were performed with coolness and despatch, though not with skill and gentleness; but a single illustration must suffice.

On one voyage a man's hand and foot were completely cut off by the flying loop of a whale-line. It was necessary to amputate the leg and to dress the stump of the arm.

Equipped only with a carving-knife, a carpenter's saw, and a fish-hook, the captain lashed the man to the carpenter's bench, after several members of the crew had fainted in attempting to hold him, and performed the necessary operations alone and unaided. The excruciating pain suffered by the patient in having two major operations performed by an unskilled hand armed with such instruments, and without the use of an anæsthetic, naturally surpasses description.[15]

Illness and incapacity, trying under all circumstances, were particularly so on a sailing vessel far out at sea. For not only did the patient suffer from the sickness or injury itself, the thoroughly inadequate means of treatment, and the nauseating condition of the forecastle, but in many instances he received but scant sympathy and attention. The captain regarded him as a burden and an expense, the mates were suspicious of malingering, and the unfeeling fellows in his watch resented doing his share of the work. The incapacitated foremast hand who was treated with real kindness and genuine sympathy was fortunate above the average.

At once the most troublesome and the most characteristic disease of the whaleman at sea was scurvy. Other maladies, of course, were plentiful. Numerous types of fever, contracted especially on the low-lying coasts of tropical regions; dysentery and allied ailments following frequent and rapid changes of climate; nervous disorders due to hardship, exposure, and dissipation; venereal disease; boils, sores, and other skin diseases caused by the constant chafing of coarse clothing wet with salt water — all these added their quotas to the sick lists. But scurvy remained the peculiar curse of the seaman. This was due primarily to the long continuance of a salt diet and the corresponding lack of fresh provisions; but poor food and water, general filth, insufficient rest, and endless monotony also played important parts.

The disease usually appeared first in the legs, as a swelling and a slight discoloration which gradually changed to a ghastly black. In the later stages it was accompanied by stiffness, extreme weakness, and melancholy, as well as a bunch-like contraction of the leg muscles. The best anti-scorbutic was pure

[15] See Davis, W. M., "Nimrod of the Sea," pp. 195–196.

lime-juice; but fresh provisions were also essential, especially potatoes and onions. When land was within reach, an earth poultice was sometimes applied by depositing the patient in a shallow grave, with only his head and shoulders exposed, for several successive days.[16]

When illness or accident resulted fatally, the unwritten law of the sea prescribed an auction sale of the effects of the deceased foremast hand. After the body, with a weight attached to the feet, had been allowed to slide overboard to the accompaniment of the simplest of burial services, the master or mate unsentimentally auctioned off the few belongings, consisting mainly of articles of clothing, to the highest bidder. The sums realized from the sale were held in trust by the captain, to be given later to the relatives or friends of the dead man. If these could not be found, the Seamen's Friend Society or some other charitable organization might become the beneficiary; or perhaps it would be necessary to turn the amount over to consuls or courts. Partly as a mark of respect to a dead shipmate, and partly because of a desire to assist his family, the etiquette of the sea demanded that the bids to the auctioneer be not only fair, but generous.

Facts fully justified the popular conception of the sailor as a person who entertained profound respect for customs, traditions, and superstitions. Many objects and actions were completely taboo; others were certain omens of approaching death or disaster; and many more were classed as "Jonahs," certain to bring bad luck on a smaller scale. One of the most-feared omens was a phenomenon variously known as St. Elmo's fires, ampizants, or composants. These were pale globular lights, sometimes as large as a man's head, which played about the tops of the masts and the tips of the spars when the atmosphere was heavily charged with electricity. It was widely believed that they represented the ghosts of men who had died aboard the vessel, and that they were warnings of the approaching death of one or more members of the crew.[17]

[16] For further descriptions of scurvy, see Wilkes, C., "Narrative of the U. S. Exploring Expedition," V, Chap. 12; Bennett, F. D., "Narrative of a Whaling Voyage Round the World," p. 193; Jarman, R., "Journal of a Voyage to the South Seas," p. 229; and Nordhoff, C., "Life on the Ocean," p. 214.

[17] For a discussion of superstitions of the sea, see "Fisheries and Fishery

Customary ceremonies, not to be overlooked with impunity, were both numerous and important. Perhaps the most widely observed was the custom of initiating the green hands into the mysteries of Neptune when first they crossed the equator. These festivities usually placed a severe tax upon the initiate's credulity, patience, and endurance. After a great deal of preliminary badgering and the narration of many preposterous yarns by the more sophisticated hands, the novice was at length presented to King Neptune as he came aboard, trailing long streamers of sea-weed and bearing his mystic trident. His majesty at once began a searching interrogation of the neophyte. If he refused to answer he was thumped under the chin; while if he talked too freely a "stopper," consisting of various disgusting and indigestible substances, was put into his mouth. Then his face was lathered with a mixture of lard and tar and he was shaved with a piece of an iron hoop. After undergoing a series of salt-water duckings throughout the ceremonies, he was sometimes "keel-hauled" as a final thrust. This involved being thrown overboard with a rope fastened around the body, by means of which the victim was drawn completely under the vessel and then hauled up on deck on the opposite side. Having passed through this sequence of trials, the green hand was declared to be duly initiated.

Aside from such rare occasions of one-sided mirth and boisterous celebration, there were also more normal means of seeking recreation and amusement. The standard play-time and rest-period on shipboard was during the second dog-watch (6 P.M. to 8 P.M.) At this time, in fine weather, the whole crew assembled on deck, ready for whatever form of diversion their combined resources might produce. Three types of persons were always in great demand: viz., the man who could play any musical instrument; the "minstrel boy" who possessed a wide repertory of songs and a passable voice; and the seaman who could spin a good yarn. The last two, in particular, were always heard with respect and interest. A good yarn, well-told and springing from a vivid imagination, had a never-

Industries of the U. S.," (Goode, G. B., Editor), V, p. 9; Davis, W. M., "Nimrod of the Sea," p. 290; and Olmsted, "Incidents of a Whaling Voyage," p. 15.

ending appeal; and there was a joyous contagion about the boisterous songs, for which the leader produced both familiar and extemporized verses and all hands roared out the choruses. Practical jokes, displaying a broad, rollicking sense of humor, often created great hilarity. From time to time, too, there would be dancing or "skylarking" — the whaleman's term for good-natured, informal wrestling or sparring. But boxing and card-playing were vigorously frowned upon by most whaling captains.

In more quiet moods, recreation included reading, writing, drawing, scrimshawing, smoking, re-examining old letters or other reminders of home and friends, and mending. The veteran tar was adept in the use of a needle; and necessity taught him to perform such prodigies of thrift in mending the various articles of his wardrobe that he was often clad in garments made up of "patch upon patch, and a patch over all."

But the two most picturesque forms of diversion, both peculiar to whaling, were "scrimshawing" and "gamming." A "scrimshawer" was one who carved and decorated by hand numerous articles made from the teeth and jaw-bone of the sperm whale. To most whalemen such slow, tedious work was a welcome means of whiling away many spare hours of a three or four years' voyage. Both utilitarian and ornamental articles were produced in forms and quantities limited only by the perseverance, skill, and ingenuity of the carvers; but cane-handles, pie-wheels, chess-men, and miniature vessels were among the most familiar products of these floating workshops.

"Gamming," on the other hand, was as brief and infrequent as "scrimshawing" was long and constant. When two friendly whalers met at sea, the captain of one vessel, with a boat's crew, went aboard the second whaler while his first mate remained behind to entertain the mate and a boat's crew from the other craft. Both parties, after weeks or months of solitary cruising, were hungry for the news, gossip, and reading matter of the other. Work was suspended and replaced by singing, dancing, and yarn-spinning; "duff" or some other delicacy was added to the menu; and a general holiday spirit pervaded both vessels. The resulting relief from weeks of dreary monotony, the rekindled zest for life, and the release of imprisoned spirits were so great that a whole-hearted "gam" formed one of the most colorful incidents of whaling life.[18]

[18] For interesting descriptions of the sailor at play see the following: Brown, J. T., in "Fisheries and Fishery Industries," VII, p. 231; Davis, W. M., "Nimrod of the Sea," p. 30; Hall, D. W., "Arctic Rovings," p. 51; Whitecar, W. B., "Four Years Aboard the Whaleship," p. 65; and Olmsted, F. A., "Incidents of a Whaling Voyage," pp. 69 ff.

CHAPTER VIII

AT SEA: ON THE WHALING GROUNDS

I — THE WHALEMAN'S NATURAL HISTORY

THE average whaleman was only slightly interested in the anatomy or physiology of his prey, save in so far as some knowledge of these subjects, inevitably gathered through experience, was essential to the successful conduct of the chase. He knew that the whale was a mammal and not a fish; he could distinguish at a glance the various kinds of whales, and understood perfectly the comparative commercial value and probable difficulty of capturing each; he was thoroughly familiar with the location and uses of all those portions of a whale's anatomy which played a part in the chase and in the process of extracting oil and bone; he could estimate to a nicety the number of barrels of oil which any given specimen would yield; he was a past master in the difficult art of large-scale dissection necessary in cutting up a whale's carcass; and he was greatly concerned with the huge dimensions and enormous bulk of those specimens which came under his observation.

But he knew very little, and cared less, about scientific description and classification. That all whales belonged to the Order Cetacea, which included two Sub-Orders, the Whalebone Whales (Mystacoceti) and the Toothed Whales (Odontoceti); that the former was composed of five Genera, of which two, the Right Whale (Balæna) and the Humpback (Megaptera) were commercially important; that the only representative of the Toothed Whales which was of significant importance to man was the Sperm Whale (Physeter Macrocephalus)—all these and other facts and terms of importance to the naturalist were quite unknown to the whaleman. His business, after all, was not to observe, describe, and classify, but to kill whales, to secure a cargo of oil and whalebone.

His activities on the whaling grounds, however, necessarily taught him some of the more elementary facts concerning structure and function. Whales are distinguished among mammals by the huge, fish-like body; by the reduction of the fore-limbs to the form of fins and the total absence of externally visible hind-limbs; by the presence of a thick layer of fat (blubber) beneath the skin, in order to retain the heat of the body; by the small eyes and ears; by the large cavities filled with oil; and by the peculiar breathing apparatus, which makes it possible to inhale and exhale steadily for some consecutive minutes and then to remain under water for periods ranging from thirty to sixty minutes.

But perhaps the most striking characteristic is that of size; for in length a full-grown whale ranges from thirty-five to eighty feet and even more. The following measurements of a sixty-barrel sperm whale (the whalemen always described the size of an animal in terms of the number of barrels of oil which it yielded) will convey some concept of the huge bulk of these largest of living creatures: length, sixty feet; circumference of largest part of body, twenty-four feet; length of jaw-bone, fourteen feet; distance from one point of tail to the other, seven feet; length of each fin, three feet six inches; thickness of blubber, five to nine inches; and number of teeth, forty-six.[1] This is only a good average size; for eighty-barrel sperm whales are not infrequent, and occasionally even larger ones are found.

The sperm whale, the most important representative of the Odontoceti, has a long, narrow lower jaw containing forty to fifty great teeth which fit into sockets in the upper jaw. This arrangement is admirably adapted to tearing and devouring its favorite food, the squid or octopus. The Mystacoceti, or Whalebone Whales, on the other hand, feed upon minute animalculæ which float on the surface of the sea in myriad masses commonly known as brit. The huge upper jaws are

[1] Figures taken from Macy, Obed, "History of Nantucket," p. 229.

CUTTING-IN A SPERM WHALE

Hoisting on the lower jaw

equipped with dozens of slabs of whalebone, with a hairy growth on the outer edge of each slab. In swimming along with the mouth open great quantities of water and brit are taken in until the cavity is entirely filled. Then the jaws are closed and the water allowed to drain out; but the hairy growth acts as a sieve which retains the brit until it can be swallowed.

The sperm whale has a single spout-hole near the forward extremity of the head, from which the thick spout, or exhalation, is thrown forward at an angle of about forty-five degrees; while the right whale, the most important species of the Mystacoceti, exhales perpendicularly through two spout-holes located on top of the body. So powerful is the exhalation that the vapor accompanying it rises several feet into the air. In both species the tail is large, strong, and capable of dealing terrific blows; though in the case of the sperm whale the jaws constitute a weapon fully as formidable.[2]

Only four species of cetacea were important enough commercially to repay systematic hunting on any scale. These comprised the following: the sperm whale, or cachalot, which was found principally in the tropical and sub-tropical waters of the Pacific, Indian, and Atlantic Oceans; the right whale, hunted over areas of great extent in the temperate zones of both the northern and southern hemispheres; the bowhead, or Arctic whale, which lived in the frigid waters near the North Pole, and was pursued chiefly in Davis Straits, Behring Straits, and the Sea of Okhotsk; and the humpback, which was occasionally sought in certain coastal waters of the South Pacific.

With the habits and instincts of these species the whaleman was quite familiar — for success in the chase demanded a thorough knowledge of the probable actions of the victim. Easily the most interesting and most spirited of the four was the sperm whale. The habits of this animal enriched the vocabu-

[2] Only those phases of the whale's structure and function which were of immediate importance to the whaleman are suggested here. For further material on this subject see the following works: Flower, W. H., and Lydekker, R., "Introduction to the Study of Mammals, Living and Extinct"; Beddard, F. E., "A Book of Whales"; Scammon, C. M., "Marine Mammals of the Northwestern Coast of North America"; Beale, T., "The Natural History of the Sperm Whale"; Scoresby, W., "An Account of the Arctic Regions"; Goode, G. B., (Editor), "Fisheries and Fishery Industries of the United States"; and Allen, G. M., "The Whalebone Whales of New England."

lary of whaling with a number of picturesque terms. Among these was the "glip," the name applied to a portion of oil emitted just before going below the surface. The whalemen believed that there was some mysterious connection between the "glip" and the whale, for when any object crossed over the water covered by this oil the cachalot became "gallied," or frightened. The whale "settled" when it simply dropped out of sight, almost instantaneously, without changing its horizontal position; it "breached" when it rose vertically out of the water with great velocity, often projecting one-half of its length into the air; it was said to be "lobtailing" when it seemed to hang perpendicularly in the water, head downward, and swept a great radius with its tail, or "flukes," striking the water with heavy concussions audible for several miles; and it "sounded" when it first lifted the forward part of the body a few feet out of water, gave a strong spout, and then dipped its nose, rounded its back in a high arch, threw its flukes into the air, and disappeared in a perpendicular descent.

The sperm whale was preëminently a globe-trotter; and when, as frequently happened, it covered great distances by swimming at a uniform speed, it was said to be "making a passage." The places visited seemed to include all the warmer waters of the world; for in numerous instances harpoons which had been driven in thousands of miles away and often many years before were found imbedded in the flesh or blubber of a dead whale. Since each vessel carefully marked its harpoon irons as a means of identification, circumstances often made it quite possible to ascertain the approximate time and place in which a pursued animal had escaped with one or more of these marks of battle. A plausible motive for such cosmopolitan wanderings, too, lay in the fact that the cachalot's chief food, the squid, was found most plentifully in the cold ocean currents which originated in the frigid zones and thence spread widely over the expanses of the several oceans. Though these long journeys were made by swimming steadily at a moderate pace, the sperm whale was capable of attaining a high rate of speed for shorter distances. Many writers refer to the terrific speed of many individuals just after being struck; and the number of fathoms of whale-line taken out in a given number of minutes (a fairly accurate test) was often truly astounding.

Spouting was referred to as "blowing." Consequently "There she blows!", or simply "B-l-o-o-w-s!", sung out from the mast-head in long-drawn accents, was the time-honored cry aboard every whaler when whales were sighted. This "blowing" was accomplished with great regularity. Any given whale exhaled a certain number of times, disappeared below the surface with the supply of fresh air which had been simultaneously inhaled, and came up again for the same number of exhalations at the end of a period ranging from forty minutes to an hour. Sometimes the sperm whales traveled in schools, in which case the sea would be dotted with spouts; while at other times single individuals, practically always older males, would be encountered.

Next in interest to the cachalot, and fully as important commercially, was the right whale. Most of the terms used in the preceding paragraphs applied also to this species; but in addition it was accustomed to indulge in the peculiar habit of "sweeping" its flukes from side to side with terrific force. This, however, formed its only means of defense; whereas the sperm whale, adding a pugnacious lower jaw, armed with rows of great teeth, to the threat of its flukes, "was dangerous at both ends." The right whale usually had a heavier coat of blubber, and therefore yielded more oil than a cachalot of equal length. Because of its fondness for cooler waters, it never crossed the equator, and in fact was seldom seen in the latitudes below twenty-five degrees; and because of this wide separation there were noticeable differences between the individuals of the same species found in the northern and southern hemispheres. In common with other cetacea, the right whale seemed to possess a remarkable sense, very disconcerting to the whalemen, which enabled it to know instantly whenever another individual was wounded within a radius of several miles.

The bowhead was found in north polar waters. It differed from the right whale in having no bonnet, or protrusion with deepset barnacles, on its nose; and also in having a smoother skin, longer head, and longer and smoother whalebone. In other respects the two were essentially alike.

Last in order, and so unimportant as to be sought only in the absence of other game, was the humpback. This species was to be found in comparatively shallow water, especially along certain coast lines in the South Pacific, and was hunted in boats which left the vessel each morning for long hours of rowing. The humpback was not only difficult to approach and highly tenacious of life, even after having been severely wounded, but in addition was likely to sink after being killed. As a result an immense amount of labor was required in attempting to hoist the carcass to the surface and to tow it back to the ship. Often, too, it was lost entirely or was devoured by sharks. These facts, together with the inferior quality of the oil which it yielded, insured its comparative safety from the harpoons of the whalemen except in dull seasons.

The bodies of the whales slain in the fishery furnished five important products. From the cachalot came sperm oil, found in a pure state in its head, and of such high quality that it required little or no refining before being used for illuminating and lubricating purposes; spermaceti, a spongy, fatty substance, inodorous and nearly tasteless, which was extracted from the oily matter of the whale's head, and was used in the manufacture of the finest grades of candles; and ambergris, an opaque, ash-colored matter found in rare instances in the intestines of diseased sperm whales.[3] The remaining species furnished whale oil, a heavier, inferior oil used as a lubricant and for the cruder forms of illumination; and whalebone, light, strong, and flexible, incomparably adapted to the manufacture of stays, corsets, hoops, whips, umbrellas, and other articles combining toughness and flexibility.

II — WHALING GROUNDS, ROUTES, AND SEASONS

THE term whaling ground was used to describe those stretches of water, never precisely defined or bounded, which were most

[3] Ambergris was used particularly in the manufacture of fine perfumery, since it possessed the property of uniting thoroughly and permanently all the other ingredients. It was also in demand in certain Oriental countries, especially Turkey, as an aphrodisiac. Owing to its rarity it was exceedingly valuable. The whole American whaling fleet produced only 1667 pounds during the years 1836 to 1880. In 1878 the *Adeline Gibbs,* of New Bedford, secured a fabulous catch of 136 pounds, worth $23,000. See Goode, G. B. (Editor), "Fisheries and Fishery Industries of the United States," II, p. 4.

largely frequented by the animals sought. Both the number and extent of these grounds changed from time to time as old ones were deserted, first by the whales and then by their pursuers, and new ones were added by exploration or accidental discovery. No systematic effort to delimit and describe them was made until 1847, when steps were taken to assemble data for the construction of a series of whale charts designed to show the favorite places of resort of both sperm and right whales during each season of the year.[4] But long before this date most of the well-known grounds had been discovered and their names had become commonplaces in the industry.

These resorts of the whalemen may be classified according to locality or according to the species of whales frequenting them. The Indian Ocean, about 1840, was credited with six separate grounds; the Atlantic with nine; and the Pacific with fifteen. Of those in the Pacific the most important were the On-Shore Ground, including the whole extent of ocean stretching along the coasts of Chile and Peru; the Off-Shore Ground, bounded by the parallels 5 and 10 degrees South Latitude and the meridians 90 and 120 degrees West Longitude; the Middle Ground, located between New Zealand and New Holland (Australia); the waters between the coast of Japan and the Bonin Islands, known as the Japan Grounds; and the northwest coast of North America. In addition there were three long areas traversing the Pacific from east to west. One followed the equator from the coast of South America to the Kingsmill Islands; another extended across the South Pacific between the parallels of 21 and 27 degrees South Latitude; and a third formed a broad belt between the 27th and 35th parallels North Latitude. Other grounds in the vicinity of the Hawaiian, Society, Samoan, Fiji, and Kingsmill Islands, and off the east coast of New Zealand, completed the list.

[4] One of the most interesting results of this survey was the verification of the fact that the right whales never entered the torrid zone. In view of the further fact that the right whales of the North Atlantic and North Pacific were exactly alike, it seemed to follow that the famous Northwest Passage sought in vain by the early explorers did in fact exist in the form of a water communication (open to whales if not to men) through the polar regions. See Lieut. M. F. Maury, U.S.N., "The Winds and Currents of the Sea," p. 34, for an account of the elaborate plans for the construction of these charts.

Later, however, other areas were opened in Behring Straits and in the frigid waters to which the Straits gave access.

The favorite localities in the Atlantic comprised the Carroll Ground between St. Helena and Africa; the north and south coasts of Brazil; the Caribbean Sea; the Gulf of Mexico; the waters in the vicinity of Greenland and Davis Straits; and other regions off the Azores and Cape Verde Islands, east of the Windward Islands, and north of the Bahama Banks. The Indian Ocean afforded favorable opportunities for the chase in the waters off the coast of Arabia, the west coast of Java, the south and northwest coasts of New Holland (Australia), and the southern and northern ends of Madagascar.

In a classification based upon the species sought, the sperm whale grounds included practically all those in the tropical and temperate waters of the world, while the right whale resorts were situated in the frigid and temperate zones. There was thus an appreciable overlapping of the two in the temperate regions. The domains of the cachalot have been listed fully in the preceding paragraphs. The northern right whale, however, was to be found also in the waters of Spitzbergen, Kamchatka, Baffin Bay, Hudson Bay, the Sea of Okhotsk, the Japan Sea, and the Arctic Ocean; and the southern right whale frequented the areas adjacent to Tristan d'Acunha, St. Paul's, Crozet, Kerguelen, and the Falkland Islands, the Cape of Good Hope, and the coasts of Chile, Patagonia, Brazil, Africa, and New Zealand.

A whaler, unlike a merchant vessel, pursued no definite course. And yet certain general routes were followed, with many deviations, in passing from one whaling ground to another. Thus sperm whalers necessarily went out either by way of Cape Horn or via the Cape of Good Hope. In the former case they were likely to leave the United States in the early summer, in order to reach the coast of Chile or Peru in time to "recruit" (putting into port in order to obtain fresh provisions and new supplies) for the season on the Off-Shore Ground which began in November. After two or three months there, one of two favorite courses was usually followed. The vessels might proceed to the Marquesas Islands, continue westward along the equator as far as the Mulgrave Islands, visit the coast of Japan, and return by way of the Northwest Coast of North America, finally reaching the Sandwich Islands in time to prepare for the next November season on the Off-Shore Grounds. Or they might sail directly to the vicinity of the Sandwich Islands, where they would pass the months of February, March, and April, then spend a long summer in cruising along the 30th parallel of North Latitude between 145 and 165 degrees West Longitude, and again arrive at Honolulu in October in time to recruit for the next Off-Shore Ground season.

But those whalemen who went out via the Cape of Good Hope commonly left their home ports in the autumn, in order to arrive on the grounds in the vicinity of New Zealand not later than March. Here they remained for six to eight weeks until the southern winter set in. Thereupon two alternate courses again presented themselves. Some captains went north and then cruised eastward between 22 and 28 degrees South Latitude until they reached the coast of South America, where they recruited at Payta or at Talcahuana Bay and began the season on the Off-Shore Grounds in November; while others proceeded slowly to the Society Islands, arriving there usually in June, continued westward to the Fiji and Samoan groups, turned south to the Middle Ground between New Zealand and Australia, and in March were again ready for the brief New Zealand season. Although the ships and barks pursuing these routes were for the most part the regular sperm whalers, many of them captured right whales also whenever a convenient opportunity presented itself. They seldom returned in less than three years; and voyages of forty to forty-five months were so common that they excited no comment.

The right whalers, on the whole, sailed on shorter and simpler schedules. Some of the earliest recognized whaling grounds were in the vicinity of Spitzbergen, Greenland, and Davis Straits, where it was possible to go and to return in a single season. The southern right whale was hunted most frequently in the high latitudes of the South Atlantic, where quantities of sea-elephant oil were obtainable as a secondary product; and these resorts also admitted of relatively short and direct voyages. Even the scattered areas of the South Pacific required cruises which were less complicated than those of the sperm industry.

But with the mid-century development of the Behring Straits and Arctic Grounds the center of gravity of the right whale fishery shifted to the North Pacific. Vessels sailed from the eastern coast of the United States in the autumn in order to round Cape Horn during the southern summer, reached Honolulu in March or April, spent one or two weeks there in recruiting, and then departed for the northern season, which kept them in the general vicinity of Behring Straits until October or November. This was known as the "regular season." When the severe cold and heavy ice of the winter months forced them to leave such high latitudes, they returned to the Sandwich Islands for several weeks' recruiting. Thereafter they proceeded to more southerly grounds, on the lookout for either sperm or right whales, until the following spring, when they again touched at Honolulu and repeated the performance of the previous year. These winter months to the southward were referred to as the "between seasons" period. In this manner one, two, or three "regular seasons," requiring respectively eighteen, thirty, or forty-two months, would be passed before returning to the home port in the United States.

Many captains who were attracted by the Sea of Okhotsk went out by way of the Cape of Good Hope; but the Cape Horn-Honolulu route remained the more popular. Honolulu formed the logical base for such long-time operations, and possessed many strategic advantages which could not be ignored. Unfortunately the element of chance was so prominent in whaling that deviations from these established lanes were frequently necessitated. In fact, at times the prearranged routes seemed honored more in the breach than in the observance. But in spite of countless variations they were commonly followed as the basic plans for the long voyages.[5]

[5] For further detailed material concerning the whaling grounds, routes, and seasons, see Wilkes, C., "Narrative of the United States Exploring Expedition," V, pp. 487 *et passim;* Goode, G. B. (Editor), "Fisheries and Fishery Industries of the United States," II, pp. 7–26; Scammon, C. M., "The Marine

III — WORKING CONDITIONS: SPERM WHALING

THE ordinary routine aboard a vessel on the whaling grounds was quite different from that observed on the passage thither. At night boat's crew watches, consisting of the four men and a boatsteerer assigned to each boat, divided the time between 6 P.M. and 6 A.M. into four short periods of three hours each; or, if there were only three boats' crews, into three periods of four hours each. Each watch was commanded by a boat-steerer, thus allowing the captain and mates the luxury of an uninterrupted night's sleep; while even the foremast hands were on deck for the space of only three or four hours, instead of the customary alternating four-hour periods of the watch-and-watch system. Soon after daylight all hands were called to reinforce the watch on deck. Sails were set, the mast-head lookouts posted, the decks scrubbed, and breakfast eaten. During the remainder of the day the members of the crew held themselves in readiness to lower the boats for whales at the first opportunity. All work of every sort was minimized; for now the all-consuming aim of life, calling for the rigid subordination of every other consideration, was the sighting and capture of the prey.

Consequently, if no whales were sighted, there was little for any given individual to do except to take his regular two-hour turn at the helm and at the mast-head, answer an occasional call to tack ship or to shorten sail in rough weather, eat dinner and supper, take in sail at sundown, and make all snug for the night. This comparatively light schedule was a welcome and necessary change from the long days and nights of severe, steady labor caused by the capture of one or more animals.

In most cases the whalemen made a systematic survey of the vaguely bounded space of water known as a whaling ground, leaving as little of its area unscanned as possible. With this end in view a vessel stood along under easy sail, keeping the same course for days at a time until the ground had been completely traversed from one side to the other.

Mammals of the North-Western Coast of North America," p. 214; and Holmes, L., "The Arctic Whaleman," pp. 267 f.

Then a similar long sweep, parallel to but a number of miles to one side of the first, was made in the opposite direction. This process, often involving an indefinite number of tacks and interminable periods of beating to windward, was repeated time after time until a great area had been carefully surveyed with surprising thoroughness. On a clear day and in calm weather a man at the mast-head of a ship could see a whale's spout at a distance of four to six miles. This meant that under normal conditions a circular area eight to twelve miles in diameter was constantly visible from any given vessel during the daylight hours.

In order to make the inspection as complete as possible all progress was halted for the night. At sundown each day the light sails were taken in, the topsails close-reefed, and the vessel brought close to the wind, with the helm hard down and the remainder of the sails so balanced that she would remain nearly stationary. In addition it was not uncommon to wear around once or twice during the night in order to counteract whatever slight progress might be made. As a result of these measures it was possible to keep a whaler in almost exactly the same spot during an entire night, ready to take up the search in the morning precisely where it had been discontinued twelve hours before.

In order to stimulate the lookouts to the greatest vigilance it was customary for the officers to offer certain prizes to the man who "raised," or reported, the first whale which was actually captured and "stowed down" in the form of oil. Such rewards usually consisted of money or tobacco, in amounts determined by the nature of the circumstances and the generosity of the captain. From five to ten dollars in cash or five to ten pounds of tobacco were normal amounts. Sometimes such offers were allowed to stand for an indefinite period, and a stipulated amount was given to every man who raised a whale which was captured.

In a few cases the foremast hands themselves agreed to provide a further sum. One such agreement read as follows: "The undersigned, hands before the mast, agree to pay the sum affixed to our respective names, on every barrel of oil raised by a subscriber, to said subscriber; the oil to be measured as stowed down."[6] The "sum affixed" was two cents per barrel for an able seaman, one and one-half cents for an ordinary seaman, and one cent for a green hand. That is, if a certain hand raised a whale which yielded sixty barrels of oil, he was to receive one dollar and twenty cents from every able seaman who had signed the agreement, ninety cents from every ordinary seaman, and sixty cents from every green hand. This arrangement, if continued for any length of time, gave rise to a rather complicated system of mutual debts which largely offset each other. In the long run, however, the more vigilant and more fortunate hands would gain, while those who were less keen-sighted or less fortunate would lose.

Allowing for differences in men and in weather, "standing mastheads" might involve hardship, monotony, or pleasure. In good weather a thoughtful or imaginative man might enjoy to the full this opportunity to escape from the roistering, foul-smelling forecastle; while in bad weather an empty-minded seaman who craved companionship would find his trick of two hours running into several eternities. Always, however, there was the insistent call for vigilance and the cramped discomfort of such an exposed position.

But Herman Melville has devoted a chapter of "Moby Dick" to a portrayal of the everlastingness, the spiritual possibilities, and the physical discomforts of "The Masthead."

In most American whalemen the mastheads are manned almost simultaneously with the vessel's leaving her port; even though she may have fifteen thousand miles, and more, to sail ere reaching her proper cruising ground. And if, after a three, four, or five year's voyage she is drawing nigh home with anything empty in her—say, an empty vial even—then, her mastheads are kept manned to the last; and not till her skysail-poles sail in among the spires of the port, does she altogether relinquish the hope of capturing one whale more . . .

The three masts are kept manned from sunrise to sunset; the seamen taking their regular turns, as at the helm, and relieving each other every two hours. In the serene weather of the tropics it is exceedingly pleasant—the masthead; nay, to a dreamy meditative man it is delightful. There you stand, a hundred feet above the silent decks, striding along the deep, as if the masts were gigantic stilts, while beneath you and between your legs, as it were, swim the hugest mon-

[6] Browne, J. R., "Etchings of a Whaling Cruise," pp. 161 f.

sters of the sea, even as ships once sailed between the boots of the famous Colossus at old Rhodes . . .

In one of those southern whalemen, on a long three or four year's voyage, as often happens, the sum of the various hours you spend at the masthead would amount to several entire months . . . Your most usual point of perch is the head of the t'-gallant-mast, where you stand upon two thin parallel sticks (almost peculiar to whalemen) called the t'-gallant cross-trees. Here, tossed about by the sea, the beginner feels about as cosy as he would standing on a bull's horns.

When, after weeks of such vigilance, whales were finally sighted, the whaleman was projected at once into the most exciting and hazardous of his manifold duties. Few methods of making a living devised by man have combined so much uncertainty, excitement, heavy labor, hardship, and danger as the pursuit and capture of a full-grown sperm whale. This huge game usually betrayed its whereabouts by "blowing," or spouting; by "white-watering," or raising miniature geysers of water and spray by falling heavily upon its side after "breaching"; or by "lobtailing," or striking the top of the water with its massive tail in a series of resounding whacks.

Each lookout was instructed to "sing out" in stentorian tones every time one of these signs was observed. This was the captain's cue to inquire from the deck, "Where away?" When the answer came, "Three points off the weather bow, sir," a series of rapid orders galvanized the whole crew into action. If the spout was a considerable distance away, the sails were trimmed to head the vessel in the proper direction; but if the distance was not too great the boats were lowered immediately and the chase was begun.

The scene of action was now transferred from the ship itself to the whaleboats. The larger whalers carried four of these small, graceful craft — one on the starboard quarter and the remaining three on the larboard quarter, waist, and bow. The starboard boat was reserved for the captain or fourth mate; while the larboard, waist, and bow boats were used by the first, second, and third mates, respectively. The whaleboats were admirably adapted to the industry's insistent demand for combined speed, strength, and seaworthiness. They were usually twenty-eight or thirty feet long, strong but light, shaped to a sharp point at each end, and adapted to rowing, sailing, or paddling. With a small mast and sail which could be set or taken down at will, and a long, heavy steering-oar, they were both fast and easily controlled.

A boat's crew consisted of six men. The boatsteerer occupied the extreme forward thwart and pulled the bow oar until close to the whale, when he stood in the bow ready to hurl the harpoon; the bowman handled an oar and assisted the boatsteerer in setting and taking in the mast; the midship oarsman pulled the longest and heaviest oar; the tub oarsman was responsible for the line in the tubs as well as for his share of rowing; the stroke oarsman rowed, bailed the boat when necessary, and assisted in handling the mast; and the boat-header, or mate, manned the steering-oar, commanded the small crew, and determined the strategy of the chase.

The carrying capacity of a whaleboat was great enough to contain, in addition to six thwarts, an assortment of paraphernalia known collectively as "whaling craft" and "whaling gear." "Craft" included the razor-edged weapons employed in attacking the prey — the harpoon for fastening to the cachalot, the long, slender lance for penetrating deep within its vitals, and the long-handled cutting-spade for severing the tendons of the flukes and for making the incisions to hold the heavy line by which the ponderous body was towed back to the vessel.

The term "gear" applied to the remaining equipment, comprising oars, paddles, a wooden bucket for bailing, a compass, hard bread, water, and other emergency articles, and two wooden tubs containing about three hundred fathoms of whaleline, carefully coiled away and ready to be taken out at great speed when the harpoon to which one end was attached was sunk in a whale. This whaleline, in its combined strength and lightness, represented the highest art of the rope-walk. It was made of the best hemp, uniformly imbued with vapor of tar; and though only two inches in circumference, it was composed of three strands, each strand containing seventeen yarns, and each yarn tested to sustain a weight of at least one hundred and twelve pounds. During the latter half of the eighteenth century the harpoon-gun and the bomb-gun were sometimes added as part of the equipment; but the majority of whalemen clung tenaciously to the hand-harpoon and the hand-lance.[7]

With all this equipment kept constantly in the boats and ready for immediate use, it required but a few moments to begin the pursuit. The mates and boatsteerers took their places at once, while the oarsmen swung the boats free from the davits and then clambered down the sides of the vessel and leaped in as they struck the water. The task now confronting each boat-header was to put his craft alongside a whale without "gallying," or frightening, the prey before the harpoons had been darted. In order to do this it was essential to cover the intervening distance as speedily and as quietly as possible, at the same time carefully avoiding the "glip," or portion of oil emitted in the wake of the whale; for it was believed that the slightest contact with this "glip" would warn the animal of impending danger. If there was a favorable breeze the sail was hoisted and the boat skimmed along with or without the use of the oars, according to the strength of the wind. If the sea was calm and the whales near enough to render the sound of the oars audible, paddles were used. But in most cases the oars alone were employed. Usually the mates were allowed to use their own judgment in conducting the chase. Each vessel, however, had its own code of signals, given by flags and sails; and whenever it seemed advisable, the movements of the boats might be directed in this manner.

With the game in sight, the atmosphere of the chase became electric with excitement and suspense. Every effort was made to reach the cachalots before they sounded; and since, in rowing, the remaining members of the crew faced backwards, the mate was the only one who could follow the incidents and estimate the varying chances of the chase. Partly because of this fact, and partly because of his position of command,

[7] A number of impracticable methods of capture were suggested from time to time. These included, among others, the use of huge nets, which were obviously unsuitable for such large, powerful creatures; the employment of bomb-lances filled with prussic acid, which were given up after a few trials because of the danger of infection from the poisoned blubber; and a scheme (never actually put into practice) for conveying heavy electric shocks through the line and harpoon into the victim's body. See "Fisheries and Fishery Industries," VII, pp. 247–250.

even the most staid and taciturn mate conceived it to be his function to urge the crew to the utmost exertion by a series of excitable comments and picturesque exhortations. By turns he entreated, besought, cajoled, urged, commanded, bullied, threatened, and cursed, employing for the most part a slang vocabulary peculiar to the whaling industry. A large share of this jargon was crude, vulgar, and obscene; but there was also much of the colorful and the picturesque about it.

Sometimes these efforts were successful, and the whale selected by a given boat was reached while still on the surface. But when, in spite of the most vigorous exertions, the animals sounded before they could be struck, there was nothing to be done except to await their reappearance. Knowing that they would seek the surface again in approximately an hour's time, and estimating as carefully as possible their direction and probable speed while under water, the boat was placed as near as might be to the spot where they were expected to rise and the event anxiously awaited. If it transpired that the location had been well chosen, the whales would rise within a comparatively short distance; and this time it would be relatively easy to reach them while they were still on the surface.

In coming to close quarters with a whale it was customary to approach either from the rear (known as "going on the flukes") or from the front (termed "taking it head and head"). This was because the animal's eyes were so situated on the sides of its huge body that it could see neither directly ahead nor directly behind, but only to the side. Hence the obvious advantage of avoiding its field of vision as far as possible. When within a short distance of the goal the mate warned the boatsteerer to stand up in the bow and to prepare for the dart. The latter braced himself and, with harpoon poised, awaited the most propitious moment. This was often delayed until the prow of the boat was only a few feet from the body of the whale, when the harpoon was sunk deep into the yielding blubber and flesh; and if time permitted a second harpoon, attached to the same line, followed the first.

Such normal and successful harpooning, however, was by no means universal. Sometimes the whale "settled" at the last moment, leaving the disappointed boatsteerer without a target for his poised weapon; sometimes the harpoon failed to penetrate far enough to secure an adequate hold; and at other times the chase might be continued for hours without obtaining a single opportunity to strike. In some instances, when the animal had become alarmed while still beyond reach, resort was had to "pitch-poling," a peculiar manner of throwing the harpoon, with tolerable accuracy, for longer distances.

But a successful cast of the harpoon, when it came, was succeeded by moments replete with activity, excitement, and danger. The most immediate necessity was to keep clear of the wild, convulsive movements of the stricken animal; and at the command, "Stern all! For your lives!" every oar was reversed in a frantic effort to keep out of the danger zone. As soon as a position of momentary safety had been attained, the mast was lowered, the mate went forward to the bow, where he was thenceforth to wield the lance, the harpooner took the steering-oar in the stern, and preparation was made for battle at close quarters.

Meanwhile the whale was following one of three courses of action. It might turn at once in an attempt to demolish the boat; it might start off at a tremendous speed, swimming along the surface and towing the boat and its occupants by means of the whaleline attached to the harpoon; or it might "turn flukes" and "sound," taking out fathom after fathom of line as it descended with startling rapidity.

The first case was comparatively rare, and was confined to those few "ugly" whales which were disposed to show fight at once. When such an individual was encountered, however, the only recourse open to the whalemen was the precarious one of attempting to evade the jaws and flukes of the cachalot, as best they might, until the latter tired of its tactics and ceased to take the initiative.

If the victim of the harpoon chose to "run" along the surface the occupants of the boat were given a wild ride, facetiously termed a "Nantucket sleigh ride," in the wake of the fleeing monster. Oftentimes a boat would be towed thus for several hours at a pace which defeated all attempts to shorten materially the length of the tow-line; and it was only after the whale became tired that the mate could be brought close enough, by means of hauling on the line, to deliver a lance thrust.

In many instances, too, the animal tried to rid itself of its pursuers by sounding, sometimes going to such depths that it took out more than the three hundred fathoms of whaleline carried by each boat. When this occurred, a second boat "bent on" with the line in its tubs, thereby taking over the pursuit; and sometimes it was necessary to press a third boat into service before the steady pull on the line was stopped by the whale's ascent. If the tubs were emptied before another boat could "bend on," it was essential to cut the line, thus losing whale, harpoon, and whaleline, in order to prevent boat and equipment from being drawn under water in the wake of the descending leviathan.

But whether the whale sounded or ran along the surface, the energies of the boatmen were concentrated for the time being on the effort to retard as much as possible the terrific rate of speed at which the line was being drawn out of the tubs. This was accomplished largely through the creation of friction. The line was led from the tubs (which were in the after part of the small craft) to a heavy post in the stern, known as the loggerhead. After making two or three turns around this obstruction, it traversed the whole length of the boat before going out through a special slit, or hole, in the bow. The friction developed by the rope in flying about the heavy post was often so great as to cause it to burst into flames; and during the early part of the chase, when the speed of the whale was greatest, it was necessary for one man to wet the timber at frequent intervals.

In commenting upon the amazing rate at which whaleboats were often drawn through the water, one trained observer, an American naval lieutenant, said: "The boat is so much buried in her rapid flight, that I have at times only been able to see the persons in her, for the water on each side was thrown so high as to conceal the hull from a distant observer, although the sea was otherwise quite smooth." [8]

The aim of the pursuers during the whole period after the

[8] Wilkes, Charles, "Narrative of the American Exploring Expedition," V, p. 496.

planting of the harpoon was to seize the first opportunity to haul up to their game again in order to ply the lance. Accordingly when the whale tired after running along the surface, or rose for a fresh supply of air after sounding, the line was hauled in, sometimes rapidly, sometimes with tantalizing slowness, until the boat was again within a few feet of the goal. This was the signal for the mate to plunge his lance several feet into the huge body, preferably piercing the lungs or puncturing some of the larger blood reservoirs known as the "life." With the lance sunk in a vulnerable spot, the officer gave it a continuous up and down motion, known as "churning." This soon tinged the spout with blood, which gradually became thicker and heavier as the wound was aggravated by fresh thrusts of the razor-edged weapon. Sometimes it was possible to keep the boat continuously within striking distance until a mortal wound had been inflicted. At other times it was necessary to keep the small craft off for long intervals spent in awaiting a favorable chance to dash in and deliver a few precarious thrusts. But sooner or later, if the pursuit was successful, the spout showed quantities of heavy, dark blood — the sure sign that the prize was about to be taken.

The final scene of the chase was enacted when the mortally wounded animal "went into its flurry." This consisted of swimming furiously in a narrowing circle, spouting great quantities of blood, and often vomiting the contents of its stomach, made up largely of fragments of its chief food, the ocean squid. At last, after describing many concentric circles in this manner, the tortured cachalot made a final lurch, threshed the sea violently for a few moments, and then rolled over on its side and back with the dorsal fin out of water. This was the end; for "finning out" was accepted by the whalemen as the final sign of death. Tradition had it, too, that the conclusion of this death-scene always came when the whale's head was turned toward the sun.

Only an actual eye-witness, however, could describe adequately the colorful details of the chase and capture of a large sperm whale. In many respects it was a royal sport, involving kaleidoscopic thrills and breath-taking excitement.

The following account, written by a mid-century whaleman, leaves no doubt on that score.

"There she blows!" was sung out from the mast-head.

"Where away?" demanded the captain.

"Three points off the lee bow, sir!"

"Raise up your wheel. Steady!"

"Steady, sir!"

"Mast-head ahoy! Do you see that whale now?"

"Ay, ay, sir! A school of sperm whales! There she blows! There she breaches!"

"Sing out! Sing out every time!"

"Ay, ay, sir! There she blows! There—there—thar' she blows—bowes—bo-o-o-s!"

"How far off?"

"Two miles and a half!"

"Thunder and lightning! so near! Call all hands! Clew up the fore-t'gallant-sail—there! belay! Hard down your wheel! Haul aback the main yard! Get your tubs in your boats. Bear a hand! Clear your falls! Stand by all to lower! All ready?"

"All ready, sir!"

"Lower away!"

Down went the boats with a splash. Each boat's crew sprang over the rail, and in an instant the larboard, starboard, and waist boats were manned. There was great rivalry in getting the start. The waist boat got off in pretty good time; and away went all three, dashing the water high over their bows. Nothing could be more exciting than the chase. The larboard boat, commanded by the mate, and the waist-boat, by the second mate, were head and head.

"Give way, my lads, give way!" shouted P——, our headsman; "we gain on them; give way! A long, steady stroke! That's the way to tell it!"

"Ay, ay!" cried Tabor, our boat-steerer. "What d'ye say, boys? Shall we lick 'em?"

"Pull! pull like vengeance!" echoed the crew; and we danced over the waves, scarcely seeming to touch them.

The chase was now truly soul-stirring. Sometimes the larboard, then the starboard, then the waist-boat took the lead. It was a severe trial of skill and muscle. After we had run two miles at this rate, the whales turned flukes, going dead to windward.

"Now for it, my lads!" cried P——. "We'll have them the next rising. Now pile it on! a long, steady pull! That's it! that's the way! Those whales belong to us! Don't give out! Half an hour more, and they're our whales!"

The other boats had veered off at either side of us, and continued the chase with renewed ardor. In about half an hour we lay on our oars to look round for the whales.

"There she blows! right ahead!" shouted Tabor, fairly dancing with delight.

"There she blows! There she blows!"

"Oh, Lord, boys, spring!" cried P——.

"Spring it is! What d'ye say, now, chummies? Shall we take those whales?"

To this general appeal every man replied by putting his weight on his oar and exerting his utmost strength. The boat flew through the water with incredible swiftness, scarcely rising to the waves. A large bull whale lay about a quarter of a mile ahead of us, lazily rolling in the trough of the sea. The larboard and starboard boats were far to leeward of us, tugging hard to get a chance at the other whales, which were now blowing in every direction.

"Give way! give way, my hearties!" cried P——, putting his weight against the aft oar. "Do you love gin? A bottle of gin to the best man! Oh, pile it on while you have breath! pile it on!"

"On with the beef, chummies! Smash every oar! double 'em up, or break 'em!"

"Every devil's imp of you, pull! No talking; lay back to it; now or never!"

On dashed the boat, cleaving its way through the rough sea as if the briny element were blue smoke. The whale, however, turned flukes before we could reach him. When he appeared again above the surface of the water, it was evident that he had milled while down, by which maneuver he gained on us nearly a mile. The chase was now almost hopeless, as he was making to windward rapidly. A heavy, black cloud was on the horizon, portending an approaching squall, and the barque was fast fading from sight. Still we were not to be baffled by discouraging circumstances of this kind, and we braced our sinews for a grand and final effort.

"Never give up, my lads!" said the headsman, in a cheering voice. "Mark my words, we'll have that whale yet. Only think he's ours, and there's no mistake about it, he will be ours. Now for a hard, steady pull! Give way!"

"Give way, sir! Give way, all!"

"There she blows! Oh, pull, my lively lads! Only a mile off! There she blows!"

The wind had by this time increased almost to a gale, and the heavy black clouds were scattering over us far and wide. Part of the squall had passed off to leeward, and entirely concealed the barque. Our situation was rather unpleasant: in a rough sea, the other boats out of sight, and each moment the wind increasing.

We continued to strain every muscle till we were hard upon the whale. Tabor sprang to the bow, and stood by with the harpoon.

"Softly, softly, my lads," said the headsman.

"Ay, ay, sir!"

"Hush-h-h! softly. Now's your time, Tabor!"

Tabor let fly the harpoon, and buried the iron.

"Give him another!"

"Ay, ay! Stern all!"

"Stern all!" thundered P——.

"Stern all!"

And, as we rapidly backed from the whale, he flung his tremendous flukes high in the air, covering us with a cloud of spray. He then sounded, making the line whiz as it passed through the chocks. When he rose to the surface again, we hauled up, and the second mate stood ready in the bow to dispatch him with lances.

"Spouting blood!" said Tabor. "He's a dead whale! He won't need much lancing." It was true enough; for, before the officer could get within dart of him, he commenced his dying struggles. The sea was crimsoned with his blood. By the time we had reached him, he was belly up. We lay upon our oars a moment to witness his last throes, and, when he had turned his head toward the sun, a loud, simultaneous cheer burst from every lip.[9]

But even the capture of their victim was, after all, but an incident in the labors of the boat's crew. There was always the possibility of the sinking and consequent loss of the valuable carcass.[10] In most instances, however, the body of the sperm whale floated without difficulty; and if there chanced to be other cachalots not too far away, the victorious crew imbedded the pole of a "waif," or flag, in the flesh as a mark of ownership, and then started out in an attempt to overtake and capture another prize.

This was true under favoring conditions; but at times further pursuit was impossible, despite the tantalizing nearness of the game. Often the vessel was entirely lost to sight during the excitement and dangers of the chase, particularly in stormy weather or upon the approach of darkness. When this oc-

[9] Browne, J. R., "Etchings of a Whaling Cruise," pp. 115–121.

[10] The danger of sinking was less for the sperm whale than for the right whale, bow-head, and hump-back, all of which showed a disappointing percentage of losses. One contemporary writer was so impressed with this phase of the industry that he suggested the use of india-rubber or bladder buoys which might be attached to the harpoons and thus keep the dead bodies afloat. This cumbersome procedure, however, was not adopted. See Cheever, H. T., "The Whale and His Captors," pp. 75 ff.

curred, it was imperative to locate and reach this refuge at the earliest possible moment. In other instances it was essential to repair damages sustained by the boat, to care for injuries incurred by its occupants, or to allow time in which to recover from the violence and consuming fatigue of the pursuit.

But if the chances for a second capture seemed favorable, the mate might safely leave the marked carcass for the time being; for the unwritten laws relating to the ownership of floating whales were strictly observed throughout the industry. Thus a carcass containing the "waif" of a vessel believed to be in the general vicinity was never disturbed by another whaler. In case a wounded whale escaped, but subsequently died and was found by some other ship, the harpoons or lances ("whaling craft") found in the body were taken to establish a definitive claim to the oil and bone. The universal rule was: "Craft claims the whale." As a result all lances and harpoons were carefully marked with the name or initials of the vessel to which they belonged. But if the whaleship possessing the rightful claim did not appear within a reasonable length of time, the finder might proceed to "cut-in." In case the owner put in an appearance while the cutting-in process was going on, he was entitled to all the blubber still in the water; but the vessel which had been working on the carcass was allowed to retain everything actually on board. The same rule held true if no evidence of ownership was discovered until after the cutting-in operations had begun.

With one or more whales killed and marked, the next task was to bridge the space intervening between the huge, unwieldy mass of flesh and blubber and the vessel. This might prove either exceedingly laborious or very easy. If the whaler chanced to be to windward of the carcass, it would beat down upon it, leaving little for the occupants of the boat to do except to rest quietly beside their prize. But if fortune had placed the larger craft to leeward, or if a calm prevailed, the boat's crew was forced to tow the tons of dead weight for the full distance. To men already tired out by the exertions of the chase, this long, heavy tugging at the oars, often requiring several hours, was the most gruelling of tasks.

Even after reaching the end of the weary pull it was necessary to attach the valuable prize securely to the vessel before any rest period could be allowed. This operation was by no means an easy one, especially when carried out after dark, as was often necessary. Two boats, one on each side of the animal, began to row from behind towards the head. Each boat held one end of a light line with a weight attached to the middle, by means of which it was sought to pass the line underneath the great body. Since the flukes, in particular, extended under water for an uncertain distance, it required both judgment and patience to bring about the end desired. Having achieved it, however, the light line was replaced first by a stout rope, then by a hawser, and finally by a heavy iron chain. This last was secured about the whale's "small," or narrow portion of the tail between the body and the expanding flukes; and by means of this chain the body was then made fast to the starboard side of the vessel, with the head toward the stern.

With the whale at length firmly secured in this position, preparations were begun at once for the laborious and exacting process of stripping off the blubber, known as "cutting-in." Every effort was made to complete this task in the shortest possible time. For not only did the ravages of sharks result in an appreciable loss of blubber, but the weight and size of such a ponderous mass of flesh alongside rendered it difficult to handle the vessel in rough weather, in pursuing other game, in keeping a definite course, and in any emergency. This was especially true when, as sometimes happened, two or more carcasses were alongside at the same time.

The first step was to rig the "cutting-stage" in such a manner that a narrow plank platform, with a railing on the inside, was provided just outside of and several feet above the body of the whale. From this platform the captain and first and second mates, armed with long-handled cutting-spades, began to attack the carcass. The master and first mate usually undertook the task of severing the head, containing the valuable oil and spermaceti of the "case" and "junk," from the remainder of the body, while the second mate began the process of "scarfing," or cutting the blubber crosswise into long strips twelve to eighteen inches wide. These strips were then peeled off and hoisted on deck by turning the body round and round in the water, just as a layer of tape may be stripped from a cane.[11]

While the blubber was being taken on board, it was necessary for one of the boatsteerers to get down on the slippery heaving carcass, which was only partly out of water and covered with oil and blood, in order to adjust the hooks used in hoisting. In the lurching, rolling, tossing motion of a heavy or even moderate sea, and with such an exceedingly treacherous footing, this duty was unpleasant and dangerous in the extreme. In addition to the possibilities of being struck by the swaying iron blubber-hook, of sliding on to one of the mates' keen-edged cutting-spades, of tumbling into the sea on one side and of being drawn between the whale and the ship on the other side, there was the ever-present danger of being torn by the multitudes of sharks attracted by the flesh of a dead cachalot. In recognition of this precarious situation the boatsteerer always had a rope fastened under his arms and a man on deck detailed to pull him out of danger whenever necessary.

When the first strip of blubber was ready for hoisting, the hook was inserted into a hole cut near the fin and the word given to heave away. Through the combined efforts of most of the foremast hands, who were heaving at the windlass, the blubber was peeled off and hoisted many feet into the air by means of a great block and tackle depending from the main-yard. When this strip had reached the utmost length allowed by the height of the block, a second hook was caught near the bottom and the blubber cut with a boarding-knife just above this. The pendent piece of blubber which resulted from this

[11] The following diagram, taken from Beane, J. F., "From Forecastle to Cabin," p. 35, will illustrate the important parts of the sperm whale which figured in the cutting-in operations.

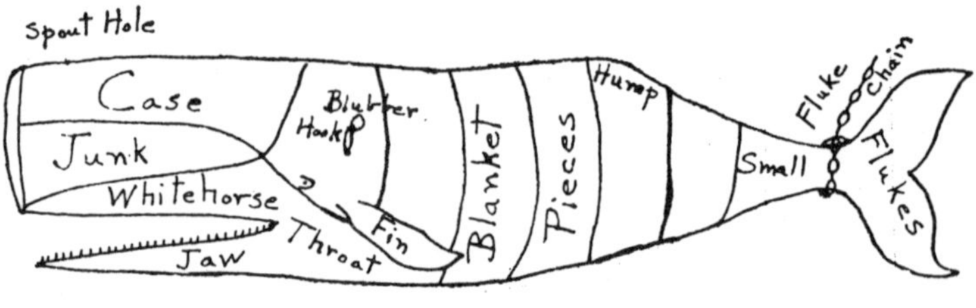

operation, known as the "blanket-piece," many feet long and twelve to eighteen inches in width, was lowered through a hatchway into the blubber-room, where it was immediately cut up into smaller "horse-pieces," about six by twelve to eighteen inches in size. This process of hoisting "blanket-pieces," to be forthwith converted into "horse-pieces" and stored in the blubber-room, was continued until all the available blubber had been stripped from the carcass.

Meanwhile the head was being completely severed from the remainder of the body and divided into three portions — the lower jaw, the "junk," and the "case." The jaw, with its heavy, white bone and huge, glistening teeth, had no commercial value; but both bone and teeth formed the raw material for the numerous objects which resulted from the long hours of "scrimshawing," and therefore an adequate supply was always kept on board. The junk and the case, on the other hand, constituted the most valuable parts of the whole animal. From the junk came spermaceti, a spongy, oily, fatty, inodorous substance used particularly in the manufacture of fine candles and of various ointments and unguents; and from the case came the finest grade of sperm oil in such a pure, liquid state that the large reservoir containing it had only to be bailed out. When all the oil and spermaceti readily procurable had been obtained, the mangled remains of the case and junk, which had been half hoisted on deck, were simply pushed overboard; and then the mutilated body, minus the head and the thick covering of blubber, was cut adrift to become a magnificent prize for countless sharks and sea-birds.

An illuminating account of these cutting-in operations was set down by a contemporary writer who had often participated in such orgies of blood and oil.

> When the whale has been towed alongside by the boats, it is firmly secured by a large rope attached to the "small" by a running noose. The fluke rope is then made fast on the forecastle, and the flukes are hauled up to the bow, or as near as they will reach, leaving the head pointed aft . . . To prevent concussion, the whale is always on the weather side. The progress of the vessel, which is usually under easy sail during the time of cutting in, keeps the whale from drifting out at right angles from the side; though, in most cases, the head is kept in its appropriate position by a small rope made fast aft.

The cutting tackle is attached to a powerful strap, or pendant, passing round the mast in the main-top by two large blocks. There are, in fact, two tackles, the falls of which pass round the windlass. To each of these tackles is attached a large blubber hook, which, upon being made fast to the blubber, are hauled up by the windlass, one only being in operation at a time, so that when the first strip of blubber, or "blanket-piece," reaches the stationary block on the pendant, the other can be made fast by a strap and bolt of wood to a hole cut below the point at which that blanket piece is to be cut off. . . . The blanket pieces are stripped off in a spiral direction, running down toward the flukes; the whole turning, at every heave of the windlass, till the whole covering of blubber is stripped off to the flukes, which are hoisted on board, and those parts containing oil cut off, and the remainder thrown overboard. The head having, in the first place, been cut off and secured to the stern, is now hauled up, with the nose down, if too large to be taken on board, and hoisted as far out of the water as may be found convenient, and the oil or liquid spermaceti bailed out with a vessel attached to a long pole, and thus taken in and saved. As there is no little risk attending this mode of getting the spermaceti, and a great deal of waste, the head is always taken on board, when not too large or heavy.

The "case," which is the name given by whalers to the head, sometimes contains from ten to fifteen barrels of oil and spermaceti. A single "blanket piece" not infrequently weighs a ton or upward. In hauling it up by the tackles, it careens the vessel over frequently to an angle of fifteen or twenty degrees, owing to its great weight, combined with that of the whale, the upper surface of which it raises several feet out of the water. When the blanket piece has reached the stationary block in the top, it is cut off by a boat-steerer, who stands by with a boarding-knife, having first, however, been secured below by the other blubber hook, which is hauled taut to prevent it from breaking away by too sudden a jerk. The upper piece then swings in, and, when it ceases its pendulating motion, is dropped down into the hold or blubber-room, where it is cut up into blocks of a foot and a half or two feet in length, and eight or ten inches in width. These blocks are called "horse-pieces." . . . The carcass of the whale, when stripped of its blubber, is cast loose, and soon sinks from the want of its buoyant covering.[12]

During these cutting-in operations the work of the entire crew was carried on at a pace and under conditions which would have become intolerable if continued over longer periods. The task in hand was pressed forward unremittingly, allowing only the barest minimum of time for food and sleep. Often-

[12] Browne, J. R., "Etchings of a Whaling Cruise," pp. 127–130.

CUTTING-IN A SPERM WHALE

The "Junk" nearly clear

times the vessel was simply permitted to drift, with no attempt to steer a course; and practically every normal activity of shipboard life was given up in a frantic endeavor to get the oil and blubber on board at the earliest possible moment. The decks of the whaler literally ran with oil; and everyone, from the captain to the latest green hand, was fairly coated with oil, blood, and grease.[13]

If several whales were captured in rapid succession, the work of cutting-in might continue uninterruptedly for an entire week, or even longer. In one instance, after an unusual run of luck which brought six whales alongside at one time, "cutting in, trying out, and clearing up the decks, occupied us for the next six days. We had an average of five hours' sleep out of the twenty-four. Working incessantly in oil, which penetrated to the skin, and kept us in a most uncomfortable condition, besides being continually saturated with salt water, produced a very disagreeable effect upon those who were not accustomed to such things, by chafing the skin, and causing painful tumors to break out over the whole body. Before I had half finished my share of the labor, I heartily wished myself in the meanest dog-kennel ashore." [14]

Such a strenuous schedule necessitated a specific allocation of tasks for all members of the crew. The captain, first, and second mates were busy on the cutting-stage; the third mate, stationed in the waist, had charge of hoisting and stowing away the blanket-pieces; the fourth mate divided his attention between the waist and the windlass; and the boatsteerers performed the most exacting tasks in the blubber-room, in the waist, and on the whale. Most of the foremast hands were employed in heaving at the windlass. But there were other assignments as well. One man was detailed to turn the grindstone for the cooper, who was interminably sharpening the cutting-spades; another handled a "scoop-net" in an effort to save the small scraps of blubber which were set adrift; a third

[13] In some cases so much oil collected on the decks that the scupper-holes were stopped up to prevent it from running into the sea. When the operations were completed it could then be soaked up and saved. It need scarcely be remarked that the mere attempt to move about on a rolling and pitching vessel with such decks was exceedingly precarious.

[14] Browne, J. R., "Etchings of a Whaling Cruise," p. 135.

guarded the boatsteerer who was down on the whale; and others were employed in cutting the great blanket-pieces as they came into the blubber-room. The only persons on board who were enabled to follow a reasonably normal program were the cook and the steward. Even these two, however, were often required to reënforce the efforts of the men at the windlass.

Following hard upon the heels of the "cutting-in" operations came the activities of "trying-out," whereby the oil was boiled out of the blubber and stowed away in huge casks in the hold. For this purpose the crew was divided into two watches which alternated for six-hour tours of duty. Here again a regular routine was followed. The horse-pieces were first taken from the blubber-room to the mincing-block, where they were cut into thin slices, but with just enough of one side left uncut to prevent the original horse-piece from falling apart. The resulting appearance of the slices was not unlike that of the leaves of a book; and for this reason they were dubbed "Bible-leaves." These fanciful volumes were thrown into two large iron try-pots, having a combined capacity of three to four hundred gallons, which were built into brick furnaces erected on deck. After boiling for some time the oil became separated from the fibers and was dipped out into a copper container, where it was allowed to cool. Thence it was transferred to casks, wherein it was kept for several days in order to insure a thorough cooling process; and thereafter it was stowed away in the hold for the remainder of the voyage.[15] The fibrous matter left in the try-pots after the oil had been extracted, known as "cracklings," was carefully saved to serve as fuel for the next trying-out period.

As in cutting-in, these trying-out activities were kept up continuously, day and night, until completed. The mates and boatsteerers superintended the loading of the pots, the feeding of the fires, and the ladling of the seething oil into the copper cooler. The remainder of the men on duty were engaged in preparing the blubber for the try-pots, in feeding the

[15] In the earlier days of the industry the same casks which had been filled on deck were re-coopered when the oil was thoroughly cool and then stowed away in the hold until the end of the cruise. Later, however, the oil was put into casks which never left the hold.

fires with cracklings, in moving the empty casks up to the copper cooler, in lashing them securely on deck after being filled, and in a multitude of minor tasks. The slippery, oil-saturated, blood-stained condition of vessel, clothing, and bodies was now greatly aggravated by the heat and smoke of the fires and by the nauseating, fetid odor of the burning cracklings and of the boiling oil.[16]

Nowhere is there half so good a description of the picturesque processes and equipment involved in trying-out as in Melville's chapter on "The Try-Works," in "Moby Dick."

> Besides her hoisted boats, an American whaler is outwardly distinguished by her try-works. She presents the curious anomaly of the most solid masonry joining with oak and hemp in constituting the completed ship. It is as if from the open field a brick-kiln were transported to her planks.
>
> The try-works are planted between the foremast and mainmast, the most roomy part of the deck. The timbers beneath are of a peculiar strength, fitted to sustain the weight of an almost solid mass of brick and mortar, some ten feet by eight square, and five in height. The foundation does not penetrate the deck, but the masonry is firmly secured to the surface by ponderous knees of iron bracing it on all sides, and screwing it down to the timbers. On the flanks it is cased with wood, and at the top completely covered by a large, sloping, battened hatchway. Removing this hatch we expose the great try-pots, two in number, and each of several barrels' capacity. When not in use, they are kept remarkably clean. Sometimes they are polished with soapstone and sand, till they shine within like silver punch-bowls. During the night-watches some cynical old sailor will crawl into them and coil himself away for a nap . . .
>
> Removing the fireboard from the front of the try-works, the bare masonry of that side is exposed, penetrated by the two iron mouths of the furnaces, directly underneath the pots. These mouths are fitted with heavy doors of iron. The intense heat of the fire is prevented from communicating itself to the deck, by means of a shallow reservoir extending under the entire enclosed surface of the works. By a tunnel inserted at the rear, this reservoir is kept replenished with water as fast as it evaporates. There are no external chimneys; they open direct from the rear wall . . .
>
> Here be it said that in a whaling voyage the first fire in the try-

[16] The odor, taste, and feel of blubber were practically always offensive to the uninitiated. The older whalemen, however, often developed a tolerance or even a liking for it. Biscuits were sometimes soaked in the hot oil; and whale meat, prepared in various ways, was occasionally eaten.

works has to be fed for a time with wood. After that no wood is used, except as a means of quick ignition to the staple fuel. In a word, after being tried out, the crisp, shriveled blubber, now called scraps or fritters, feed the flames. Like a plethoric burning martyr, or a self-consuming misanthrope, once ignited, the whale supplies his own fuel and burns by his own body. Would that he consumed his own smoke! for his smoke is horrible to inhale, and inhale it you must, and not only that, but you must live in it for the time. It has an unspeakable, wild, Hindoo odor about it, such as may lurk in the vicinity of funereal pyres . . .

By midnight the works were in full operation. We were clear from the carcass; sail had been made; the wind was freshening; the wild ocean darkness was intense. But that darkness was licked up by the fierce flames, which at intervals forked forth from the sooty flues, and illuminated every lofty rope in the rigging, as with the famed Greek fire. The burning ship drove on, as if remorselessly commissioned to some vengeful deed . . .

The hatch, removed from the top of the works, now afforded a wide hearth in front of them. Standing on this were the Tartarean shapes of the pagan harpooneers, always the whale ship's stokers. With huge pronged poles they pitched hissing masses of blubber into the scalding pots, or stirred up the fires beneath, till the snaky flames darted, curling, out of the doors to catch them by the feet. The smoke rolled away in sullen heaps. To every pitch of the ship there was a pitch of the boiling oil, which seemed all eagerness to leap into their faces. Opposite the mouth of the works, on the farther side of the wide wooden hearth, was the windlass. This served for a sea-sofa. Here lounged the watch, when not otherwise employed, looking into the red heat of the fire, till their eyes felt scorched in their heads. Their tawny features, now all begrimed with smoke and sweat, their matted beards, and the contrasting barbaric brilliancy of their teeth, all these were strangely revealed in the capricious emblazonings of the works.

The weirdness and flaming barbarity of such midnight scenes could not fail to leave an indelible suggestion of the infernal. "It requires but little stretch of the imagination to suppose the smoke, the hissing boilers, the savage-looking crew, and the waves of flame that burst now and then from the flues of the furnace, part of the paraphernalia of a scene in the lower regions . . . I know of nothing to which this part of the whaling business can be more appropriately compared than to Dante's pictures of the infernal regions . . . A trying-out scene has something peculiarly wild and savage in it; a kind

Trying out

Showing the brick try-works erected on deck for boiling out the oil

of indescribable uncouthness, which renders it difficult to describe with anything like accuracy. There is a murderous appearance about the blood-stained decks, and the huge masses of flesh and blubber lying here and there, and a ferocity in the looks of the men, heightened by the red, fierce glare of the fires, which inspire in the mind of the novice feelings of mingled disgust and awe."[17]

But after several days and nights of such feverish, barbarous activity even the largest catch was converted into oil and stowed away. Fortunately all signs of conflict could then be washed away without leaving any permanent stains on either vessel or clothing. The thorough cleansing administered to every part of the craft removed all traces of oil, blood, grease, and dirt so effectually that within two or three days the whaler had fully resumed its wonted appearance, except for the blackening of sails, masts, and yards by the smoke from the fires. This last, however, was a condition common to all whaleships except those fresh from their home ports; so that even the experienced eye of another whaleman would have found it difficult to decide, as the result of a casual inspection, whether the last whale had been cut-in and tried-out three days, three weeks, or three months beforehand.

Again Melville gives us a graphic description, this time in the chapter on "Stowing Down and Clearing Up."

In the sperm fishery, this is perhaps one of the most remarkable incidents in all the business of whaling. One day the planks stream with freshets of blood and oil; on the sacred quarter-deck enormous masses of the whale's head are profanely piled; great rusty casks lie about, as in a brewery yard; the smoke from the try-works has besooted all the bulwarks; the mariners go about suffused with unctuousness; the entire ship seems a great leviathan himself; while on all hands the din is deafening.

But a day or two after, you look about you, and prick your ears in this self-same ship; and were it not for the tell-tale boats and try-works, you would all but swear you trod some silent merchant vessel, with a most scrupulously neat commander. The unmanufactured sperm oil possesses a singular cleansing virtue. This is the reason

[17] Browne, J. R., "Etchings of a Whaling Cruise," p. 63. Another good description of "trying-out" operations may be found in Nordhoff, Charles, "Life on the Ocean," pp. 128–132.

why the decks never look so white as just after what they call an affair of oil. Besides, from the ashes of the burned scraps of the whale, a potent lye is readily made; and whenever any adhesiveness from the back of the whale remains clinging to the side, that lye quickly exterminates it. Hands go diligently along the bulwarks, and with buckets of water and rags restore them to their full tidiness. The soot is brushed from the lower rigging. All the numerous implements which have been in use are likewise faithfully cleansed and put away. The great hatch is scrubbed and placed upon the try-works, completely hiding the pots; every cask is out of sight; all tackles are coiled in unseen nooks; and when by the combined and simultaneous industry of almost the entire ship's company, the whole of this conscientious duty is at last concluded, then the crew themselves proceed to their own ablutions; shift themselves from top to toe; and finally issue to the immaculate deck, fresh and all aglow . . .

But mark: aloft there, at the three mastheads, stand three men intent on spying out more whales, which, if caught, infallibly will again soil the oaken furniture, and drop at least one small grease spot somewhere. Yes; and many is the time, when, after the severest uninterrupted labors, which know no night; continuing straight through for ninety-six hours; when from the boat, where they have swelled their wrists with all day rowing on the Line — they only step to the deck to carry vast chains, and heave the heavy windlass, and cut and slash, yea, and in their very sweatings to be smoked and burned anew by the combined fires of the equatorial sun and the equatorial try-works; when, on the heel of all this, they have finally bestirred themselves to cleanse the ship, and make a spotless dairy room of it; many is the time the poor fellows, just buttoning the necks of their clean frocks, are startled by the cry, "There she blows!" and away they fly to fight another whale, and go through the whole weary thing again.

Such ill-timed smiles of fortune meant that the life of the crews during the months on the whaling grounds was a curious combination of consuming activity and monotonous idleness. At highly irregular intervals a run of luck in raising and capturing whales would involve long periods of fatiguing, unremitting toil. Seemingly endless hours, and sometimes whole days, were spent at the oars, either in the pursuit of game or in towing dead bodies back to the vessel. Upon arriving there it was necessary to take part in cutting-in the most recent capture or to stand a long watch at the try-pots. The meals, inadequate before, now had to be swallowed hastily, in spite of ravenous appetites; and the short and broken hours of sleep were barely sufficient to prevent utter collapse.

On the other hand, when no whales were sighted for weeks or months in succession, the very tedium and monotony of the enforced idleness became almost insufferable. The scanty stock of reading matter was soon exhausted; the entire repertory of songs and yarns was known to all; the mending gave out; card playing was forbidden or waned in interest; scrimshawing could not be pursued interminably; and even sleep could not be courted both night and day. As a result of such an existence most of the men permitted themselves to sink into a state of tedious and quarrelsome boredom, in which the mind atrophied while the body performed mechanically those functions which were demanded of it. The log-books of hundreds of voyages furnish ample evidence of the unvarying sameness and ennui of these long periods of inaction. Such evidence commonly assumes the form of the meager, automaton-made entries extending over page after page; but occasionally it is embodied in direct statement.

An unusually interesting and illuminating record is that of the whaleship *Acushnet*. The first mate of this vessel, in keeping the log during a voyage which began in July, 1845, exhibited a quaint combination of ennui, cynicism, humor, and pained resignation which is both revealing and delightful. Thus on Sunday, October 26, 1845, he wrote: "Your humble servant employ'd in kiling time." On the following Friday the wind was "from Oh! I don't know where or about the same place the whales are I guess and thats where the Devill can't find them." The entry for Tuesday, December 16, 1845, contained this remark: "Busy Doing nothing — nothing to do it with." On Friday, January 23, 1846, he was "Doing nothing special. Dull as you please." One week later they were "Cruising for the Lord knows what. Oil I believe though." On Wednesday, February 4, 1846, it was the "Same old Story. Calculated to see whales made a misscallculation." And on the Fourth of July of the same year he was "Employed Eating & Drinking, fretting, Playing Backgammon & sleeping Mending a pair of pants." [18]

[18] This log-book, a part of the Daniel B. Fearing Collection, is now in the

Most whaling log-books contained an interesting pictorial record of all whales encountered during the course of a voyage. Small ink drawings of various portions of a whale's anatomy were placed in the margin of the log according to a scheme which was as significant to the initiated as it was meaningless to the casual reader. If a victim had been killed and tried-out, a representation of the whole animal, with a space for the initials of the victorious boat and the number of barrels of oil obtained, was given in the margin opposite the appropriate date. If one or more mammals had been raised but not captured, a vertical pair of flukes, suggestive of the act of sounding, was shown for each individual. If a prize had been killed, but subsequently lost through sinking, the word "sunk" was printed beside the pair of flukes; and certain other refinements of pictorial representation were sometimes employed. Some mates of artistic pretensions sketched their figures by hand; but in most instances the regular, standardized outlines betrayed the use of stamping-pad outfits.

Under such circumstances whales were pursued whenever they chanced to be encountered, Sundays and holidays included. After weeks of barren search and unrequited hopes even the foremast hands were eager to forego their Sabbath pseudo-holiday if game were sighted on that day. If, on the other hand, the crew had been tired and overworked for a long period, it was the captain, representing the owners, who insisted upon sending out the boats at every possible opportunity. For every potential prize which was sighted but unpursued and uncaptured increased by an appreciable margin the length of already overlong voyages. Consequently a Sunday on the whaling grounds was in actual practice a work-day whenever there was anything to be done.

There were, it is true, a few so-called "Sunday whalers," whose pious captains refused to lower for whales on Sunday. Now and then, too, a faint and isolated voice of protest was raised against a practice which the rigid, ecclesiastical thought

Treasure Room of Widener Library, Harvard University. Two of the largest and most accessible collections of whaling log-books, however, are to be found in the Whaling Museum and in the Public Library at New Bedford, Massachusetts.

of the day considered a desecration of the Sabbath. But the ministers of the gospel in the whaling ports said little or nothing about Sunday whaling — a fact which one contemporary writer attributed to apathy, a fear of offending wealthy parishioners, and a disinclination to mingle in practical affairs.[19] Partly as a result of this general lack of interest in the question, the percentage of "Sunday whalers" was negligible; and for the most part it was accepted as a matter of course that whales were to be pursued upon every possible occasion, unless a vessel was homeward bound and fairly bursting with oil.

To the officers and boatsteerers this was, on the whole, a satisfactory and even desirable condition of affairs. For them the pursuit and capture of a whale was a gloriously thrilling sport, to which undoubted dangers merely added zest. For in addition to maneuvering the boat and wielding the lance and harpoon, they were in a position to plan the strategy and to observe the varying fortunes of the chase.

But to the men at the oars it was "a good deal as though they were being conveyed to the center of a field of battle, blindfolded, and seated on a car, with their backs to the enemy." [20] At best they could follow the course of the action only imperfectly and at second-hand. Their part was to obey the excited orders of the mate and to send their frail craft into the danger zone without knowing when or how the crises of the battle were to be met. And yet they must manifest no outward sign of fright or panic; for in spite of the nature of the conditions, which were such as to tax the courage of the steadiest as well as of the most reckless of men, no quality was more universally and heartily despised by both officers and foremast hands than the fear of "going on to a whale." Any form of weakness or villainy might be condoned more readily than timidity in pressing the pursuit. To many men, however, such heroic stoicism was unattainable; and in spite of the certainty of incurring contempt and ostracism, they never succeeded in overcoming a natural, understandable, and, to a novice, well-nigh inevitable feeling of mingled fear and despair.

[19] See Cheever, H. T., (Reverend), "The Whale and His Captors," pp. 232 *et passim.*

[20] See Nordhoff, Charles, "Life on the Ocean," p. 225.

IV—WORKING CONDITIONS: RIGHT WHALING IN TEMPERATE WATERS

The equipment, methods of procedure, and general working conditions pertaining to right whaling, when carried on in temperate waters, were essentially similar to those existing in the sperm fishery. In fact, many vessels made a practice of capturing both species during the course of a single voyage, and returned home with a cargo containing not only sperm oil, but whale oil and whalebone as well.

There were, however, some significant differences in the habits of the two species which the activities of the whalemen reflected. The right whale, in addition to an insistent preference for the cooler waters of higher latitudes, often displayed a fondness for the more shallow areas of sounds, bays, or even harbors; whereas the cachalot seldom forsook the deeper portions of the ocean. The former was not only more plentiful, but was also slower in action and less likely to show fight when attacked. Consequently the pursuit of a specimen of this species usually involved less difficulty, excitement, and danger than came to those who sought the cachalot. The right whale possessed one effective weapon in its massive flukes, which could be swept from side to side with great speed and terrific momentum. It also made use of one curious method of passive resistance known as "slack blubber," whereby a portion of the body was doubled into a bow-shape which rendered it impenetrable to the harpoon. But an experienced harpooner seldom allowed himself to be foiled by "slack blubber"; and ordinarily it was possible to evade the flukes by "taking it head and head." Having reached this relatively favorable location, any rushes which might be made at the boat could be stopped by the expedient of pricking an extremely sensitive spot on the end of the whale's nose.

If the right whale was easier to capture, however, the processes of cutting-in and trying-out involved more labor than in the case of the sperm whale. Since the former possessed a heavier coat of blubber, any given individual yielded more oil, on the average, than a cachalot of equal length. A large

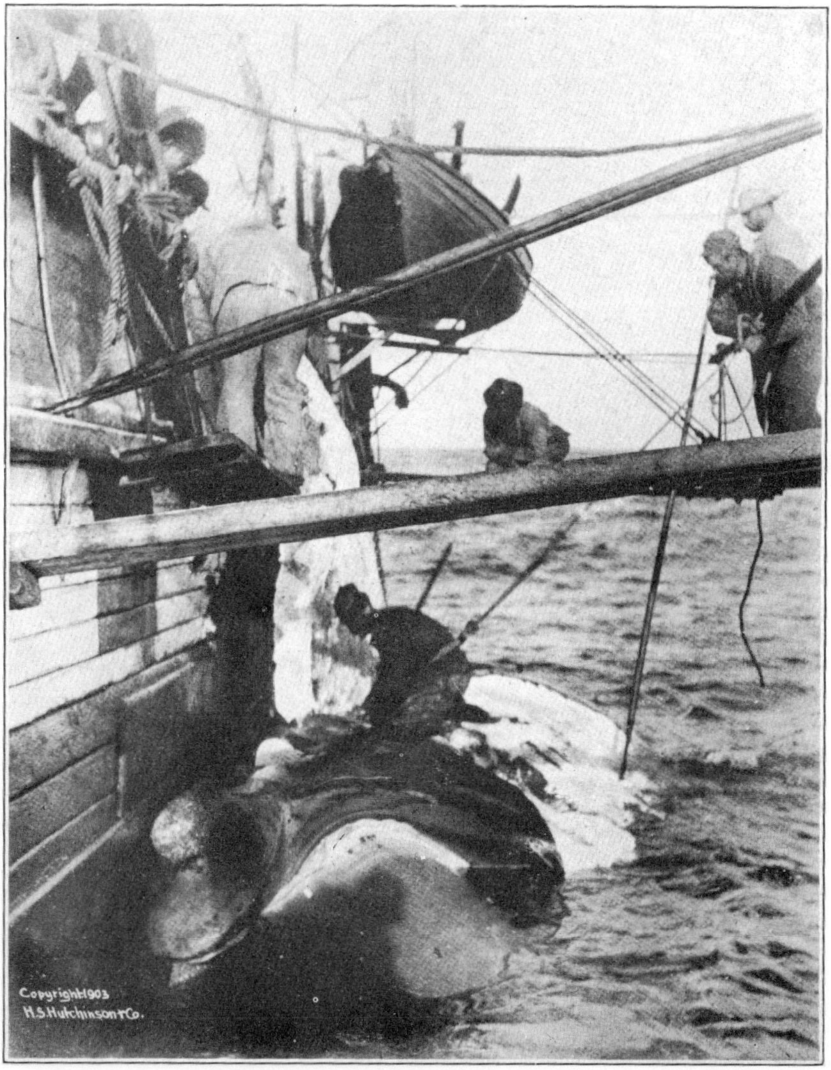

Cutting-in a Right Whale

Showing the Bonnet and a Blanket Piece

specimen yielded from sixty to one hundred barrels of sperm oil; whereas the amount of whale oil secured from a large right whale varied from one hundred to two hundred barrels. And not only was there more oil, but it was heavier and thicker, and all of it had to be tried-out from the blubber; for the right whale had no great reservoir filled with oil in a liquid state, such as that contained in the "case."

Another laborious task confronting the right whaleman was to extract and clean the hundreds of pounds of whalebone which became more and more valuable as the nineteenth century progressed. This bone, found in the right whale's head in the form of dozens of long, heavy slabs, had to be extracted from the ponderous jaws and painstakingly scraped and cleaned before the pieces could be tied together into bundles of convenient size and stowed away in the hold. And since a single large prize yielded from one to two thousand pounds of bone, this was an operation requiring time and patience.

V — WORKING CONDITIONS: ARCTIC WHALING

WHALING in and near the frigid zones was carried on under climatic disadvantages which greatly increased the ordinary hardships and hazards of the industry. The severe cold was a constant enemy, to be fought with heavy, burdensome clothing of such weight that the strenuous duties of shipboard life were rendered even more onerous than usual; and the long periods of inactivity spent at the mast-head and at the helm frequently resulted in frozen hands and feet. The coating of ice and snow deposited on decks and rigging rendered it both difficult and dangerous to handle the vessel, especially in rough weather. Blizzards and snowstorms were likely to overtake the boats when away from the larger craft and to blot it from view. Icebergs, ice-floes, and field-ice proved to be both menacing dangers to the whalemen and friendly allies to their prey; for it was aggravatingly common to have a whale escape by sounding under a large area of ice, often carrying valuable lines and other equipment in its wake. And for those captains who dared to continue their operations until late in the season there was the imminent risk of having their vessels crushed in the ice, or of finding themselves frozen in for the winter, hundreds of miles away from the nearest outpost of civilization.

As if to compensate for such severe climatic drawbacks, however, the right and bowhead whales hunted in these frigid waters were rather more timid and sluggish, on the whole, than those found in warmer seas. This was particularly true of the bowhead, a relatively stupid and slow-moving animal which was less difficult to approach and to kill than any other species of whale.

But this greater ease of capture was not so marked as to permit any significant modification of vessels or equipment. Some appreciable differences in materials, provisions, and working and living conditions were necessitated by the cold and the ice; but aside from these changes, the methods, processes, and general background of Arctic whaling were essentially similar to those of the other branches of the industry.[21]

[21] The classic account of Arctic and Northern whaling was written by an Englishman as early as 1820. See Scoresby, William, "An Account of the Arctic Regions." The best American material was gathered by Clark, A. H., in "Fisheries and Fishery Industries of the United States," V, pp. 73–102; and by Holmes, L., in "The Arctic Whaleman."

CHAPTER IX

HAZARDS AND COMPENSATIONS

THE dangers of whaling were legion. Whalemen were exposed not only to the recurrent risks of storm, fog, and reef which beset every vessel at sea, but also to countless perils arising within the industry itself. In cutting-in it was necessary to guard against sharks, boat-spades, and the treacherous footing afforded by the oil-covered decks of a rolling vessel; in trying-out there was the constant possibility that the flames or the boiling oil would escape from confinement and set fire to the ship; while in a gale the heavy casks of oil might break loose from their lashings and charge about the decks to the imminent peril of both vessel and crew.

The average whaler, too, was commonly isolated and friendless. Always far from the regular lanes of commerce, her captain was essentially an explorer, often years in advance of those first forerunners of civilization, the trading-post and the missionary. Consequently succor was seldom to be expected in any emergency. Unfrequented seas, with their reefs, shoals, dangerous currents, rocky coasts, and strange lands and peoples, had to be navigated with incomplete charts and maps or with none at all. If shipwreck ensued, there was a choice between long days and nights at sea in an open boat, possibly without food and drink, and the prospect of longer weeks or months on uninhabited islands or in the company of unfriendly savages.

But the greatest risks came only when the boats were lowered in actual pursuit of whales. Although the very frequency of such lowerings forced the whalemen to become expert in handling their well-constructed boats under all conditions, gruesome accidents of the chase were all too common. When at close quarters with such gigantic game, the slightest mishap was a death threat. Inexperience, faulty judgment, and carelessness were mortally costly; terrifying disasters occurred in spite of skill, courage, and hardihood; and the deliberate attacks of "ugly" whales, and the convulsive movements of massive bodies seeking to escape, were alike dangerous to the pursuers.

These boat accidents assumed various forms. Frequently a man was whisked away to a watery grave by a flying loop of a "foul line" which was being taken out at lightning speed. The few who were released by the slackening of the line and rose to the surface with ghastly wounds were among the favored of the gods. At times one or more men would be picked out of a boat by a sweep of the flukes which missed the craft itself. But when the flukes registered a square hit, or when the small vessel was caught in the jaws of a sperm whale, the result was a "stove boat" — a catastrophe in which the occupants were often wounded or drowned. Upon other occasions the speed and strength of a "running" whale carried a boat out of sight of the larger vessel; and if then a storm or darkness set in before the whaler was rediscovered, the situation was precarious indeed. Sometimes the following morning brought rescue; but often it was only the first of days that ran into weeks of maddening drifting, hunger, and thirst. In many cases boats so lost were never heard of again. Others were picked up only after one or more members of the small crews had died, and, in a few extreme instances, had been devoured by the survivors.

Only a detailed recital of specific happenings, however, can suggest adequately the dramatic perils, the cruel sufferings which characterized the industry. The desperately earnest nature of the chase was epitomized in the slogan, "A dead whale or a stove boat." And very often it was the latter alternative which prevailed. For a "stove boat" was one of the most common major accidents suffered on the whaling grounds.

The facts of one harrowing encounter which took place off the coast of Chile on December 24, 1847, are given in the logbook of the whaler *Acushnet*. In spite of a rough sea three boats were lowered in pursuit of two whales, a cow and a calf.

After a long period of maneuvering one of the boats was partially destroyed, and a second one, which was fast to the cow, was forced to cut its line in order to go to the rescue of the men in the water. Meantime everyone had lost sight of the third boat. Surmising that some calamity had befallen it also, the ship followed its supposed course and just before nightfall picked up a Sandwich Islander who was swimming with the aid of a piece of plank. From him it was learned that his boat had been completely demolished while attacking the second whale, and the mate killed at once. The five remaining men clung to the wreckage until they were knocked off by another onslaught of the wounded animal. At this time a second man was drowned; and the quartet of survivors, each clinging to a fragment of the boat, began to swim to leeward in an attempt to get beyond reach of the threshing flukes. The height of the waves, however, made swimming exceedingly difficult, and one after another they dropped out of sight until only the Kanaka was left. This man, with typical Hawaiian skill, was able to keep himself afloat until rescued.

A certain Captain A. C. Baker suffered a remarkable series of misfortunes in having boats destroyed under him. On October 13, 1864, while third mate of the bark *Awashonks*, then off the coast of Patagonia, he lowered in pursuit of two whales, and after an exciting chase harpooned one of them. Upon attempting to pull in to lance the victim, however, the boat was caught by the sweeping flukes and demolished. The mate and one oarsman were entangled in the wreckage on the forward fragment, and after a few moments of unconsciousness awoke to find themselves still fast to the whale and being jerked about in the most hazardous manner. At length they succeeded in cutting the line, whereupon they were picked up and taken to the bark. There it was found that Mr. Baker, in addition to many other wounds, had sustained a broken leg. About nine months later, still on crutches and just emerging from a long period of convalescence, he lowered in pursuit of a school of sperm whales and again had his boat knocked to pieces. And during the following voyage, which he made as second mate of the bark *Stafford*, a line was caught while the whale to which it was fast was sounding, and the entire boat carried far under water. Upon this occasion, however, he fortunately escaped serious injury; and throughout the remainder of the cruise his efforts proved more successful.

At times certain "ugly" cachalots seemed to display a deliberate desire for revenge. One animal, upon being harpooned, contrived to catch the line in its open jaws and to swim at great speed in a circle so small that the boat was almost capsized. The line was finally severed by the friction and sawing of the great teeth; and the liberated whale then apparently took to flight and "turned flukes" about a mile away. Fifteen minutes later, however, it came up without warning directly under the boat, capsized and smashed it, threw the crew into the sea, and then made off again, with the harpoon still firmly imbedded in its flesh. Upon another occasion a sperm whale, after having been attacked, sounded and remained under water for a long period. Then suddenly it "breached," caught the boat squarely between its jaws, and crumpled it into bits, giving the men barely time enough to dive into the sea.

In June, 1842, a fearful adventure befell a boat's crew from the whaleship *Russell*, of Dartmouth. While the *Russell* was cruising on one of the Pacific grounds this boat succeeded in fastening to a victim which ran along the surface until out of sight of the vessel. At length, while the mate was endeavoring to pull in for a lance thrust, the cachalot overturned the craft with a blow of its flukes. One man was drowned, but the others regained the capsized boat and clung to the keel. The sea was so rough, however, that even such a small craft could not be righted; and it was not until they had been in the water for some thirty-six hours that the wind abated sufficiently to allow them to turn the boat over, bail it out, and clamber aboard. In spite of hunger, thirst, and exposure, they finally reached Tacamas, where they were providentially found by their shipmates a few days later.

A pugnacious whale was seldom encountered; but when such an individual was attacked, it constituted a formidable adversary indeed. One right whale struck and wrecked a boat in "breaching," and afterwards swept the fragments with its flukes. Two men were killed at once; but the captain managed to keep himself afloat. Upon seeing him in the water, the now infuriated animal attempted to fall upon him with its gigantic body or to strike him with its flukes. Time after time the master was forced to dive in order to escape the sweeps of the massive tail; for neither the vessel nor the remaining boats could get into position to extend assistance. Finally, however, when the swimmer was completely exhausted, the mate succeeded in making a rush and in hauling him out of danger. Another fighting leviathan, encountered by a Captain Huntting off the Rio de la Plata, destroyed in rapid succession four boats sent against it, and finally escaped with several harpoons imbedded in its flesh and twelve hundred fathoms of whale-line trailing in its wake. Though not a single man was killed, the four boats' crews were so unnerved by their harrowing experience that they were unfit for further service, forcing the captain to put into Buenos Aires to ship a new crew, as well as to refit his vessel.

In coming to close quarters a blow from the flukes or jaws sometimes struck the man standing in the bow without touching the boat itself. Captain Edmund Gardner, a well-known whaleman of New Bedford, was so wounded while serving as an officer on the whaler *Winslow*. While attempting to lance a large sperm whale, he received a blow from the animal's lower jaw which was all but fatal. His skull was laid bare, his jawbone broken, and his arms and shoulders severely lacerated. All sail was made at once for the nearest port, which chanced to be Payta. But upon arriving there six days later it was found that the only available surgeon was fifty miles away, and consequently another thirty-six hours elapsed before the wounds could be adequately dressed. In spite of such tardy treatment, however, the patient was later able to resume the voyage and to recover completely.

Another source of mishap was the fouling of the line as it was being taken out by a running or sounding whale. One of the earliest records of an American whaling accident described such an event. The log-book of a certain Peter Folger, kept during the course of a voyage made in 1761, contained this entry: "July ye 29 we stowed away our whale. We saw 2 sloops to the Easterd, and we saw divers sparmocities and we struck one and made her spout Blood. She went down and their came a Snarl in the Toe-line and catched John Meyrick and oversot the boat and we never saw him afterwards. We saved the whale." [1]

During the nineteenth century similar occurrences were frequent. While cruising in the Japan Sea in July, 1846, Captain J. R. Corwin, commanding the whaleship *Portland*, of Sag Harbor, was caught by a flying loop and dragged far under water. By a great stroke of good fortune he was able to cut the line in front of him during a momentary slackness which came when the whale suddenly changed direction; and as a result he rose to the surface in time to be rescued by the boat from which he had disappeared with such magic swiftness. Only a few years later another boat-header saved himself in identical circumstances. In 1853, however, when a third man was similarly snatched out of his boat, he was released only after the combined tension and friction of the line had severed one hand from the remainder of the body.

But even without such calamities as a foul line or a stove boat, the sheer vitality, speed, and strength of a harpooned whale often led its pursuers into the most extraordinary difficulties and adventures. At times a victim sounded to such depths that more than a mile of whaleline was taken out before the ascent began. Then it was necessary that several boats contribute to the lengthening line, for a single set of tubs seldom contained more than three hundred fathoms. If, on the other hand, the animal chose to run along the surface instead of sounding, the six men were drawn along mile after mile at a rate of speed which in such a small craft seemed truly precarious.

Pursuits and combats which continued for periods of eight to twelve hours were well-known; and frequently the body of a hunted whale was pierced by dozens of harpoon and lance thrusts before it succumbed to the attack. On May 28, 1817, the whaler *Royal Bounty* began an amazing adventure in which a huge leviathan dragged from one to six boats and from three hundred to sixteen hundred fathoms of line through the water

[1] This log-book entry is reproduced conveniently in Macy, W. F., and Hussey, R. B., "The Nantucket Scrap Basket," p. 18.

for more than twenty-four hours. After several unsuccessful attempts, a line was finally transferred to the ship itself; and for an appreciable period even this tremendous weight was towed at a rate of two knots per hour. The annals of whaling afford many other examples of the incredible strength and tenacity of life often exhibited by both sperm and right whales.

Because a boat was often towed by a running whale for many miles before it was possible to haul up close enough to deliver a lance thrust, the pursuers were frequently overtaken by darkness and forced to spend the night at sea in an open boat. Often it was not until daybreak that the mutual efforts of vessel and boat to locate each other were successful, although the search was continued throughout the night. Seldom was the quest so happily ended as in the case of two boats which had fastened to a large sperm whale late in the afternoon and were drawn out of sight before sunset. While the ship was sailing to windward in an effort to overtake them a night of such pitch blackness set in that for the time being all further efforts seemed hopeless. Only a few hours later, however, both boats and the captured whale were discovered by means of a lantern held aloft on an oar.

Far less fortunate was the second mate of another Pacific whaler who fastened to a cachalot and killed it just before dark, after a long chase which had taken him far away from his ship. In securing the animal a cutting-spade slipped and all but amputated one hand. But that was not all: one of the South Sea Islanders, crazed through fear of the darkness, threatened momentarily to overturn the boat in his terror. As a last resort the mate was compelled to quiet the cowering Kanaka by keeping his pistol pointed at him. And throughout the night he sat awake, with the excruciating pain of his gaping wound in one hand and a pistol in the other. Nor was it until broad daylight that the boat's crew was discovered by the vessel and taken aboard.

In other cases, however, discovery and rescue did not come even with the morning; and then followed harrowing days and weeks of experiences to be numbered among the most terrifying of maritime adventures. A well-known instance was that of a boat's crew from the bark *Janet*, which in 1849 was cruising about a thousand miles off the coast of Peru. After having been thrown into the water while attacking a cachalot, Captain Hosmer and his five companions vanished from the sight of the bark and the remaining boats. In a small, leaky craft, which they finally succeeded in righting, and without food or water, they were adrift from June 23 until July 13, when Captain Hosmer and one other survivor reached Cocus Island, where they were found two days later by a boat's crew from the whaler *Leonidas*. During this long period of horror the more hardy were driven to subsist on the flesh and blood of those who died, one after another, from starvation and exposure; and in one instance lots were cast to determine which member of the crew should be killed in order to furnish food for those remaining.

Similar experiences, save only for the final eating of the dead, befell the members of another and larger group. While cruising "on the line" (near the equator), three boats from the bark *Harriet* were forced to spend an entire night beside a sperm whale which they had killed just before nightfall. In the morning it was blowing a gale; and they lay by the body for three days, waiting for the gale to blow itself out and hoping against an empty horizon that the bark would find them in cruising about the vicinity. In their disappointment they stood to the westward day after day, drinking rain-water and subsisting upon a shark and several flying-fish which they were fortunate enough to catch. On the eleventh day they were finally picked up by the bark *Hanseat*, of Hamburg; and only a few days later, by a great stroke of good fortune, they fell in with their own vessel, which had given them up as lost and was making its way to Hawaii in order to refit and to ship a new crew. As late as 1881 a boat from the schooner *Edward Lee*, of Provincetown, likewise drifted about for eleven days without sighting a single sail. With no water and with only a little raw meat, cut from the captured whale which was the cause of their difficulties, the members of the crew suffered agonies of hunger, thirst, and exposure before they were rescued by a friendly vessel which took them into Pensacola, Florida.

But even such experiences were, amazingly, not the worst.

Practically every boat which was picked up after days or weeks of drifting could be matched by another which was never heard of again. Through every whaling port drifted the shadowy memories of such losses. And while the sufferings and adventures which preceded death can only be surmised, one must conclude that they were at least as harrowing as those of the men who were eventually saved.

At other times an unsuccessful encounter with a whale involved a less tragic sequel, in that only one or more harpoon irons were carried away by the escaping animal. Since every harpoon was stamped with the name or initials of the vessel to which it belonged, the subsequent capture of the same individual, with this curious means of identification still intact, furnished the material for many a strange tale of the forecastle. Thus in 1802 the ship *Lion*, of Nantucket, Peter Paddack, Master, lost an "iron" in a whale which could not be lanced; but in 1815 the same Captain Paddack, then in command of the *Lady Adams*, captured the identical mammal and recovered his long-lost harpoon. A similar incident occurred when Captain Bunker, of the ship *Howard*, of New Bedford, who had lost an iron in a whale, recovered it five years later. An even longer period elapsed between escape and recapture when in 1851 the ship *Catawba*, of Nantucket, planted an iron which was found by the *Andrew Hicks*, of Westport, twenty years later. Perhaps the strangest case of all, however, was that of a right whale which was killed in the North Pacific by the *Cornelius Howland*, of New Bedford. This animal, captured in 1870, contained an iron belonging to the bark *Ansel Gibbs*, which for at least ten years before 1870 had been whaling only in the North Atlantic. In many other instances, too, recaptured harpoons were found at points far distant from the original battlegrounds.

The destructive activities of the game reached a climax, however, when whaling vessels themselves were attacked. Though such instances occurred only rarely, they included several of the most fearful disasters in the history of the industry. One of the earliest experiences in point was that of the whaleship *Union*, of Nantucket, which was struck by a large sperm whale on October 1, 1807. So heavy was the blow delivered by the infuriated animal that the vessel soon settled and compelled the crew to take to the boats. And only after a week at sea, with an inadequate supply of food and water, did they succeed in landing on the island of Flores, in the Azores. Almost a full century later, on March 17, 1902, when American whaling was rapidly approaching extinction, the crew of the bark *Kathleen* underwent a similar experience. Again the vessel sustained a heavy blow from a cachalot while cruising in mid-Atlantic, and again there was scarcely time to save a little bread and water and to take to the boats before the bark disappeared from sight. Thereafter, however, fortune was kind. For three of the boats were sighted during the following morning by a Scotch steamer; and the fourth made the Barbadoes, distant about a thousand miles, after nine days at sea.

But the two classic cases in which whales turned and destroyed the vessels pursuing them concern the *Essex* and the *Ann Alexander*. The adventures of the crew of the *Essex*, of Nantucket, George Pollard, Master, form a tale astonishing even in an industry replete with ghastly legends. On November 20, 1819, while cruising in Latitude 40 minutes South and Longitude 119 degrees West, three boats were lowered after sperm whales. Soon the mate's boat was damaged so seriously that he was forced to return to the vessel. As if awaiting this event as a signal, a huge cachalot bore swiftly down upon the craft and struck a tremendous blow, followed at once by a second staggering onslaught. The two impacts were so powerful that the ship rapidly filled with water and fell over on her beam-ends. After a consultation with the other two boats, which had returned without securing their game, it was decided to obtain water and provisions from the wreck, strengthen the three small craft as much as possible, and attempt to make for the coast of Chile or Peru.

Then began one of the most ill-fated journeys on record. The three boats, containing fourteen white men and six blacks, all on the most meager allowance of bread and water, half-sailed and half-drifted from November 20 until December 20 without encountering the slightest cause for hope. On the last-mentioned date they sighted land, but only to find a small, uninhabited rock, known as Ducie's Island. Here were such scanty supplies of food and water that they were compelled to sail on, after leaving three men who preferred the possibility of starvation on the island to the interminable torture of suspense in the boats. In spite of much rough weather, intense heat, and increasing hunger, thirst, and weakness, the three leaky, weather-beaten boats managed to stay together until January 12, 1820, when the mate's craft separated from the others in a storm. On January 28 the two remaining groups parted company under similar circumstances; and thereafter the isolated parties suffered even more excruciatingly than before.

After two men in the mate's boat had died and had been eaten by the remainder, the three survivors were finally picked up on February 19, 1820, and taken into Valparaiso. In the captain's boat, meantime, two men died and one was shot after drawing lots. Their bodies kept the master and one hand alive until they were sighted by the whaler *Dauphin*, of Nantucket. This was also late in February. The second mate's crew was never heard of again, and undoubtedly perished at sea; while the five emaciated wretches who were rescued had been forced to spend three full months in open boats.

The loss of the *Ann Alexander*, though occurring under similar circumstances, involved far less suffering on the part of the crew. On August 20, 1851, an "ugly" sperm whale, harpooned on the Off-Shore Ground, turned and crushed two boats in succession. After rescuing the crews, the captain pressed the attack in the vessel itself, barely avoiding several dangerous rushes. Just before sunset, however, the cachalot succeeded in delivering such a heavy blow that the ship became water-logged and the crew was forced to spend the night in the boats. The next day was spent in securing provisions from the wreck and in preparing for a long journey; but fortunately they were picked up within twenty-four hours and taken into Payta. Five months later, by a curious coincidence, the same whale, with two harpoons in its blubber and pieces of the ship's timbers imbedded in its head, was killed by the *Rebecca Sims*, of New Bedford.

But not all the hazards of whaling life were due to actual combat. Other dangers, too, there were in abundance. Fire was a significant risk aboard vessels which were compelled to keep the contents of the try-pots at the boiling point for many hours in succession. Thus, on January 31, 1835, the ship *Lydia* caught fire while cruising in the Pacific, and burned so fiercely that she had to be abandoned in less than an hour. Luckily the three boats were sighted within a short time; but the vessel was a total loss. The *Lady Adams*, last spoken off the coast of Japan in July, 1823, is believed to have burned soon afterwards, as an English whaler reported seeing a great light in the probable location of the vessel. Another conflagration provides an interesting sidelight upon the institution of American slavery. For the ship *Cassander*, of Providence, was deliberately set on fire by two negroes who were afraid of being sold into slavery. This was on May 1, 1848, in the middle of the South Atlantic. The crew was forced to take to the boats; and, in spite of an inadequate supply of food and water, reached the coast of Brazil after ten days at sea.

The losses due to storm and reef were even greater. Gales, hurricanes, and typhoons swept away so many vessels that their number is legion. One of the most tragic whaling wrecks was that of the ship *Henry*, of Nantucket, which, after a long Pacific voyage, was completely dismasted on July 27, 1813, when only a few days from her home port. Three men were drowned at once; but the remainder clung to the bowsprit, the only portion of the vessel above water. From this precarious perch one after another dropped into the sea until only five were left. Heavy seas washed over them repeatedly, and only the smallest rations of bread and water could be obtained from the derelict. But this quintet lived on the bowsprit for forty days, until at length they were taken off by an English vessel.

The strange and uncharted waters so often explored by the whalers increased immeasurably the dangers resulting from reefs and shoals. One of these uncharted reefs, struck on March 4, 1854, by the ship *Canton*, of New Bedford, caused another amazing journey in small boats. This obstruction, situated in the Pacific and almost on the equator, was near a small, barren island. There the crew lived for four weeks, subsisting upon food from the wreck and enduring the most enervating heat. At the end of that time the supplies of food and water were so low that they were forced to put to sea in four boats, with an allowance of one-half pint of water and one-half biscuit per man per day. For forty-five days they drifted without seeing either a sail or a sign of land. Then they landed on one of the Ladrone Islands; but since no fresh water was found, another start was made, this time for Tinian. There the Spanish governor insisted that they were pirates and refused to allow them to come ashore. At length, however, he was prevailed upon for a small supply of bread and water; and with this scant encouragement they set out again, finally reaching Guam and safety only after another month and a half of semi-starvation.

The ship *Junius* was likewise lost on a reef in the Mozambique Channel in October, 1851; and on February 11, 1823, the whaler *Two Brothers* sank within thirty minutes after striking heavily on some hidden rocks in Latitude 24 degrees 9 minutes North and Longitude 167 degrees 30 minutes West. In the former case the crew, with only four salt hams as food, spent six days and nights at sea before reaching land; while the twenty-one men of the *Two Brothers,* crowded into two boats, were fortunately picked up within a few hours by another whaler.

Another curious accident befell the ship *Minerva Second,* of New Bedford, while cruising in the South Pacific in 1856. A reef was touched so lightly that no evidence of serious damage could be found. Several months later, however, a heavy leak began quite unaccountably; and the crew, working furiously at the pumps, was barely able to prevent the vessel from foundering before reaching Sydney. There it was discovered that the contact with the reef had loosened a portion of the copper sheathing, and that the few boards thus exposed had been so eaten by worms that at length the water was allowed to rush into the hold.

A further series of dangers awaited those whalemen who ventured into high latitudes in pursuit of the right whale and the bowhead. The extreme cold and the utter barrenness of the Arctic regions were sufficiently serious handicaps; but the greatest perils centered about the existence of vast quantities of ice. Whether in the form of icebergs, ice-floes, field-ice, or shore-ice, it was a constant menace. And perhaps the most dreaded fate of all was to be frozen in through a long Arctic winter.

One of the few vessels which ever returned from such a captivity was the English whaleship *Diana*, which left the Shetland Islands in 1866. When about to start for home after a season's whaling, the ice closed in and both vessel and crew were imprisoned for the entire winter. Cold, hunger, scurvy, and dysentery were responsible for such ravages that when the ice finally broke up in April, 1867, only five members of the original crew of fifty were able to aid in navigating the ship. Ten had died, and the remaining thirty-five were so weakened by cold, starvation, and disease as to be unable to walk. The vessel itself was reduced to a mere shell, for boats, spars, and every loose piece of timber had been used as fuel. Consequently the partially dismantled hulk, laden with weakened and diseased like a hospital ship, was barely able to creep back to civilization.

A still more gruesome tale is one which, in its tragic, ghostly outlines, can be matched only in the "Ancient Mariner." In August, 1775, the master of a Greenland whaler sighted a strange vessel in a lane which had opened in an ice-field. Upon going aboard he discovered that it was abandoned and drifting. A closer inspection revealed, too, the frozen bodies of the master and his wife, the mate, and a number of seamen. The log-book, which was well preserved, contained this final entry: "Nov. 11, 1762. We have now been inclosed in the ice seventeen days. The fire went out yesterday, and our master has been trying ever since to kindle it again without success. His wife died this morning. There is no relief." Later, upon returning to England, it was discovered that the vessel had sailed in 1762 and had never been heard of again. Thus it seemed to be established that this floating sepulcher, with its corpses embalmed by the cold, had been held icebound, through some curious trick of fate, for thirteen years.[2]

In 1871 the entire American Arctic fleet, with the exception of five vessels, was similarly caught in the ice at the end of the season. Fortunately, however, the nature of the loss was reversed; for this time it was the crews, and not the ships, which

[2] See Cheever, H. T., "The Whale and His Captors," pp. 175–80.

were saved. Thirty-three whalers had to be abandoned to the mercy of the winter storms and crushing ice-fields. But the five members of the fleet which chanced to be in clear water transported the crews of all the deserted vessels to Honolulu without the loss of a single life.[3]

But neither the attacks of whales nor the forces of nature were responsible for all the dangers of whaling. Human nature, in the rôles of hostile savages and of mutineers, was also a contributing factor. Savage treachery probably reached its climax in the tragedy of the *Sharon*. On Sunday, November 6, 1842, this vessel lowered all of its boats in pursuit of whales, leaving only Captain Norris, a white boy, and three Kingsmill Islanders on board. During the day one of the natives stole up from behind and decapitated the captain with a cutting-spade. The boy, however, took refuge in the rigging, where his agility frustrated the attempts of the natives to follow him. When the boats returned their crews were prevented from boarding the ship by the formidable array of whaling weapons which the islanders had collected. But plans to sail away from the boats were defeated by the boy, who slid from mast to mast on the stays and cut the ropes and sails in such a manner that little headway could be made. During the night the third mate swam to the ship, climbed up the rudder, and squirmed through the cabin window. His discovery, while still engaged in loading the captain's firearms, precipitated a terrible hand-to-hand conflict in which one native was killed and the mate and a second islander severely wounded. The third assailant hid in the hold, where he was subsequently found and carried into the port of Sydney in irons.

The relations between the whalemen and the South Sea Islanders ranged from consistent friendliness to open combat. In some regions the natives maintained an air of light-hearted welcome and respect, often in spite of insults and indignities; but in other localities they were openly or covertly hostile. Thus in 1835, while touching at the Marshall Islands, the

[3] The best accounts of this catastrophe are contained in the Old Dartmouth Historical Sketches, Number 45, p. 43; and in Starbuck, Alexander, "History of the American Whale Fishery," Section F.

ship *Awashonks*, of Falmouth, was attacked by a group of natives who had come aboard on a "friendly" visit. The result was a bloody fray in which the decks were cleared only after several members of the crew and a number of attackers had been killed.

But savage hostility was most feared in case of shipwreck. Seamen who contrived to escape from a wreck, only to be cast upon unfriendly shores, were sometimes led to consider fortune's smile a sneer. For their reception often resulted in death, torture, or a period of captivity which continued for years. The mere titles of the occasional records of such experiences strongly suggest the strange, barbaric sufferings. Witness one account, published in Boston in 1836, which was entitled: "A Narrative of the Shipwreck, Captivity, and Sufferings of Horace Holden and Benj. H. Nute; who were cast away in the American ship *Mentor*, on the Pelew Islands, in the year 1832."

With mutiny the list of whaling misadventures described a full circle. Such uprisings were brought about either through half-justifiable resentment wrought by inhuman treatment and unbearable conditions, or through the machinations of the criminal and irresponsible individuals often found in whaling crews. During one voyage, for example, the captain and mates had been unreasonable in their demands, which were stubbornly resisted by the crew. At length the men became mutinous. Upon arriving at Payta, they drove overboard a number of Spanish soldiers who had been sent on board to furl the sails and to deal with the mutineers, and then left the ship in a body. So strongly did the seamen from all the other vessels in the harbor sympathize with the deserters that the authorities dared not arrest them; and in the end the captain was compelled to ship a new crew.

The mutiny on board the ship *Junior*, which took place on Christmas night, 1857, was far more tragic and bloodthirsty. Shortly after midnight five members of the crew, headed by a certain Cyrus Plummer, crept into the cabin and attacked the sleeping occupants with guns and whaling weapons. All of the officers were brutally murdered except the first mate, who, though seriously wounded, managed to escape into the hold. The mutineers now attempted to take charge of the ship; but finding it impossible to steer a straight course, they were compelled to offer the mate his safety in return for his services as navigator. With this assistance they finally neared a little-known shore in the South Pacific, where the ten mutineers, taking with them a large quantity of supplies, put off in two boats. Before disappearing from view, however, they forced the mate to swear that he would sail away at once and not seek to molest them; and in return they wrote the full and true account of the whole episode in the log-book, in order to free the remainder of the crew from all suspicion of complicity in the crime.

The classic case of mutiny in the annals of whaling, however, was that which occurred on board the ship *Globe*, of Nantucket, in January, 1824. Again several mutineers, headed by a boatsteerer named Comstock, attacked and killed the sleeping captain and mates. Comstock, who seems to have been actuated by personal grievances as well as by the general dissatisfaction of the crew, thereupon took command of the vessel, with another hand named Payne acting as mate. Those members of the crew who had not openly joined the mutineers were forced to an apparently willing acquiescence; and with affairs in this state the ship reached the Mulgrave Islands. Here preparations were made for an extended sojourn on shore. Payne and Comstock quarreled, however, and the latter was murdered. Soon thereafter six members of the crew, who had never voluntarily joined in the mutinous acts, succeeded in escaping with the vessel. In spite of their short-handed condition they finally reached Valparaiso, where the case was officially reported. Meantime Payne and his associates were subjecting the natives to such indignities that the islanders turned and massacred all of the white men except two, who were protected because they had shown some kindness to several families. These two men, William Lay and Cyrus Hussey, lived with their rescuers on terms of mingled captivity and adoption for almost two years, until they were finally discovered by an American naval vessel in December, 1825.

Such a record of occupational disaster and adversity, running the gamut from stove boats and foul lines to shipwreck and mutiny, contributed to a manner of life which bred ill health, disease, and incapacity.[4] Many whalemen, more fortunate than their comrades, lived to a ripe old age; but they did so in spite of their whaling careers. Countless circumstances of whaling life, both at sea and ashore, carried disease, injury, or lessened vitality in their wake. Well-known habits of dissipation in many ports made venereal disease a commonplace on shipboard. Coarse, salty, monotonous food and brackish water, coupled with a minimum of fresh provisions and months of confinement on shipboard, led to scurvy and organic disorders. Shore leave, with its fresh food and hectic indulgence in drunkenness and vice, relieved the scurvy but only aggravated the disturbance of kidneys and liver. And, in turn, the organic functions were still further disorganized by a return to the whaling grounds, where days of comparative idleness alternated with periods of superhuman exertion under gruelling conditions. Bruises and fractures resulted from accidents in the boats, falls from aloft, slipping on the oily decks in heavy weather, or blows delivered by the mates. And tuberculosis, dropsy, and other chronic affections arose through long and repeated exposure to cold, heat, and water.

An authoritative description of the peculiar ailments of whalemen was given in an official consular despatch written at Paita, Peru, on August 10, 1863. The State Department

[4] For accounts of the specific incidents related in the preceding pages, as well as of many other whaling mishaps, see especially the following works: Starbuck, A., "History of the American Whale Fishery"; Brown, J. T., writing in Vol. VII of "Fisheries and Fishery Industries of the U. S."; Davis, W. M., "Nimrod of the Sea"; Macy, Obed, "History of Nantucket"; Ricketson, D., "History of New Bedford"; Cheever, H. T., "The Whale and His Captors"; and Nordhoff, Charles, "Life on the Ocean." The same happenings, together with others, were also repeated in many contemporary accounts of whaling life. Several of the most spectacular disasters, too, were recounted in narratives written by the survivors. William Lay and Cyrus Hussey published "A Narrative of the Mutiny on Board the Ship *Globe,* of Nantucket, in the Pacific Ocean, Jan. 1824, and the Journal of a Residence of Two Years on the Mulgrave Islands." Owen Chase, the first mate of the *Essex,* wrote a similar "Narrative of The Most Extraordinary and Distressing Shipwreck of the Whaleship *Essex,* of Nantucket." Jenkins, T. H., published a pamphlet entitled "Bark *Kathleen* Sunk by a Whale." And only recently the diary of the surgeon of the *Diana,* edited by the author's son, C. E. Smith, appeared under the title, "From the Deep of the Sea: An Epic of the Arctic."

had inquired, rather sharply, why the consular expenses for sick and destitute seamen were proportionately greater at Paita, frequented largely by American whalers from the Off-Shore Ground, than they were at such a port as Liverpool, crowded with merchant shipping. In his reply Consul C. F. Winslow, who had had a formal medical education, pointed out that the long cruises and general conditions of whaling life gave rise to a set of diseases which were distinct from those suffered by merchant seamen. "During these long confinements on shipboard, subsisting upon coarse salt beef, fat pork, and hard biscuit, without a system of constant and uniform labor (hard labor going by snatches of luck and short spells in whaleships) many chronic complaints are engendered—such as scurvy, gastritis, affections of the liver and kidneys; and the habits of whalemen, when in port, add another category of complaints which, generally badly treated by mercury at sea, leave the system racked with neuralgia, rheumatism, carious bones, and a class of diseases which it takes a long time, and great care, to eradicate."

Such cases, "even occurring in officers," and either ignored or subjected to the crudest treatment, were kept on board for weeks and months until, in due course of time, the vessel again made port. If the patient had become incapacitated, he was then either discharged with three months' extra pay, as the law required, or he was "stealthily put ashore in some hut to save this extra pay, and to come upon the Consul's hands afterwards." Whaling hands were often forced to desert, the consul added, because their captains refused to admit that they were sick enough to merit a discharge.[5]

But in spite of its triple threat against life, health, and character, whaling had its compensations. Gamming, scrimshawing, and "skylarking" deserve to be ranked, perhaps, as but incidental and inadequate recreations in a life of unutterable boredom and hardship. But in the opportunities for travel and in the excitement of the chase many whalemen, at least,

[5] These statements were taken directly from the original manuscript, which is incorporated in the files of the Consular Letters in the Library of the Department of State at Washington. Reference to these Letters is by port and date.

were enabled to find real satisfaction. The world-encircling journeys of single whalers on single voyages offered intimate contact with strange climes and stranger peoples. The luxuriant South Seas, the voluptuous Orient, and the barren wastes of Arctic or Antarctic all came within the view of a whaling hand. And with them came, too, their widely differing inhabitants, from Fiji Islander to Eskimo. Granted that the occasions for observation were woefully restricted, and that the average foremast hand knew more about foreign brothels and grog-shops than about lands and peoples, it still remained true that many an insistent case of wanderlust found its relief and indeed its cure in a whaling forecastle. And in an industry which took so much from its human material, and gave so little, such satisfactions were not to be overlooked.

An even greater compensation, however, was the savage joy of the chase which whaling offered to some (though by no means to all) of its followers. To pursue the world's largest creatures with short-range weapons and relatively simple implements could not fail to appeal to an age-old hunting instinct. And when such pursuit was carried on in a watery environment both foreign and dangerous, there were untold possibilities for the development of a royal sport. Regarded purely as a form of sport, whaling has perhaps never had an equal: even the hunting of tigers and of elephants had less to offer to the ardent sportsman. Any one of a half-dozen major and common exigencies of the chase, from a stove boat to a "Nantucket sleigh-ride," afforded as much of thrill and of dangerous excitement as mere man could well expect to survive. Few persons have ever lived to describe an intensity of mingled fear and joy greater than that which came with the pursuit and capture of an eighty-barrel "ugly" cachalot.

In fact, whale-hunting was cast in such heroic proportions that only the elect could appreciate it as a form of sport. Many whalemen never succeeded in overcoming a terrifying and paralyzing fear of "going on to a whale." And with what excellent reason! The more hardy and adventurous hands, however, and particularly the harpooners and boat-headers who were responsible for the immediate strategy of the chase, often secured the most intense pleasure from the hunt. One whaleman, for instance, said, "The excitement on a whaling voyage had for me its pleasures. Yes, I think I may say its thrilling interest, and I longed once more to be on the deep blue sea in pursuit of the monsters." [6]

The intense excitement and thrilling interest of the actual pursuit, with its shifting hopes and fears and imminent dangers, could not be better portrayed than in an early account which was originally written for the *Nantucket Inquirer*. This recital, couched in the contemporary slang and technical vocabulary of the industry, presented with a fine flavor the kaleidoscopic changes of action and emotion which characterized the chase from the moment when a spout was discovered by the lookout until the whale, after death, turned "fin up." The description is reproduced in its entirety.[7]

The man at the mast-head upon the look-out, having discovered whales, vociferates with all his might, "There She Blows!" The captain immediately exclaims — "Where away?" and "How far off?" When being answered as to their being to windward, to leeward, right ahead, or astern, he goes aloft himself to determine that they are sperm whales, and which way bound. We will suppose that they are three points off the larboard bow, distant about 4 miles, and heading along the same course as the ship. Now the captain cries, "Keep her off two points," which being done, "Steady-steady as she goes," is his next order. "The weather braces a small pull." "Loose top-gallant sails, bear-a-hand." Scarcely a man is to be found on deck after these orders are executed, except the helmsman; all are eagerly jumping aloft to catch sight of the whales previous to their going down; and hope and fear are alternately expressed in the faces of each, as the fish are seen to glide through the water rapidly, and in a straight course, or occasionally to play upon the surface — to lobtail it, is the technical term. The ship nearing the whales, the next order is, "See the lines in the boats!" "Swing the cranes!" The after-oarsman now fills his boat-keg with water, puts some bread under the stern sheets, and sees that a bucket is in the boat. We will imagine that the whales are now sounding, and that the captain, having run

[6] "Scraps From the Log Book of George Lightcraft, Who Was More Than Twenty Years a Sailor," etc., p. 38.

[7] This picturesque account was published by Chase, Washington, in a small volume with the cumbersome title, "A Voyage From the United States to South America, Performed During the Years 1821, 1822, and 1823. Embracing a Description of the City of Rio Janeiro — and of an Eighteen Months Cruise in a Nantucket Whaleship." The work was published at Newburyport, Massachusetts, in 1823; but the author states that this description, found on pp, 79–80, was originally composed for the *Nantucket Inquirer*.

down with the ship as near as he thinks advisable, has ordered the main-topsail to be backed. All hands are now straining their optics to discover the whales when they first blow. They are at length seen a short distance from the ship. "Stand by the boats, there," cries the captain; and each man, knowing his station, is found at his respective boat, eager for the chase. "Lower away" — the boats are precipitated into the water, and the crews are at their oars in a twinkling. After pushing from the ship, it takes some 2 or 3 minutes for the harpooner to adjust his craft, he then seats him on his thwart, and takes his oar. Now the officer who heads the boat, speaks, "Line your oars, boys, and pull ahead — (a lapse of two or three minutes) — pull ahead, I tell you, why don't ye — Oh, how they lay, heads and points, look at 'em — pull ahead, I tell ye — long and strong, head boat, I say — (an interval of about 60 seconds) — every man do his best — lay back, I tell ye (fiercely) — why don't ye spring — don't let that boat pass ye (despondingly) — spring, I tell ye (authoritatively) there, there they be, round and round with 'em, for God's sake, pull ahead (entreatingly) — (lapse of a few seconds) — everything — everything I've got in my chest I'll give ye, do spring boys, let's go on first. Now then, back to the thwarts, give her the touch; I feel ye (encouragingly) — five seas off, but five seas off, spring! — 3-oar side best; pull all, pull every son of you (boisterously) — I'll give you all my tobacco, everything I've got — look at her, O, what a hump, and slow as night — don't you look round (passionately) — I tell you she don't blow, she only whiffs it out — at the end of your thwarts, pull, and we'll be on this rising — she's an 80-barrel whale; there she mills; by jingo she's heading to leeward; a large fellow separate from the school (shoal)— why the harry don't you pull — now do boys, do your best, won't you (soothingly) — I tell you we are jam on to her! One minute more! Half a minute! — O, boys, if you want to see your sweethearts, if you want to see Nantucket (with emotion), pull ahead — spring, b—t ye, that whale will shorten our voyage six months — I tell you we gain her fast — now's the time — mills still-heading to leeward — slap on to her in a moment — harpooner stand by — all my tobacco — all my clothes — everything that I possess — pull — O, what a whale (softly) — I've hove my soul out — harpooner — harpooner — harpooner . . . — one minute more, lay back; spring half a minute more; all my tobacco, a double share of grog — we are in her wake — (whispers) make no noise with your oars — STAND UP HARPOONER — pull the rest — GIVE IT HER SOLID!

. . . Stern, stern I tell ye (loudly) — stern all, stern like the devil — stern, and get clear of the whale — harpooner come aft — wet the line there, you 2nd oarsman, we are fast — there she's up, there she spouts, now haul me on — stern, stern I tell ye — lay to leeward of the whale — that's a good one (straightens his lance) — lay the head of the boat off — I've hon'd my lance d—n her — give me a chance, won't ye; do haul me on, will ye — there's the FLAG, She Spouts Blood — stern, I tell ye — lie — give us a set upon her — thick as tar, there she clotters — stern, she's going in her flurry — stern all — there, she's fin up; pass the spade forward, let's haul up to her, get harness on, and tow her alongside.

It is only in Herman Melville's immortal "Moby Dick," however, that the quintessence, the very soul of whaling is laid bare. Melville was both a whaleman and a writer who at times neared genius; and in consequence his epic of whaling life, while admittedly a work of fiction, contains many descriptive passages which combine accuracy of detail with a graphic portrayal of the spirit of the fishery far more successfully than any piece of historical scholarship. Witness the following description of "The First Lowering."

"There she blows! there! there! there! she blows! she blows!"

"Where — away?"

"On the lee-beam, about two miles off! a school of them!"

Instantly all was commotion.

The sperm whale blows as a clock ticks, with the same undeviating and reliable uniformity. And thereby whalemen distinguish this fish from other tribes of his genus.

"There go flukes!" was now the cry from Tashtego; and the whales disappeared.

"Quick, steward!" cried Ahab. "Time! time!"

Dough-Boy hurried below, glanced at the watch, and reported the exact minute to Ahab.

The ship was now kept away from the wind, and she went gently rolling before it. Tashtego reporting that the whales had gone down heading to leeward, we confidently looked to see them again directly in advance of our bows. For that singular craft at times evinced by the sperm whale when, sounding with his head in one direction, he nevertheless, while concealed beneath the surface, mills round, and swiftly swims off in the opposite quarter — this deceitfulness of his could not now be in action; for there was no reason to suppose that the fish seen by Tashtego had been in any way alarmed, or indeed knew at all of our vicinity. One of the men selected for shipkeepers — that is, those not appointed to the boats, by this time relieved the Indian at the mainmast head. The sailors at the fore and mizzen had come down; the line tubs were fixed in their places; the cranes were thrust out; the mainyard was backed, and the three boats swung over the sea like three samphire baskets over high cliffs. Outside of the bulwarks their eager crews with one hand clung to the rail, while one foot was expectantly poised on the gunwale. So look the long line of man-of-war's men about to throw themselves on board an enemy's ship . . .

"Lower away then; d'ye hear?" shouting across the deck. "Lower away there, I say."

The men sprang over the rail; the sheaves whirled round in the blocks; with a wallow, the three boats dropped into the sea; while, with a dexterous, off-hand daring, unknown in any other vocation, the sailors, goat-like, leaped down the rolling ship's side into the tossed boats below . . .

. . . "Pull, pull, my fine hearts alive; pull, my children; pull, my little ones," drawlingly and soothingly sighed Stubb to his crew, some of whom still showed signs of uneasiness. "Why don't you break your backbones, my boys?" . . . "So, so; there you are now; that's the stroke for a thousand pounds; that's the stroke to sweep the stakes! Hurrah for the gold cup of sperm oil, my heroes! Three cheers, men . . . all hearts alive! Easy, easy; don't be in a hurry . . . don't be in a hurry. Why don't you snap your oars, you rascals? Bite something, you dogs! So, so, so then; . . . softly, softly! That's it . . . that's it! long and strong. Give way there, give way! The devil fetch ye, ye ragamuffin rapscallions; ye are all asleep. Stop snoring, ye sleepers, and pull. Pull, will ye? pull, can't ye? pull, won't ye? Why in the name of gudgeons and ginger-cakes don't ye pull? . . . pull and break something! pull, and start your eyes out! Here!" whipping out the sharp knife from his girdle; "Every mother's son of ye draw his knife, and pull with the blade between his teeth. That's it . . . that's it. Now ye do something; that looks like it, my steel-bits. Start her . . . start her, my silver spoons! Start her, marling-spikes!" . . .

Meanwhile, all the boats tore on. The repeated specific allusions of Flask to "that whale," as he called the fictitious monster which he declared to be incessantly tantalising his boat's bow with his tail — these allusions of his were at times so vivid and lifelike, that they would cause some one or two of his men to snatch a fearful look over the shoulder. But this was against all rule; for the oarsmen must put out their eyes, and ram a skewer through their necks; usage pronouncing that they must have no organs but ears, and no limbs but arms, in these critical moments.

It was a sight full of quick wonder and awe! The vast swells of the omnipotent sea; the surging, hollow roar they made, as they rolled along the eight gunwales, like gigantic bowls in a boundless bowling-green; the brief suspended agony of the boat, as it would tip for an instant on the knife-like edge of the sharper waves, that almost seemed threatening to cut it in two; the sudden profound dip into the watery glens and hollows; the keen spurrings and goadings to gain the top of the opposite hill; the headlong, sled-like slide down its other side; — all these, with the cries of the headsmen and harpooners, and the shuddering gasps of the oarsmen, with the wondrous sight of the ivory *Pequod* bearing down upon her boats with out-stretched sails, like a wild hen after her screaming brood; all this was thrilling. Not the raw recruit, marching from the bosom of his wife into the fever heat of his first battle; not the dead man's ghost encountering the first unknown phantom in the other world; — neither of these can feel stranger and stronger emotions than that man does, who for the first time finds himself pulling into the charmed, churned circle of the hunted sperm whale.

The dancing white water made by the chase was now becoming more and more visible, owing to the increasing darkness of the dun cloud-shadows flung upon the sea. The jets of vapor no longer blended, but tilted everywhere to right and left; the whales seemed separating their wakes. The boats were pulled more apart; Starbuck giving chase to three whales running dead to leeward. Our sail was now set, and, with the still rising wind, we rushed along; the boat going with such madness through the water, that the lee oars could scarcely be worked rapidly enough to escape being torn from the row-locks.

Soon we were running through a suffusing wide veil of mist; neither ship nor boat to be seen.

"Give way, men," whispered Starbuck, drawing still further aft the sheet of his sail; "there is time to kill fish yet before the squall comes. There's white water again! — close to! Spring!"

Soon after, two cries in quick succession on each side of us denoted that the other boats had got fast; but hardly were they overheard, when with a lightning-like hurtling whisper Starbuck said: "Stand up!" and Queequeg, harpoon in hand, sprang to his feet.

Though not one of the oarsmen was then facing the life and death peril so close to them ahead, yet with their eyes on the intense countenance of the mate in the stern of the boat, they knew that the imminent instant had come; they heard, too, an enormous wallowing sound as of fifty elephants stirring in their litter. Meanwhile the boat was still booming through the mist, the waves curling and hissing around us like the erected crests of enraged serpents.

"That's his hump. There, there, give it to him!" whispered Starbuck.

A short rushing sound leaped out of the boat; it was the darted iron of Queequeg. Then all in one welded commotion came an invisible push from astern, while forward the boat seemed striking on a ledge; the sail collapsed and exploded; a gush of scalding vapor shot up near by; something rolled and tumbled like an earthquake beneath us. The whole crew were half suffocated as they were tossed helter-skelter into the white curdling cream of the squall. Squall, whale, and harpoon had all blended together; and the whale, merely grazed by the iron, escaped.

And in another chapter, entitled "Stubb Kills a Whale," Melville gives a description of a second chase, this time successful, which formed an admirable supplement to "The First Lowering."

. . . Close under our lee, not forty fathoms off, a gigantic Sperm Whale lay rolling in the water like the capsized hull of a frigate, his broad, glossy back of an Ethiopian hue, glistening in the sun's rays like a mirror. But lazily undulating in the trough of the sea, and ever and anon tranquilly spouting his vapory jet, the whale looked like a portly burgher smoking his pipe of a warm afternoon . . .

"Clear away the boats! Luff!" cried Ahab. And obeying his own order, he dashed the helm down before the helmsman could handle the spokes.

The sudden exclamation of the crew must have alarmed the whale; and ere the boats were down, majestically turning, he swam away to the leeward, but with such a steady tranquillity, and making so few ripples as he swam, that thinking after all he might not as yet be alarmed, Ahab gave orders that not an oar should be used, and no man must speak but in whispers. So seated like Ontario Indians on the gunwales of the boats, we swiftly but silently paddled along; the calm not admitting of the noiseless sails being set . . .

"There go flukes!" was the cry, an announcement immediately followed by Stubb's producing his match and igniting his pipe, for now a respite was granted. After the full interval of his sounding had elapsed, the whale rose again, and being now in advance of the smoker's boat, and much nearer to it than to any of the others, Stubb counted upon the honor of the capture. It was obvious now, that the whale had at length become aware of his pursuers. All silence or cautiousness was therefore no longer of use. Paddles were dropped, and oars came loudly into play. And still puffing at his pipe, Stubb cheered on his crew to the assault . . .

"Start her, start her, my men! Don't hurry yourselves; take plenty of time — but start her; start her like thunderclaps, that's all," cried Stubb, spluttering out the smoke as he spoke. "Start her, now; give 'em the long and strong stroke, Tashtego. Start her, Tash, my boy — start her, all; but keep cool, keep cool — cucumbers is the word — easy, easy — only start her like grim death and grinning devils, and raise the buried dead perpendicular out of their graves, boys — that's all. Start her!" . . .

And thus with oars and yells the keels cut the sea. Meanwhile, Stubb retaining his place in the van, still encouraged his men to the onset, all the while puffing the smoke from his mouth. Like desperadoes they tugged and they strained, till the welcome cry was heard — "Stand up, Tashtego! — give it to him!" The harpoon was hurled. "Stern all!" The oarsmen backed water; the same moment something went hot and hissing along every one of their wrists. It was the magical line. An instant before, Stubb had swiftly caught two additional turns with it round the loggerhead, whence, by reason of its increased rapid circlings, a hempen blue smoke now jetted up and mingled with the steady fumes from his pipe . . .

"Wet the line! wet the line!" cried Stubb to the tub oarsman (him seated by the tub) who, snatching off his hat, dashed the seawater into it. More turns were taken, so that the line began holding its place. The boat now flew through the boiling water like a shark — all fins. Stubb and Tashtego here changed places — stem for stern — a staggering business truly in that rocking commotion.

From the vibrating line extending the entire length of the upper part of the boat, and from its now being more tight than a harpstring, you would have thought the craft had two keels — one cleaving the water, the other the air — as the boat churned on through both opposing elements at once. A continual cascade played at the bows; a ceaseless whirling eddy in her wake; and, at the slightest motion from within, even but of a finger, the vibrating, cracking craft canted over her spasmodic gunwale into the sea. Thus they rushed: each man with might and main clinging to his seat, to prevent being tossed to the foam; and the tall form of Tashtego at the steering-oar crouching almost double, in order to bring down his center of gravity. Whole Atlantics and Pacifics seemed passed as they shot on their way, till at length the whale somewhat slackened his flight.

"Haul in — haul in!" cried Stubb to the bowsman, and, facing round towards the whale, all hands began pulling the boat up to him, while yet the boat was being towed on. Soon ranging up by his flank, Stubb, firmly planting his knee in the clumsy cleat, darted dart after dart into the flying fish; at the word of command, the boat alternately sterning out of the way of the whale's horrible wallow, and then ranging up for another fling.

The red tide now poured from all sides of the monster like brooks down a hill. His tormented body rolled not in brine but in blood, which bubbled and seethed for furlongs behind in their wake. The slanting sun playing upon this crimson pond in the sea, sent back its reflection into every face, so that they all glowed to each other like red men. And all the while, jet after jet of white smoke was agonizingly shot from the spiracle of the whale, and vehement puff after puff from the mouth of the excited headsman; as at every dart, hauling in upon his crooked lance (by the line attached to it), Stubb straightened it again and again by a few rapid blows against the gunwale, then again and again sent it into the whale.

"Pull up — pull up!" he now cried to the bowsman, as the waning whale relaxed in his wrath. "Pull up! close to!" and the boat ranged along the fish's flank. When reaching far over the bow, Stubb slowly churned his long sharp lance into the fish, and kept it there, carefully churning and churning, as if cautiously seeking to feel after some gold watch that the whale might have swallowed, and which he was fearful of breaking ere he could hook it out. But that gold watch he sought was the innermost life of the fish. And now it is struck; for, starting from this trance into that unspeakable thing called his "flurry," the monster horribly wallowed in his blood, overwrapped himself in impenetrable, mad, boiling spray, so that the imperilled craft instantly dropping astern, had much ado blindly to struggle out from that frenzied twilight into the clear air of the day.

And now abating in his flurry the whale once more rolled out into view; surging from side to side; spasmodically dilating and contracting his spout-hole, with sharp, cracking, agonized respirations. At last, gush after gush of clotted red gore, as if it had been the purple lees of red wine, shot into the frighted air; and falling back again, ran dripping down his motionless flanks into the sea. His heart had burst!

"He's dead, Mr. Stubb," said Daggoo.

"Yes; both pipes smoked out!" and withdrawing his own from his mouth, Stubb scattered the dead ashes over the water; and, for a moment stood thoughtfully eyeing the vast corpse he had made.

Such accounts, reflecting the hectic atmosphere and breathtaking fortunes of the chase, can leave little doubt concerning the thrilling excitement experienced by the men in the boats. On the other hand, the very intensity and scope of whaling dangers and vicissitudes lent an air of apparent exaggeration to their recital which sometimes raised up scoffers and disbelievers. For instance, the *London Punch,* after reading the story of the loss of the ship *Ann Alexander* which appeared in the *New Bedford Mercury,* was moved to write a scoffing poem which was as delightful as it was unwarranted by the facts.[8] In spite of its many stanzas this poem, which was published in the issue of December 6, 1851, is worthy of reproduction in full; for beyond doubt it represented in a light and supercilious vein the reactions of many less clever persons to the supposed exaggerations of tales of whaling disasters.

[8] See the preceding pages of this chapter for a brief account of the essential facts regarding the loss of the *Ann Alexander.*

WHALER *Ann Alexander*, OF NEW BEDFORD

Attacked by a whale in 1850

BARK *Canton*, OF NEW BEDFORD

Built at Baltimore in 1835

Once again truth was so much stranger than fiction that it often failed to find credence!

THE WONDERFUL WHALERS

(See the *Bedford* [U. S.] *Mercury*)

Fathers of the oratory,
List to my surprising tale,
Hearken to a wondrous story
More than very like a whale;
Each mesmeric marvel-monger,
Lend to me your ears likewise;
If for miracles you hunger,
You shall ope both mouth and eyes.

In the ship *Ann Alexander,*
Cruising in pursuit of whales,
Bold John S. Deblois, commander,
With a crew so gallant, sails.
In the South Pacific Ocean,
Reaching to the Off-Shore Ground,
'Mong the waves in wild commotion,
Several monstrous whales they found.

These two boats did follow after,
Larboard boat and starboard too,
And with shouts of glee and laughter,
The leviathans pursue;
When the larboard boat, commanded
By the stout first mate, did soon
In a whale, with force strong-handed,
Deeply plunge a sharp harpoon.

Off the mighty monster started;
Pain and anguish gave him cause;
Suddenly he backwards darted,
Seized the boat between his jaws;
Into smithereens he cracked it;
Or, as witnesses declare,
Who beheld the thing transacted,
Bits no bigger than a chair!

In the starboard boat the captain
Quickly to the rescue struck,

And, although the bark was snapped in
Pieces, saved the crew — by luck.
Now the good *Ann Alexander*
To their aid the waist-boat sent;
Half the band then having manned her,
At the whale again they went.

Soon the ocean-giant nearing,
They prepared to give him fight,
Little thinking, never fearing,
That the beast again would bite.
But without their host they reckoned;
At their boat he also flew;
Like the first he served the second,
Snapped it into pieces too.

Sure his jaws, together clapping,
Had the gallant seamen crushed;
But, when they perceived him snapping,
Straight into the sea they rushed.
To afford the help they needed,
Bold Deblois repaired again;
Once more, also, he succeeded,
In the aim to save his men.

Tired, perhaps, of sport renewing,
To their ship this time they hied,
When, behold the whale pursuing,
With his jaws extended wide.
Gloating with revenge, he sought 'em;
But, with blubber pierced, and gored,
He was crippled, or had caught 'em;
But they all got safe on board.

Risk the heroes little cared for;
Speedily they set their sail
In the ship herself — prepared for
One more tussle with the whale.
Now they reached him — plunged a lance in
The infuriate monster's head;
Then — of course they had no chance in
Close encounter — onward sped.

For the ship they saw him making,
But the chase he soon gave o'er,
Which the animal forsaking,

Down on him again they bore!
Fifty rods below the water,
There they saw the monster lie;
So, despairing him to slaughter,
They resolved no more to try.

At this time, Deblois was standing
Sternly on the larboard bow,
Ready, with harpoon his hand in,
To inflict a deadly blow:
Up he saw the monster rising,
With velocity and power,
At the rate of speed surprising
Of full fifteen knots an hour.

In an instant — Heaven defend us! —
Lo, the whale had, near the keel
Struck, with such a force tremendous
That it made the vessel reel;
And her bottom knocked a hole in,
Into which the water poured;
And the sea so fierce did roll in,
That the billows rushed and roared!

Yet the ship was saved from sinking,
Though so riddled by the whale,
And Deblois and his unshrinking
Crew survive to tell the tale.
Strong are those daring fellows,
Doubtless, the harpoon to throw;
And — to judge from what they tell us —
Stronger still to draw the bow!

Other would-be poets, instead of scoffing, were attracted by the romantic and sentimental phases of the fishery; and again they voiced a certain section of contemporary public opinion. Poetry was perhaps too flattering a word for the fugitive and amateurish literary efforts which dealt with whaling: mere versification was a more accurately descriptive term.

A certain Dr. John Osborn, born on Cape Cod in 1713 and educated at Harvard College, set the standard in this field by composing "A Whaling Song" of such ponderous dullness that its eighteen verses seem doubly long.[9] Nor were Dr. Osborn's scattered successors able to display any marked improvement. American whaling was responsible for one masterpiece of prose composition — Herman Melville's immortal "Moby Dick"; but it was able to inspire no poetry worthy of the name.[10] The following bits of whaling verse, mostly obvious adaptations of well-known English poems, will convey some concept of the caliber of the efforts which were made. As poetry their value is nil; but as reflections of certain current tendencies to romanticize the pleasures, adventures, and compensations of whaling life, they have a slight significance.[11]

The whale he shall still be dear to me,
When the midnight lamp grows dim;
For the student's book, and his favorite nook,
Are illumined by aid of him. — *Sailor's Song.*

Ye gentlemen of England, that live at home at ease,
Ah, little do ye think upon the dangers of the seas. — *Ocean Song.*

Spout! spout! spout!
The waves are purling all about,
Every billow on its head
Strangely wears a crest of red.
See her lash the foaming main
In her flurry and her pain.

Take good heed, my hearts of oak,
Lest her flukes, as she lies,
Swiftly hurl you to the skies.
But lo! her giant strength is broke.
Slow she turns, as a mass of lead;
The mighty mountain whale is dead. — *Whaler's Song.*

[9] Both this "Whaling Song" and some remarks about its author may be found in Ricketson, D., "History of New Bedford," p. 68.

[10] Melville published, in connection with "Moby Dick," a curious assortment of literary odds and ends which referred to whaling. But these references, gleaned from many nooks and crannies of the world of books, possess little value except as a sort of whaling curiosity shop.

[11] These selections were taken from Cheever, H. T. (Reverend), "The Whale and His Captors; or, The Whaleman's Adventures, and The Whale's Biography, as Gathered on the Homeward Cruise of the *Commodore Preble.*" The bits of verse are placed at the beginning of various chapters; and it is to be inferred, at least, that the author was personally responsible for many of them.

Oh, the whale is free, of the boundless sea;
He lives for a thousand years;
He sinks to rest on the billow's breast,
Nor the roughest tempest fears.
The howling blast, as it rushes past,
Is music to lull him to sleep:
And he scatters the spray in his boisterous play,
As he dashes — the King of the deep. — *Sea Song.*

The whaling industry itself, however, nurtured certain rough-and-ready versifiers, the "minstrel boys," who extemporized verse after verse of the chantey songs while the crews rolled out the boisterous and endlessly-repeated choruses. These spontaneous outbursts of accentuated rhythm, always robust and often uncouth, seldom attained any set form. Instead, in true minstrel fashion, they were transmitted by word of mouth; and each recipient reserved the right to adapt the matter to please his own fancies. Now and again, however, one of the widely known songs was committed to writing, and thus saved from oblivion. Of this group "Captain Bunker," termed by a mid-century writer "a good specimen of sea-spun poetry," is perhaps as characteristic as any other.[12]

Captain Bunker

Our captain stood upon the deck,
A spyglass in his hand,
A viewing of those gallant whales,
That blowed at every strand.
"Get your tubs in your boats, my boys,
And by your braces stand,
And we'll have one of those gallant whales,
Hand, boys, over hand!"

Chorus

So be cheery, my lads! let your hearts never fail
While the bold harpooner is a striking of the whale.

"Overhaul, overhaul!
Your davit-tackles fall,
Till you land your boats in the sea

[12] This "poem" is reproduced in Browne, J. R., "Etchings of a Whaling Cruise," p. 77.

One and all!"
Our waist-boat got down,
And of course she got the start:
"Lay me on, Captain Bunker,
I'm h—l for a long dart."

Chorus

Our first mate he struck,
And the whale he went down;
The captain he stood by
All ready for to bend on;
Which caused the whale to vomit,
And the blood for to spout;
In less than ten minutes,
He rolled both fins out!

Chorus

CHAPTER X

EARNINGS AND THE LAY

THE method of wage payment in the whaling industry was a singular one. The whaleman was not paid by the day, week, or month, nor was he allowed a certain sum for every barrel of oil or for every pound of bone captured. Instead, his earnings consisted of a specified fractional share, known as a lay, of the total net proceeds of a voyage. This system of lays, which probably originated in the Dutch Greenland fishery during the early seventeenth century, was employed to the virtual exclusion of other types of remuneration; and throughout the middle decades of the nineteenth century, the golden era of American whaling, the assigned fraction ranged from 1/8 or 1/10 in the case of a few favored captains to 1/250 for a young and inexperienced cabin-boy. Between these two extremes there was at any given time an approximation toward a going rate for each of the different ranks on board a whaler. The captains, mates, boatsteerers, and coopers received "short lays" ranging from 1/8 to 1/100 of the net proceeds; the able and ordinary seamen, stewards, cooks, and blacksmiths were entitled to shares which varied from 1/100 to 1/160; the green hands and boys had to be content with "long lays" which fluctuated from 1/160 to 1/200; and instances of fractions as small as 1/250, or even 1/350, were not unknown.[1]

Within each rank, too, the intense individualism of the wage bargain brought about variations which were traceable to such

[1] These figures were secured through an analysis of hundreds of individual accounts found in the large collection of original manuscript account-books now in the New Bedford Public Library. Unless otherwise stated, all figures given throughout this chapter, together with all material relating to specific vessels, have been obtained from the same sources.

factors as experience, skill, courage, length of voyage, policy of agents or owners, and the port from which the vessel was sailing. Still other fluctuations affected all ranks, though commonly in minor degree. Such changes were due to the scarcity or oversupply of hands; the relationship between the prospect of return from a whaling voyage and the rate of wages prevailing in the merchant marine and among unskilled shore workers; and the quantity, quality, and price of whaling products.

Obviously the amount of earnings represented by a given lay could not be definitely calculated until the close of an entire voyage, which often extended over a period of forty months and frequently over forty-five to forty-eight months. To determine the exact amount of wages due to the various members of a crew therefore involved a large amount of bookkeeping. The gross value of a cargo was computed at the current market prices for the various qualities of oil and bone, regardless of whether it was sold at once or held for future sale by the owners. The net proceeds were then obtained by subtracting from this gross value a number of charges which included pilotage, wharfage, gauging, commissions for the sale and handling of the cargo, coopering, the expenses of watchmen, and, upon occasion, insurance premiums, as well as special guarantees to cover a sale made on credit. Each man's earnings then consisted of that fractional share of this net sum which was specified in his lay.

Thus at the close of the fourth whaling voyage of the ship *Benjamin Tucker*, of New Bedford, in 1851, the cargo consisted of 73,707 gallons of whale oil, 5,348 gallons of sperm oil, and 30,012 pounds of whalebone. At the current market prices of 43¢ per gallon for whale oil, $1.25 per gallon for sperm oil, and 31¢ per pound for whalebone, the gross value of this cargo was found to be $47,682.73. Subtracting from this a series of charges which amounted to $2,362.73, the net proceeds were fixed at $45,320.00. A green hand with a lay of 1/200 was then entitled to $226.60 as his gross earnings for the entire voyage; while a seaman with a lay of 1/160 was credited with $283.25.

But the amount of a man's lay was only one of many items

which entered into the determination of the amount of money which he actually received at the close of a cruise. For throughout all the slowly passing months of a voyage both captain and agents kept an account showing all deductions, advances, and credits for each member of the crew. The final settlement with each man then consisted of balancing his account after computing and including the amount of his lay as the dominant credit factor. Normally there was a balance in his favor, which was paid in cash or by means of an order on the owners. But it was not uncommon for a seaman to find himself actually in debt to the agents of the vessel in which he had worked for a period of two to four years!

In such instances there was no settled practice; for the disposition of the matter depended upon the policy and attitude of owners and agents. In those rare cases in which a man was able to pay, the sum was collected. But if, as was almost universally true, the debtor was penniless, there were several alternatives. Where the amount was small and the owners generous, the debt might be cancelled; and if he was peculiarly unfortunate, an individual might be given a few dollars in cash in spite of his debit balance. But most often the persuasive agents and the conspiring landsharks induced the hand to embark upon another voyage for the same firm. If this plan proved to be successful, the amount due was carried over as the first debit item in the new account; if unsuccessful, the seaman escaped without payment and the owners were forced to accept the loss.

The items on the credit side of a typical whaleman's account were scanty. In truth, the only significant entry was the amount of his lay. A few dollars which he had earned as a bounty for sighting whales which were subsequently captured, and an occasional order against the lay of some other member of the crew — these completed the scanty list of his rights to receive actual money. In addition, however, there were several other things which he had obtained in kind. These included meals, a bunk in the forecastle, and all necessary equipment except his clothing and bedding, which he was expected to provide for himself.

In contrast to such a clean credit sheet, the debit side of the ledger was discouraging. Some of these charges were made by the owners in the home port, while others originated with the captain during the voyage. The owners' list was made up of the sum advanced for the man's original outfit of clothing, secured just before sailing; interest on this sum; a charge of five to ten dollars for loading the vessel, and a similar amount for unloading; a contribution of one to three dollars for defraying the expenses of a medicine-chest; and, in some instances, special payments to a man's family or dependents.

A second set of debits was derived from the captain's records. The most significant of these deductions comprised the slop-chest account, containing a notation of all the supplies which any given man had purchased on credit from the ship's store, or slop-chest; the amounts of cash which had been advanced to him, mainly to provide spending-money while in port; and the interest charges, calculated at generous if not exorbitant rates, on the sums advanced. In addition there were frequent special charges against individuals. If a deserter had been unfortunate enough to be caught and returned to his vessel he was made to bear the expenses of his recapture. Members of the crew were held responsible for stolen or deliberately damaged articles whenever the guilt could be fixed. And as a particular favor a master occasionally honored an order which some man in dire need of funds had drawn against his lay.

Not all individual accounts contained this full list of subtractions; but the exigencies of seafaring life were such that virtually all whalemen were compelled to ask for advances of cash and equipment and to patronize the slop-chest. These twin necessities were responsible for well-nigh universal debit items which quite overshadowed those arising from all other causes, and constituted formidable deductions from the gross lays.

Except for some agricultural pursuits and certain branches of the fishing industry, both past and present, the type of labor contract represented by the lay is essentially without a parallel.[2]

[2] Even the various forms of product-sharing practised in fishing and in agriculture bear a resemblance to the lay which is more apparent than real. For whereas in principle these schemes are not unlike the lay system, in the necessary modifications of circumstances the differences in degree become so great that they are quite as significant as differences in kind. Thus the fishing and

There are, it is true, other arrangements which resemble its general principle of dividing the net proceeds of an undertaking, in some prearranged proportion, between one set of persons charged with the functions of management and ownership, and another set whose obligation it is to provide the necessary kind and amount of labor. But with this general principle the resemblance ceases. For it is impossible to classify the lay system under any one of the commonly recognized methods of remuneration. Obviously it bears no relation to the familiar time wage, or the piece wage, or the numerous variants of the task and bonus system. Nor is there any resemblance between it and the sliding scale, the arbitrary bonus, or labor copartnership. Profit-sharing and gain-sharing are suggested, but hardly more. Neither is there any hint of co-operation, for the foremast hands possessed not even the most indirect part in the functions of ownership or of management.

Perhaps payment by means of the lay may be best described as a system of product-sharing accompanied by an enforced, rather than voluntary, division of risks. By placing the remuneration of the whaleman wholly upon the basis of an assigned fraction of the net proceeds of an entire voyage, the whaling merchants forced their crews to share to the full the risks involved in the enterprise, without allowing them the slightest part in the determination of the degree of those risks. The hazards of the chase and the dangers of wind and wave were, of course, in the last extremity beyond human control. But there remained a considerable range of risk-determining factors which were amenable to conscious regulation. Such were the selection of captain, mates, and seamen, the soundness of the vessel, the adequacy of the equipment, the choice of whaling-grounds, the strategy of the chase, and the business judgment displayed in disposing of the product.

In the decisions regarding these matters, however, the mem-

agricultural plans involve relatively small-scale, short-time, direct-bargaining projects in which the number of persons is restricted to a mere handful. On the other hand whaling displayed, among other significant peculiarities, long periods of two to four years between settlements, heavy investments, impersonal relationships between owners and large groups of seamen, an elaborately organized system of exploitation, and a divergence of bargaining power so wide that the crews were reduced to the position of sweated labor.

bers of the crew had no voice. Instead, they contributed their labor throughout the voyage, often under conditions of hardship, privation, and imminent danger, and in addition assumed their full share of the risks of the enterprise, financial as well as physical. The owners, on the other hand, undertook the responsibilities of management, the assumption of business, physical, and labor risks, and the provision of enough capital to furnish vessel, food, equipment, and the advances of cash and supplies necessitated by the penniless condition of the foremast hands. Food and bunk space, such as they were, were guaranteed to all members of a crew as long as they remained on board.

The earnings of a whaleman thus constituted a reward not only for the performance of labor under peculiarly trying conditions, but also for the assumption of personal, business, and physical risks. For the size of his lay, representing wages, depended directly upon the business risks centering about price fluctuations as well as upon the physical risks of storm, fire, stranding, and poor luck on the whaling grounds. Even after oil had been taken it was at the crew's risk as long as it was in the vessel. Consequently wages varied with profits, and were payable only in direct proportion to the amount of profits. If the profits of a given voyage were great, the earnings of the crew were correspondingly high; if profits were small, earnings were low; while if there were no profits at all, wages, too, disappeared entirely or were reduced to a nominal sum.

Obviously, then, this lay system partially shifted the most distinctive and onerous entrepreneurial function, the bearing of industrial risk, from the entrepreneur to the worker. Instead of the usual situation in which the entrepreneur contracted to pay a definite rate of wages to his workmen and assumed the risks of an industry, a special condition was created under which the whaling merchant materially lightened his financial burdens by proportioning his wages bill to the amount of his profits. And obviously the more his risks were decreased in this manner, the more were those of his workers increased. For their wages depended not upon the mere performance of certain stipulated kinds and amounts of labor, as in the ordinary industry, but upon the financial success of the undertaking in which their employer had engaged them to take part.

The question here is whether such a change in the allocation of the risks of the industry was accompanied by a corresponding change in the relative proportions of profits and wages. Did the whaling entrepreneurs, having divested themselves appreciably of the most important functions for which business profits constitute a reward, suffer also a reduction in the percentage of the net proceeds available as profits? Did the laborers, having assumed in part the burdens of business risk, secure an appropriate increase in the percentage of the net proceeds going to wages?

Unfortunately the available material does not lend itself to a neat and conclusive answer. It may be stated with some assurance, however, that the total wages bill for the industry, as indicated by the gross lays, did not increase relatively to the total amount of profits in the same degree as did the risks of the workers in proportion to the risks assumed by the profit-takers. Evidence tending to support such a viewpoint may be found in the fact that approximately seventy per cent of the net proceeds of the average voyage went to owners and agents, leaving only thirty per cent for the crew, including captain and mates; in the slow but steady decrease of the fractional lay throughout the first half of the nineteenth century; in the common occurrence of ludicrously small money lays; and in the notorious system of extortionate charges for advances of goods and of cash which reduced still further the real wages.[3]

It is necessary to recall also that then, as now, the factor of industrial risk was ignored or underestimated by workers in making their wage demands. The resulting tendency toward the under-payment of risk-bearing by laborers was greatly strengthened in whaling by the lack of bargaining power, either

[3] Warrant for these statements was found in the collection of manuscript account-books now in the New Bedford Public Library. The percentages for profits and wages were corroborated, too, by two contemporary writers. See Williams, J. R., in an article published in the *North American Review* for January, 1834, XXXVIII, p. 105; and Grinnell, Joseph, "Speech on the Tariff, With Statistical Tables of the Whale Fishery," p. 9. The latter was published in booklet form in 1844.

individual or collective, on the part of the foremast hands.

It seems reasonably clear, then, that under the lay system the whaleman received little or no extra reward for the enforced assumption of an appreciable share of the business risks of the industry. In view of the unusual extent and intensity of those risks it is not surprising to find that they were diffused as widely as possible, instead of being focussed sharply upon the entrepreneur. Risk-sharing by means of product-sharing was a natural and logical method of remuneration, in particular, during the simple beginnings of American whaling. When worker and entrepreneur were intimately associated, if not entirely merged, in making short voyages in small vessels, each party to the division of product was able to bargain on terms of equality with the other.

But with the later expansion of the industry came a specialization which might have been expected to make management and financial risk-bearing the peculiar functions of the entrepreneur. The owners and agents who remained at home were now sharply differentiated from the workers who went to sea; the thousands of men in the crews made it possible for the outfitters, boarding-house keepers, shipping-masters, and other species of landsharks to intrench themselves strongly in the whaling ports; and the older conditions of simplicity and intimacy were displaced by specialization of function, intricacies of structure, and problems of organization. In spite of these developments, however, the entrepreneurs steadily refused to allow any crystallization of risk-bearing — a refusal which was made possible by the force of custom and tradition (always strong in seafaring communities), the deteriorating character of the crews, and the greater bargaining power of the owners. As a result the continuance of the lay system and the fall of the fractional lay caused the whaleman to retain his full share of business risk even while relinquishing his claim to a corresponding proportion of the net proceeds.

But if the lay remained not only supreme, but virtually unattacked, its applied severity was modified in two directions. An authorized discharge, given during the course of a cruise, implied the calculation and receipt of the net earnings then due, thus obviating the necessity of waiting until the end of the entire voyage for payment. And at times the logic of circumstances was responsible for a resort to regular monthly wages. If extra hands were needed to work a full ship which was homeward bound, for instance, the impossibility of adding further captures to an already overladen vessel rendered a lay both valueless and meaningless, since it would have represented a share in zero. More common, however, were situations in which it was probable that oil and bone would be taken on a small scale, rather than in significant quantities, while certain hands were on board. In consequence the masters were sometimes compelled to offer a regular monthly guarantee in addition to the ordinary fractional lay. Such combined time and share payments applied to "seasoners," who were wanted only for a short space of time; to hands who were shipped for the purpose of helping to work a vessel from one port to another; and to men aboard ships which were carrying freight (usually supplies or products to or from other whalers) as well as seeking to add to their own catches.

The necessity for supplementing the lay under such circumstances appears clearly in a comparison of the earnings derived from fractional shares and from regular wages; for it was not unusual to have the latter outdistance the former by a wide margin. Thus the ship *William C. Nye*, during the course of a voyage extending from 1851 to 1854, carried a number of hands whose lays yielded but negligible sums. Eight men, shipped at lays of 1/130 plus twelve dollars per month, received sixty dollars each in wages and only $5.05 each as lays; three others likewise secured $5.05 each as lays, but fifty dollars each as wages; while a "short lay" of 1/75 yielded another hand only $8.76, in contrast to $73.20 resulting from wages. Such extreme divergences, of course, were not universal. At times, as on board the *Canton* in 1855, a run of luck brought to certain hands lays which equaled or exceeded their monthly wages. But the great majority of hands who sailed under this double system of remuneration found that their guaranteed wages were not only more dependable, but also appreciably larger than their lays.

In principle (and for the most part in practice as well), the sums due to these temporary hands, as well as the amounts earned by the regular members of the crew, were payable at the time and place of discharge. For the recruits such discharge normally came at the end of the stipulated period of service: for the others, shipped for the duration of an entire voyage, it commonly followed sickness, accident, disability, or a consular order, elicited through representations of particularly brutal treatment or shocking conditions.

In both instances the same methods of payment were employed. The amount of the lay was calculated by taking the required fraction of the oil and bone which had been captured while the given man was on board. This quantity was then translated into dollars and cents at the current prices for oil and bone — or rather by using the latest price quotations (usually weeks and sometimes months behind time) which had reached the port at which payment was being made. The difference between this gross money lay and the total of his debit items constituted the man's net earnings, which he might receive in one of three main ways. If small, the sum was paid in cash. If an appreciable amount was called for, it was often paid, in part at least, by means of an order on the owners. And not infrequently dollars and cents were reconverted into gallons of oil and the obligation liquidated by taking the required number of gallons from the hold and sending them ashore with the discharged hand. The accounts of the ship *Canton,* for example, show that during the years 1842 to 1853 many of her hands left the vessel with such payments in kind, ranging from 74 gallons to 393 gallons of sperm or whale oil per man.

But whether payment was made at the time of discharge or, as in the normal case, at the conclusion of a cruise, the great majority of whalemen failed to secure the full value of the sums called for by their fractional shares. So completely did the whaling merchants insist upon dividing the risks of the industry with their crews that they refused to pay outright to anyone a single cent of his earnings until vessel and cargo were safe in port. Consequently the only way a seaman might obtain even the smallest sum in cash was to secure it in the form of a loan extended by the owners through the captain; and such loans bore interest at rates which were by no means nominal. The prices charged for both original outfit and slop-chest articles, too, were notoriously high and often extortionate. And at the time of final settlement the prices of all purchases, the principal and interest of all loans, and other charges, including some of doubtful justification, appeared amongst the items to be deducted from the amount of the lay. It is apparent that in so far as these charges were either unjustifiable or exorbitant they represented deductions from total wages which were far from giving the crews full value received. Thus a large proportion of the average whaleman's gross earnings came to him in the form of goods and cash which could be secured only upon the most unfavorable terms. Purchasing power was provided in advance of actual payment, but was also greatly reduced.

Such a paring down of real wages accounted largely for the fact that the lay system failed miserably to furnish an incentive toward more energetic work by foremast hands. Since wage payment by means of the lay rendered the earnings of the whaleman directly dependent upon securing the greatest possible catch in the shortest possible time, it would seem that under it he should have been stimulated to hearty endeavor. The absence of such a looked-for result must be attributed in part to the unambitious character of the crews. But it was due even more to the many conditions which brought about a wide gap between the immediate efforts of any individual and the amount of his earnings.

These conditions included the attitude and actions of the remaining members of the forecastle; the whaling skill, navigating ability, and trading shrewdness of the captain; the business judgment of the agents and owners; the varying fortunes of the chase; the nature and amount of the debit items accumulated during the course of a voyage; and the fluctuations in the market prices of oil and bone. Any one or more of these circumstances, over which the individual foremast hand was able to exercise little or no control, might well offset any conscientious contribution to the success of a cruise. This fact, coupled with the apathy, indifference, and suspicion manifested on most whalers, caused the average hand to labor only under the immediate spur of a brutalizing discipline, and to be as inefficient and as careless as the relentless and arbitrary supervision of the mates would allow him to be.

But in spite of this lack of incentive, and the abuses, hardships, and low earnings which were largely responsible for it, the labor supply, considered from the quantitative standpoint, was adequate to meet the demands of the industry. Hands, such as they were, were not lacking: somehow the vessels were manned. And this remained true despite a continuous and kaleidoscopic change in the personnel of the crews. The steady stream of men pouring into the forecastles proved sufficient to counteract the continuous labor leakage caused by death, illness, incapacity, discharge, and desertion. For the majority of whaling hands completed only a single voyage or a fraction thereof, depending upon the generosity or niggardliness of fortune in presenting a chance for early escape.

Among the veterans who did sign for successive cruises, however, were to be found both the worst and the best elements in the industry. The former included those weak-willed and dissipated individuals who were reshipped as a result of the vicious system of organized exploitation which kept returned seamen intoxicated and in debt while ashore between voyages; while the latter comprised those men, largely of New England birth, whose ability and energy enabled them to secure a higher rank and a shorter lay with each successive cruise. These men, advancing steadily through all the gradations of the whaling hierarchy, regarded the pursuit of whales as a life-long means of livelihood, endowed with all the essential attributes of a regular trade.

Nor was there any adequate reason for refusing to classify whaling as a trade, although both the abuses and the romance of the industry tended to obscure the more technical aspects of shipboard life. The multifarious work to be performed aboard both whalers and whaleboats was carefully defined into a number of well-recognized tasks; and each one of these task-groups was carried on through methods which had been definitely organized and standardized. The successful execution of the work in each department, too, demanded skill, experience, judgment, and strength, in addition to a knowledge of a code of working rules and a familiarity with the various tools to be employed. Characteristic "tools of the trade" existed in abundance. The lance and the harpoon were as typical of the whaleman as the trowel of the bricklayer or the saw and hammer of the carpenter. Every green hand necessarily spent the early months of his first cruise in a real, though informal and unofficial, period of apprenticeship.

Furthermore, the duties and privileges of each one of the strictly-defined ranks aboard a whaleship depended upon an unwritten code observed with punctilious nicety. Succession from one rank to the next was in accordance with a definite system of advancement, based primarily upon the attainment of a minimum degree of proficiency. In fact, an analogy which is far from fanciful may be drawn between the foremast hands, boatsteerers, and mates and masters of the American whalers, and the apprentices, journeymen, and master workmen of the mediæval craft guilds. In addition, too, there is the striking and obvious similarity between the same three whaling ranks and the seamen, petty officers, and commissioned officers of the navy, as well as the privates, noncommissioned officers, and commissioned officers of the army.

The slow and cautious progress from one shipboard post to another was well illustrated in the record of a single whaling lifetime, typical of hundreds of others spent in the industry. In early youth a voyage was made as cabin boy at a lay of 1/215. During successive cruises thereafter, each of two to four years' duration, the same man sailed as foremast hand at a share of 1/165; as boatsteerer at 1/65; as third mate at 1/58; as second mate at 1/28; as first mate at 1/23; and finally as master at 1/16. Before retiring from the sea a second voyage was made as captain, this time at a lay of 1/14.[4] In other words, it was necessary for this representative whaleman to complete six entire cruises, involving some fifteen to twenty years at sea, before he was entrusted with the command of a whaler.

But if the earnings of an individual successful whaleman increased with each successive voyage, the lay of the average foremast hand, stated in fractional terms, became steadily smaller throughout the first half of the nineteenth century.

[4] See Brown, J. T., in "Fisheries and Fishery Industries of the U. S." (Goode, G. B., Editor), VII, p. 291.

That is, his wage bargain entitled him, as time went on, to a smaller fractional share of the voyages for which he shipped: so that while during the first decade of the century he received, if an ordinary seaman, a lay which was approximately 1/75 of the net proceeds, by 1850 this fraction had fallen to about 1/150. Thereafter, however, it remained virtually stationary until 1885, when the industry had shrunk to proportions no longer seriously significant. This double process of early decline and subsequent fixation applied to the lays of all ranks aboard a whaler except those of captain and first and second mates.

The juxtaposition of the lays paid from time to time on board representative vessels serves to throw into bold relief this course of the fractional share throughout the period in question.

The lays paid by the *Lydia* and the *Lion* possess a particular meaning, since they are among the earliest complete records of American whaling voyages which have been preserved.[5] When the *Lydia* sailed during the course of the 8th month, 1795, "for a whaling voyage to Delago Bay and elsewhere, via the Cape of Good Hope," such long-distance Pacific whaling was still young and hence redolent of mystery and of daring. Under the terms of the agreement drawn up to govern this cruise, the men were to perform the regular duties of their respective "qualities," to obey all lawful commands of the master, and to go ashore only after securing special permission from the commanding officer. Any unauthorized absence of more than forty-eight hours was to be construed as desertion. The owners, on their part, promised to pay to each one of the fifteen members of the crew the share of the "neat proceeds" which was set opposite his name on the crew list.

By a fortunate chance the figures pertaining to a cruise of the *Lion*, begun in 1805 and concluded in 1807, comprise not only the fractional lays, but the money lays as well. The sale of the cargo, made up of 37,358 gallons of body oil, 16,868 gallons of head matter, and 150 gallons of black oil, yielded

[5] There are, of course, records and references pertaining to whaling lays and earnings before this date; but almost without exception they are so fragmentary or so curiously arranged as to be of little value in indicating the individual lays.

FRACTIONAL LAYS PAID ON REPRESENTATIVE WHALERS, 1795–1885[6]

	Lydia *1795*	*Lion* *1805*	*Minerva* *1836*	*Brighton* *1847*	*Adeline* *1856*	*James Maury* *1864*	*Marcella* *1876*	*New Bedford Normal* *1880–1885*
Captain	1/16	1/18	1/16	1/16	1/12	1/16	1/14	1/12 to 1/17
First Mate	1/25	1/27	1/29	1/26	1/20	1/23	1/21	1/20 to 1/25
Second Mate	1/37	1/37	1/37	1/44	1/35	1/32	1/35	1/30 to 1/40
Third Mate	...	...	...	1/60	1/52	1/52	1/55	1/45 to 1/60
Fourth Mate	...	...	...	...	1/55	...	...	1/60 to 1/65
Boatsteerer	1/45	1/48	1/75	1/70 to 1/100	1/85 to 1/95	1/88 to 1/100	1/70 to 1/80	1/75 to 1/90
Cooper	...	1/60	1/34	1/52	1/60	1/60	1/48	1/50 to 1/55
Seaman	1/54 to 1/60	1/75 to 1/90	1/110 to 1/130	1/145 to 1/180	1/155 to 1/175	1/170	1/130 to 1/150	1/140 to 1/160
Steward	...	...	...	1/145	1/140	1/150	1/125	1/100 to 1/150
Cook	...	...	...	1/160	1/160	1/150 and 1/2 "slush" *	1/130 and 1/2 "slush" *	1/140 to 1/160 and 1/2 "slush" *
Carpenter	...	...	...	1/165	1/180	...	...	...
Blacksmith	...	...	...	...	1/185	...	...	...
Green Hand	...	...	1/125 to 1/150	1/165 to 1/215	1/190 to 1/200	1/200	1/170 to 1/180	1/180 to 1/190
Boy	1/90	1/120	...	...	1/350	1/195 to 1/220	...	...

[6] The figures for the *Minerva, Brighton, Adeline, James Maury,* and *Mar-*

* It was customary for whaling cooks to save the grease, or "slush," which resulted from cooking. Upon the vessel's return this grease was sold to soap manufacturers; and a portion of the sum so realized was added to the cook's earnings in order to induce a careful husbanding of this thrifty by-product.

a grand total of $37,661.02. After allowing for certain incidental expenses, $24,252.74 was apportioned amongst the owners, while the remaining $13,045.53 went to the crew. The captain received $2052.13; the first mate, $1381.41; the second mate, $1008.06; the two boatsteerers, each $777.05; the cooper, $621.64; the boy, $310.82; and the seamen, including several "blacks" and temporary hands, from $108.36 to $497.31 each.

The turn of the century which marked the interval between the voyages of the *Lydia* and the *Lion* saw the beginning of a drop in the fractional lay which was not arrested for some fifty years. The seamen who sailed on the *Lydia* in 1795 received shares which were greater than those paid during any subsequent period in the history of the industry. Only a decade later, when the *Lion* left the shallow waters of Nantucket Harbor and stood out to sea, the downward tendency of the lay was unmistakable in the forecastle, though it was barely discernible in the steerage and quite invisible in the cabin. During the following years the decline was slowly accelerated until the *Minerva*, shipping hands in 1836, was enabled to obtain seamen at half the rates prevailing in 1795, and boatsteerers for something more than half. Captain and mates, however, still exacted the same fractions. The distance between the forecastle and the cabin had been materially increased.

Another full decade was required to bring the descending lay to a halt. The wages called for by the crew-account of the *Brighton* when she sailed in 1847, and the almost identical shares distributed by the *Fabius* when she reached port early in January, 1849, were markedly lower than those offered by the *Minerva*. Masters and mates had again succeeded in maintaining their positions; but all other ranks had been forced to accept reductions. The *Brighton* paid her green hands from

cella were taken from the manuscript account-books now in the New Bedford Public Library. Those for the *Lydia* were also taken from the original manuscript, now in the museum of the Old Dartmouth Historical Society. An account of the voyage of the *Lion* is given in an article by James Freeman, called "Notes on Nantucket, August 1st, 1807," in the Massachusetts Historical Society Collections, Series 2, Volume III, pp. 19–38. The lays normally paid at New Bedford during the early eighties were set down by Brown, J. T., in "Fisheries and Fishery Industries of the United States" (Goode, G. B., Editor), VII, p. 292.

1/165 to 1/215, her seamen from 1/145 to 1/180, and her boatsteerers from 1/70 to 1/100.

With these figures, however, the lay at length reached a plane upon which it was destined to remain. Aside from the normal variations between individuals of the same rank and the minor fluctuations which always abounded in the industry, there was no significant change in the fractional shares going to the various ranks in a whaler's crew during the next thirty-five years. The gap between the earnings of officers and men, which had been opening steadily throughout the period of stationary lays for the former and falling lays for the latter, had now reached its widest extent, and thenceforth kept the two groups separated by approximately the same distance.

Several main tendencies may be distinguished in attempting to account for both the half-century fall of the fractional lay and its subsequent long period of virtual fixation. The gradually deteriorating character and efficiency of the crews was one significant factor. It was natural and perhaps inevitable that the lay should follow the downward path of the caliber of the foremast hands. As an outgrowth of this same movement, too, arose the temptation to exploit to the full the economic helplessness of the inferior crews. The use of heavier and larger vessels and of more elaborate equipment necessitated the employment of more and more capital in the industry; and this in turn created both a need and a desire to enlarge the proportion of the net proceeds available as profits. By 1850, however, the major transitions had been effected. Thereafter, speaking in broad terms, the crews became no worse, the vessels no larger or heavier, and the equipment no more costly. Nor were there any significant changes in the opposite directions. As a result the size of the lay, which but reflected these fundamentally unchanging conditions, remained itself unchanged.

The fall of the fractional lay was also interpreted, at times, as the mere counterpart to the increasing size of the cargoes secured by the larger vessels of the later years. It was held to be necessary, as a means of preventing an automatic and unintended rise in whalemen's earnings, to vary the fractional lay in inverse proportion to the quantity of oil and bone secured. Thus a "long lay" of only 1/150, if applied to a cargo of 80,000 gallons of oil and 20,000 pounds of bone, would yield the same number of dollars, gallons of oil, or pounds of bone, as a "short lay" of 1/75 of a catch of 40,000 gallons of oil and 10,000 pounds of bone. The continued payment of the larger shares of the earlier periods, in spite of the increasing size of the later cargoes, would have entailed a heavy increase in wages; and the fall in the fractional share, instead of constituting a decrease in earnings, merely served to prevent an increase. The apparent drop in earnings, when properly understood, was really the maintenance of the status quo.

Such an analysis, however, quite overlooked the fact that the greater catches were secured, in general, only as the result of correspondingly longer cruises. In reckoning whaling earnings the time element was all-important. A lay of 1/150 in a large vessel was equal, it is true, to an amount double that represented by 1/150 of a cargo half the size of the first; but if the former also required a voyage twice as long there was exactly the same earning capacity per year or month in both cases. Although actual conditions did not, of course, fit the exactitude of this hypothetical illustration, it was unmistakably true that the larger whalers of the later years were converted into "full ships" only as the result of longer and longer cruises.[7] If the greater cargoes tended to increase the monetary value of a fractional lay, this effect was largely or wholly offset by the more protracted voyages.

Other factors also served to disturb the relationship between the number of barrels of oil and the earnings of the men who filled them. Poor luck on the whaling-grounds, accident, disaster, or mutiny might force a large vessel to return with a cargo much smaller than that secured by a less roomy whaler. The fluctuations in the market prices of oil and bone, as well as the variations in quality, might rob a large catch of much of its value or confer unexpected importance upon a modest arrival. The caprices of the sea might entail the total or partial loss of hard-won products. And even if a large cargo were safely landed and sold at a satisfactory price, the number of dollars due to the crew was often shamefully cut down by

[7] See Appendix E for detailed figures concerning the average length of voyage for a period of years.

the extortionate prices of "slops" and outfit and by the over-generous rates of interest charged upon advances.

Against this background of fluctuating fortunes and of capricious phenomena the whaleman's earnings finally emerged as the result of the interaction of four major variables, — the fractional lay, the size of the cargo, the length of the voyage, and the prices of oil and bone. As such it was difficult, but not impossible, to isolate the effect of a change in the fractional lay alone. If the drop in the lay had been accompanied only by an increase in the average size of the cargo, there would have been no necessary reason to interpret it as a fall in earnings.[8] But since there was a simultaneous increase in the average length of voyage which was fully comparable to the growth in the size of the vessels, it became apparent that there was an actual fall in money wages when reckoned on a monthly or yearly basis. For even if the smaller shares had been fully and uniformly offset by larger vessels, the result would have been only the maintenance of the same earnings per voyage; and the receipt of the same sum for a forty-months' cruise which had formerly been forthcoming at the end of a thirty-months' voyage obviously marked a sharp decline in monthly wages.

This was true, in varying degrees, of all ranks aboard a whaler except those of captain and first and second mates. That the fractional lays of these three highest grades escaped the general lowering process and remained practically stationary meant that their earnings increased both absolutely and relatively — a fact which is to be attributed to the necessity of

[8] At any given time, for instance, the fractional lay tended to vary inversely with the size of the vessel. This relationship appears clearly in the following table, which shows the lays which prevailed in New Bedford and in Provincetown about 1880. The figures were taken from Brown, J. T., in "Fisheries and Fishery Industries," Vol. VII, p. 292.

	New Bedford		Provincetown	
	4-boat Bark	*3-boat Bark*	*3-boat Schooner*	*2-boat Schooner*
Captain	1/12 to 1/17	1/10 to 1/16	1/10	1/10
1st Mate	1/20 to 1/25	1/18 to 1/23	1/16	1/12
2nd Mate	1/30 to 1/40	1/28 to 1/35	1/21	...
Boatsteerer	1/75 to 1/90	1/65 to 1/75	1/43 to 1/45	1/20 to 1/22
Seaman	1/140 to 1/160	1/140 to 1/160	1/125 to 1/140	1/75 to 1/100
Green Hand	1/180 to 1/190	1/170 to 1/180	1/160	1/120

keeping capable and responsible men in the positions of command in order to cope with the increasing difficulties of supervision which accompanied the deterioration of the crews.

But in point of fact, what were the average monthly earnings of American whalemen during the mid-century decades of greatest whaling prosperity? An accurate and precise answer to this question, unfortunately, would involve a statistical cleaning of the Augean stables. Not only were there notoriously wide variations between individual seamen, voyages, vessels, and seasons, but the contemporary literature and accounting systems were couched entirely in terms of the fractional lay rather than of daily or monthly wages. Great quantities of records and account-books have been defaced, lost, or destroyed; and many of those which still remain have been mutilated or contain only fragmentary entries. But average earnings based upon considerable bodies of representative material may be discovered; and if accepted as being broadly suggestive rather than precisely conclusive, these averages possess both validity and significance.

The accounts of three typical and well-known whalers yield the following results.[9]

Average Length of Voyage and Average Earnings of 83 Foremast Hands Carried by the Ship *James Maury* During Six Consecutive Voyages, 1845–1868.

Average Length of Voyage	1218 Days
Average Lay per Voyage	$321.21
Average Earnings per Month	$ 7.92
Average Earnings per Day	$ .264

Average Length of Voyage and Average Earnings of 70 Foremast Hands Carried by the Bark *Marcella* During Four Consecutive Voyages, 1845–1856.

Average Length of Voyage	935 Days
Average Lay per Voyage	$97.60
Average Earnings per Month	$ 3.12
Average Earnings per Day	$ .104

[9] The figures entering into these averages were obtained through an examination of the original manuscript account-books, now in the New Bedford Public Library. Almost two hundred accounts, taken from thirteen voyages made by three representative vessels during the greatest period of American whaling, are included.

Average Length of Voyage and Average Earnings of 39 Foremast Hands Carried by the Bark *Minerva* During Three Consecutive Voyages, 1836–1841.

Average Length of Voyage	614 Days
Average Lay per Voyage	$94.51
Average Earnings per Month	$ 4.62
Average Earnings per Day	$ .154

Average Length of Voyage and Average Earnings of 192 Foremast Hands Carried by the Vessels *James Maury*, *Marcella*, and *Minerva* During 13 Voyages Made Between 1836 and 1868.

Average Length of Voyage	992 Days
Average Lay per Voyage	$200.09
Average Earnings per Month	$ 6.06
Average Earnings per Day	$ .202

Substantial confirmation of these figures is found in an estimate made in 1860 by W. B. Whitecar, who had but recently returned from four years of blubber-hunting. The average catch of a four-year sperm whale voyage was placed at 1200 barrels of oil. At a lay of 1/200, the average green hand was thus entitled to six barrels of oil. When translated into money at the current price of $45 per barrel, this was equivalent to $270, or $5.63 per month for 48 months! And if the more experienced foremast hand, with an average lay of 1/160, had been considered, a similar calculation would have placed his monthly wage at only $7.03.[10]

The most extensive contemporary calculation of whalemen's earnings, however, was made in England. Mr. Charles Enderby, head of the greatest whaling firm in British history, stated in 1846 that for some years the average wages of foremast hands had not exceeded £1.9.0 ($7.05) per month. Then, delving into the inherited account-books of vessels which

[10] Whitecar, W. B., Jr., "Four Years Aboard the Whaleship," p. 411. The estimate of 48 months as the *average* length of a sperm whaling voyage is open to question. Contemporary figures compiled by the *Whalemen's Shipping List* show that for the period 1842–1857, inclusive, the average length of voyage for all American sperm whalers was 42 months and 5 days. It is true that by 1860 four-year cruises were common; but it is quite improbable that the average length, though undoubtedly increasing, had gone up to four full years. If the more conservative figure of 42 months be employed in connection with Whitecar's other estimates, the average monthly wages prove to have been $6.43 for a green hand and $8.03 for seamen.

had carried the Enderby house-flag into every corner of the world, he compiled these figures:[11]

Average Monthly Earnings of Seamen in the English Sperm Whale Fishery, 1801–1845.

	£	s
1801–1810	2	11
1811–1815	3	0
1816–1820	3	2
1821–1825	1	14
1826–1830	2	3
1831–1835	1	16
1836–1842	2	2
1843	1	7
1844	1	9
1845	1	7

In sum, during the middle decades of the nineteenth century, the average wage of the whaling foremast hand would seem to have centered about a range of three to eight dollars per month. These amounts, however, were in addition to food and bunk space, which were provided without charge on board all whaling vessels. On the other hand, a large proportion of a man's earnings was required to pay for outfit and slop-chest articles and to meet heavy interest charges upon advances. Since prices were commonly exorbitant and interest rates more than generous, real wages were largely reduced by such payment in kind. These two opposing qualifications, together with the lack of comprehensive wage data, render it impossible to fix upon any definite figure for real wages which would avoid a misleading air of exactness. Similarly the average money wage must be regarded as a rough approximation rather than as precise truth.

Even after allowing for a generous margin of error, how-

[11] Enderby, Charles, "A Proposal for Re-Establishing the British Southern Whale Fishery," London, 1847. While this material applies only to the English branch of the industry, it serves in part, at least, as a means of filling the gap caused by the absence of any collected American figures of equal scope and authenticity. Mr. Enderby, because of family tradition and long personal experience, was peculiarly well fitted for the task of securing and interpreting such a mass of data. The differences between English and American whaling were not great enough to obscure a fundamental similarity which rendered the conditions prevailing under the Union Jack strongly suggestive of those existing under the Stars and Stripes.

SPERM WHALING—"THE CONFLICT"

From a lithograph by Prang and Mayer, Boston, 1859, after a drawing by J. Cole

ever, there seems to be small doubt that merchant seamen received appreciably higher earnings than whalemen. The wages paid by merchantmen at the port of Boston during the forties, as reported in the *Boston Shipping List*, ranged from $9 to $16 per month, depending upon the destination and the probable length of voyage. At the same time whaling hands, sailing from New Bedford, were making voyages which yielded them only $3 to $8 per month. And again Mr. Enderby provided corroborative evidence by calculating that the average earnings of sperm whalemen were not more than $7.05 per month, whereas merchant seamen were receiving from $10.95 to $17.03 per month.[12]

Unskilled shore labor, too, exhibited a clear-cut advantage over whaling. The landlubber who remained at home in order to fetch and carry earned much more than his adventurous cousin who found neither fame nor fortune in a whaling forecastle. Three different sets of figures, one applying to Massachusetts and the remaining two to the United States, provide an adequate background for this comparison.[13]

[12] Enderby, Charles, *op. cit.* As is indicated by these figures, there was little direct competition between whaling and the merchant service in the hiring of foremast hands. The seamen on the faster and more graceful merchantmen professed great contempt for "spouters" and "blubber-hunters"; and a real whaleman never thought of shifting his allegiance. Many men, too, went to sea for reasons other than the amount of wages. The device of the lay, with its tantalizing possibility of a lucky voyage, served to obscure the average earnings and to prevent a clear-cut comparison between wages at Boston and New Bedford. For the most part each industry possessed and recruited its own labor supply. Only in busy seasons, when there was some competition for those hands who were willing to ship in either service, was the size of whaling lays affected by merchantman wages.

[13] These figures were secured from the "Historical Review of Wages and Prices," published in August, 1885, as Part IV of the Sixteenth Annual Report of the Massachusetts Bureau of Statistics of Labor; from the Report on Wholesale Prices, Wages, and Transportation, Senate Report No. 1394, Second Session, Fifty-Second Congress (commonly known as the Aldrich Report); and from certain data regarding the wages of common laborers, 1829–1880, printed as part of the Tenth U. S. Census. The mass of invaluable but unwieldy material contained in the Aldrich Report and in the Tenth Census has been rearranged by Dr. Edith Abbott in a study published in the *Journal of Political Economy,* Vol. XIII, No. 3. The wage data include these occupations: laborers, yard hands, watchmen, teamsters, quarrymen, coal-heavers, helpers, and unskilled factory operatives. Unfortunately figures showing average wages with board and room were not included for unskilled laborers. From evidence contained in the wage tables for certain other occupations, however, it would seem that for any given grade of labor the wages with board were from 33% to 50% less than those without board.

Average Daily Wages (Without Room and Board) of Unskilled Labor in Massachusetts, 1831–1860, and in the United States, 1841–1860.

	Massachusetts (Mass. Bureau of Labor Statistics)	United States (Aldrich Report-Abbott)	United States (Tenth Census-Abbott)
Decade Ending 1840	$ .872	——	——
Decade Ending 1850	$ .852	$ .879	$ .837
Decade Ending 1860	$ .975	$ .985	$ .976
1840–1860, Inclusive	$ .913	$ .932	$ .906

Contrast these amounts, paid on shore between 1840 and 1860, with the money lays received by whaling foremast hands during the same period:

Average Length of Voyage and Average Daily Earnings of 156 Foremast Hands Carried by the Vessels *James Maury*, *Marcella*, and *Minerva* during Ten Voyages Made Between 1840 and 1860.

Average Length of Voyage	937 Days
Average Lay per Voyage	$180.54
Average Earnings per Day	$ .193

In round terms, the average whaleman was receiving about twenty cents per day, plus food and bunk space, at a time when the average unskilled shore worker was being paid about ninety cents per day without room and board. Since wages paid in addition to board and room were from 33% to 50% lower than ordinary money wages, these same shore laborers would have received from forty-five cents to sixty cents per day if they had been living with their employers. That is, when average earnings were reckoned on a comparable basis, the lowest grade of landlubber could sell his untrained strength for an amount two to three times as great as that obained by the occupant of a whaling forecastle.[14]

[14] This general conclusion is subject to qualification by a series of factors which do not lend themselves readily to numerical analysis. In extenuation of the low whaling earnings were the desperately low caliber of the crews, the lack of interested efficiency, the large percentage of untrained and exploitable Portuguese and South Sea Islanders, and the fact that the higher ranks of shipboard life were omitted from the calculations. On the other hand, green hands, "seasoners," and all abnormal cases were also omitted, so that the average figures represent the earnings of only the more steady foremast hands.

Such conditions inevitably raise the question: Why did anyone sail before the mast in a whaleship? The answer, multifold in detail, must be sought in a double set of factors which clustered about the unscrupulous practices of agents and owners and the non-pecuniary motives of many whalemen.

The ordinary shipping-agent, dealing commonly with prospective green hands who were young and unsophisticated, was notoriously shameless in his methods of persuasion. Misrepresentation, chicanery, fraud, mendacity in many forms — these constituted his regular stock in trade. And these tools, handled with an easy and enthusiastic loquacity which drew alluring pictures of stirring adventure and of exotic pleasures, were sufficient to draw many credulous youths from farm and shop. When the spell was finally broken by a few hours at sea in a real whaling forecastle, the disillusioned victims found themselves hopelessly committed to a cruise of several years' duration.

Nor did the commitments cease, in many cases, at the close of a single voyage. For every whaling hand was an actual or prospective pawn in a game of organized and commercialized exploitation which sought to keep him constantly within its toils. From the moment he stepped ashore in the home port he was fawned upon, pandered to, given intoxicants, and encouraged to indulge in every form of vice and dissipation. Then, when thoroughly befuddled and drunkenly heedless, he was cheated, robbed, and alternately cajoled, threatened, and deceived into signing an agreement to ship for another voyage. "Like the flying fish, which escapes from the albacore in its native element, only to be pounced upon by the man-of-war bird waiting to devour it, the sailor no sooner escapes the perils of the deep, than he is the object of instant attack from those who live by preying and feasting on his misery, on shore. On coming to anchor, he exhibits the spectacle of a helpless victim, bound hand and foot, and passed from the ship to the crimp, from the crimp to the long-room, from the long-room to the

Food and living conditions, too, were incomparably worse at sea than on land. The discipline, brutality, monotony, and danger of whaling life had only weak and emaciated counterparts on shore. And last, but by no means least, was the system of exploitation which took from the average seaman the bulk of the meager earnings which he did receive.

brothel, and from the brothel to a ship again — watched and guarded at every stage, and his fetters unrelaxed — glad to escape, though with injured health, and the loss of all his earnings, to take refuge amidst the perils of the sea from the greater perils of the land."[15] Thus did the industry entice and retain a significant proportion of its labor supply.

Another important element in that supply, actuated by non-pecuniary motives, entered the forecastles voluntarily. In the larger whaling ports, where tradition, ambition, and social esteem combined to make whaling the most acceptable means of livelihood, and where the families of the leading merchants constituted a virtual aristocracy, every boy of spirit sought a berth on an outgoing whaler. Other youths, coming from inland towns and farms as well as from the seaboard, were driven to offer themselves to the shipping-agents by a real or fancied love of the sea, by the lure of foreign travel, or by a dominant spirit of adventure and of wanderlust. And some dare-devil hands, fascinated by the fortunes of the chase and by the sporting element involved in the capture of such gigantic and spirited game, found it impossible to break away from the industry, in spite of the accompanying monotony, brutality, and danger.

Still other types of men found it convenient to ship on a whaler without inquiring too closely into the amount of earnings. Escaped convicts, unapprehended criminals, and that ever mysterious personage, "the man with a past," were alike attracted by the prospect of spending several years in out-of-the-way corners of the globe. Certain lascivious wretches chose a whaling voyage as the most convenient route to the carnal paradise of the South Seas. And the well-known slogan, "No Sundays off soundings," held out welcome promises to those unprincipled and maladjusted individuals who were sprinting away from the Ten Commandments.

As a result of these voluntary enlistments and of the unscrupulous recruiting practices, the number of foremast hands remained sufficient to meet the needs of the industry. In

[15] Harris, John, "Zebulon; or, The Moral Claims of Seamen Stated and Enforced," p. 63. This work (the first American edition of which was published at Boston in 1837) won for its author a prize of £50.

keeping up the supply of man-power the agents were aided and abetted, too, by the few bonanza voyages whose profitable lays served to obscure the low figure for average earnings. Still further succor came from the Cape Verde Portuguese and from the South Sea Islanders, who regarded even a lay of $1/200$ as a princely sum, and who crowded into the forecastles in increasing numbers.

The quantity of hands, then, was kept up in spite of the low earnings; but the quality deteriorated woefully. More and more the best young men of New England heeded Horace Greeley's advice and went West; and their places in the whaling forecastles were taken by Portuguese half-breeds, by Kanakas, by convicts and criminals, and by human scourings gathered from all parts of the world.

CHAPTER XI

DEBITS AND CREDITS

WHEN Albert Johnson, boatsteerer, and Josiah H. Jackson, green hand, were paid off at the close of the first whaling voyage of the ship *Brighton*, of New Bedford, their accounts had assumed the following form: [1]

Dr. Albert Johnson Boatsteerer to Ship *Brighton* & owners 1 W V Cr.

1842			1844		
8 mo. 2	For his share fitting ship $10 Less Labour $5	$5.00	8 mo. 6	By his share in Nett Proceeds of Ship Brightons Cargo	$394.90
	For his share Medicine Chest	1.50			
20	For paid his order to Joshua Richmond for Outfits on Owners for	100.70			
1844					
7 mo. 30	For his proportion disch. Cargo of ship	7.00			
	For Int. & Ins. on Amt. Advanced	16.11			
	For Bill Slops rec. on the voyage	22.89			
		153.20			
8 mo. 7	For Check pd. him balance on acc.	241.70			
		$394.90			$394.90

[1] These accounts were taken directly from the original manuscript account-books of the *Brighton*, now in the New Bedford Free Public Library. Unless otherwise noted, all figures cited throughout this chapter have been derived from the collection of whaling account-books made by this library. Illustrative material has been chosen from the records of some 64 voyages made by 20 vessels during the years 1836 to 1879, inclusive.

Dr. Josiah S. Jackson Green Hand to Ship *Brighton* & owners 1 W V Cr.

1842			1844		
8 mo. 2	For his Share fitting Ship	$10.00	8 mo. 6	By his share in Nett Proceeds of Ship Brightons Cargo	$219.39
	For his Share Medicine Chest	1.50			
24	For paid John Cummings bill Outfits	57.18			
1844					
7 mo. 30	For paid Thomas Coles bill board	9.75			
	For Int. & Ins. on Amt. Advanced	10.72			
	For Bill Slops etc. on voyage	17.77			
	For prop. disch. & cleaning ship	7.00			
	For Cash on acc.	5.00			
		118.92			
8 mo. 6	For Cash for balance acc.	100.47			
		$219.39			$219.39

In such accounts as these, multiplied many hundred-fold, is the raw material of which the wages system of the entire industry was constructed. Forecastle earnings, as well as counting-room finances, depended upon debits and credits, upon long series of cumulative book-entries punctuated by occasional and endlessly deferred settlements. Throughout the leaden-weighted months of every whaling voyage each member of the crew was steadily acquiring one debit entry after another, both in the books of the master at sea and in those of the agents on shore. But not until the cruise finally terminated, when the cargo had been safely landed and its value carefully calculated, did the single credit item of the lay appear, like a giant opposed to a crowd of pigmies. If, as ordinarily occurred, the giant outweighed the crowd, his demands were appeased by a cash offering of the balance due from owner to seamen. If, however, the crowd of little debits overcame the giant, it was a difficult and ungracious task for the owner to extract a cash balance from an employee who had gone unpaid for many weary months.

Debits and credits were of the essence of the labor aspects of whaling; and a proper understanding of the industry as a whole demands a survey of the customs and traditions behind the myriad book entries. Nor will a hasty and superficial survey suffice. For the policies and practices of whaling accounting illuminate many secluded corners and unexplored nooks.

When classified according to origin, the whaleman's debts fell into two categories: seagoing debits and landlubber debits. The first, including the slop-chest account, cash advances, interest and insurance charges, and incidental items, originated with the master at sea. The second, comprising advances to outfitters, charges for a medicine-chest and for fitting and discharging a vessel, and payments to third parties, followed the activities of the agents in the home port.

At once the most common and significant of seagoing debits was the cumulative charge for articles bought on credit from the slop-chest. This not-too-elegant term was the universal American equivalent for the ship's store. For owing to the long and uninterrupted periods spent at sea a whaling vessel had to carry a stock of goods from which the hands might purchase articles currently needed.

Such an arrangement was rendered all the more imperative by the lack of money and of foresight which prevented the whalemen from equipping themselves adequately before sailing; and also by the outfit which was sold to every foremast hand before a voyage, for most of the articles included in these outfits were of such inferior quality that it was necessary to replace them from the slop-chest long before a full cargo had been secured. Add to these demands the further replacements necessitated by accidents, losses, and the contingencies of whaling voyages, and it becomes evident that the slop-chest played a vital rôle in shipboard life.

Sales on a credit basis rendered this rôle even more significant. Since the men had to have certain supplies, and since they possessed neither cash nor a ready way to get it, there was no alternative to an extension of credit. But inasmuch as most hands could not be trusted to repay their debts voluntarily, provision was made for compulsory subtraction from the gross lay at the end of the cruise. Such an arrangement would have been workable and businesslike if made between equals. But when carried out under the shipboard conditions of brutal discipline, unquestioning obedience, and frequent clashing of personalities, it presented ready opportunities for sharp practices and vengeful distrust.

The slop-chest, however, was used not only for supplying the crew, but also for "recruiting ship" — a term describing the practice of taking aboard whatever fresh provisions and other supplies might be needed during the months and years spent away from the home port. Only occasionaly did the purchase of these provisions demand specie. Instead, it was customary to fill the slop-chest with goods to be bartered directly for desired commodities. Barter was the general procedure in the little-frequented harbors of the South Seas, where fresh provisions were secured in exchange for cotton cloth, powder, tobacco, knives, beads, and cheap ornaments and novelties. In the larger civilized ports, where barter often presented serious difficulties, it was common to sell certain slop-chest supplies to established merchants and then to purchase needed articles through other dealers.

This strong predilection for barter influenced most whaling vessels to leave their home ports with surprisingly small sums in specie. One hundred dollars, or thereabouts, was often considered an adequate provision for thirty to forty months at sea. One New Bedford whaler once set sail with only $97.60 in the captain's strong-box. These relatively small amounts, however, often proved to be quite ample; for with the exception of official port charges and occasional incidental expenses, all payments were made from the slop-chest or by robbing the cargo of oil and bone.

This method of financing a voyage combined several advantages. It was convenient; it was safe, since it avoided the obvious risks and temptations which would have accompanied the presence of large sums of money on shipboard; and it afforded an opportunity for securing a handsome profit. For if the captain were a shrewd bargainer (and the average Yankee whaling master was nothing if not a keen tradesman) the rate of exchange emerging from the bartering process was likely to be decidedly favorable to the owners of the vessel.

In short, the slop-chest was invaluable to whaling. But what of its contents? For the most part, it was stocked with assorted articles of clothing, with much cheap cloth in the bolt, and with endless pounds of tobacco. But there were also sheath knives and belts, needles and thread, combs, spoons, tin pots and pans, cups, blankets, cotton handkerchiefs, axes, axe handles, beads, brogans, yarn, bed comforters, neck comforters, round top hats, palm leaf hats, and whatnot. On the whole, a rather narrow range of prosaic commodities. Just such a collection of goods, in fact, as would suffice to meet the restricted needs and even more restricted purchasing power of a whaler's crew, and at the same time make some appeal to the untutored natives from whom provisions would have to be secured. The exact number and variety of articles on any given vessel varied, of course, with the nature of the cruise, the parts of the world to be visited, and the fancy and shrewdness of the owners. In essentials, however, one slop-chest was much like another; and the following list, given in its entirety, is a representative sample.[2] Especially noteworthy are the differences between cost prices and sale prices; the great yardage of cheap cloth (six different kinds, comprising 4855 yards); and the enormous quantity of tobacco (4329 pounds).

Invoice of Slop Clothing and Merchandise put on board ship *Canton,* George White, Master, (bound to the Indian Ocean and elsewhere on a Whaling Voyage) for the supply of the Crew and to Recruit the Ship.

In Cask No. 1

Charge Crew		*Cost*	
$1.50	50 prs. Pumps	@ 78¢	39.00
1.50	35 prs. Thick Shoes	85¢	29.75
1.75	27 prs. Thick Shoes	$1.00	27.00
6.00	8 Cadet Salt Monkey Jackets	3.25	26.00
5.00	12 Black Salt Reefing Jackets	2.50	30.00
3.00	15 prs. Cadet Salt Trowsers	1.50	22.50
2.00	15 prs. Black Salt Trowsers	1.12½	16.87
2.00	18 Heavy Stripe Kersey Shirts	1.12½	20.25
2.00	18 Blue Twill Flannel Shirts	1.10	19.80
1.50	18 Stripe Kersey Undershirts	.62½	11.25
1.50	12 pr. Stripe Kersey Drawers	.62½	7.50
1.50	12 pr. Mixed Kersey Drawers	.54	6.48
2.50	2 Reefing Jackets	1.25	2.50
1.00	1 Calico Shirt		.50
1.00	21 Stripe Cotton Shirts	.45	9.45
.50	2 8/12 Doz. Jack Knives	2.00	5.33
.25	5/12 Doz. Sheath Knives	1.75	.74

[2] This list of slop-chest articles was taken aboard by the ship *Canton* at New Bedford on November 30, 1858. It was found in an original Invoice-Book (now in the New Bedford Public Library) of Chas. R. Tucker & Co., of New Bedford. This invoice-book, covering the period 1857 to 1875, contains some 30 complete invoices of goods carried on as many different voyages "to recruit the ship and for sale to the crew." The list here reproduced is typical of the others.

Charge Crew		*Cost*		
.20	14 pr. Suspenders	.10	1.40	
3.00	1 pair Boots		1.00	
.10	11 Brittania Spoons	.06	.66	
.20	1 4/12 Doz. Sheaths & Belts	1.25	1.67	
.25	4 Tin Pans	.12½	.50	
.25	2 Doz. Shields Caps	1.75	3.50	283.65

IN CASK NO. 2

2.00 each	3 pr. Mixed Blankets	2.00	6.00	
2.00 each	5 pr. White Blankets	1.87	9.37	
2.00	6 Bed Comforters	.85	5.10	
2.00	12 Guernsey Frocks	1.00	12.00	
.75	2 Doz. Double Socks	4.50	9.00	
.40	1 Doz. Double Mittens		2.75	
.25	2 Doz. Neck Comforters	2.75	5.50	
1.25	12 Calico Shirts	.50	6.00	
.25 each	2 doz. Palms	1.50	3.00	
.10	1 4/12 Doz. Sheaths & Belts	1.50	2.00	
.10	1 Doz. Dressing Combs		.50	
.20	1 Doz. Ivory Combs		1.10	
.50	2 Doz. Rodgers Knives	3.00	6.00	
.40	3 Doz. Shepherds Knives	2.00	6.00	
.10	2 Doz. Spoons	.42	.84	
.20	2 Doz. Suspenders	1.75	3.50	
	4 lbs. English Linen Thread	.70	2.80	
	20 papers Needles	.03	.60	
	39 skeins Yarn	.09½	3.70	
.20	3 Doz. Cotton Handkerchiefs	1.50	4.50	
.25	2 Doz. Sheath Knives	1.12½	2.25	
1.00	48 pr. Best Denim Trowsers	.54	25.92	
2.00	24 pr. Pantaloon Trowsers	.75	18.00	
.50	1½ Doz. Round Top Hats	3.00	4.50	
.25	2 Doz. Palm Leaf Hats	1.25	2.50	
1.00	36 Striped Cotton Shirts	.45	16.20	
.25	2 Doz. Tin Pots	1.50	3.00	
1.00	36 Denim Frocks	.46	16.56	
.25	2 Doz. Tin Pans	1.50	3.00	182.19
2.00 (In Box)	60 Pairs Thick Brogans	1.10		66.00
1.50 (In Box)	29 Pairs Brogans	.60		17.40
(In Box)	1 Doz. Handle Axes			11.50
(In Box)	2 Doz. Axe Handles	1.50		3.00
(In Box)	1 Doz. Axes			11.00
	1 Case Bleached Cotton 650 Yds.	.09		58.50
	1 Case Blue Denims 1012 Yds.	.11½		116.38
	2 Case Blue Drills 1643¼ Yds.	.10¾		176.65
	1 Bale Brown Drillings 800 Yds.	.10		80.00

IN BOX

.40¢	10 Boxes G. R. Wilkins Tobacco	1175 lbs. @ .20	235.00
.40¢	11 Boxes Daniel Rives Tobacco	1193 lbs. @ .15	178.95
.40¢	10 Boxes Virginia Tobacco	1176 lbs. @ .15	176.40
.40¢	7 Boxes Tobacco	785 lbs. @ .15	117.75

IN A CASE

				Cost
10 Pieces Louisiana Twill	375 Yds. @	.15½	58.12	
10 Pieces Washington Jeans	375 Yds. @	.15½	58.12	116.24
4 Doz. Axe Handles		1.87½		7.50
2½ Doz. Handles		1.25		3.12
1 Chain Cable 90 Fathoms	5600 lbs. @	.05		280.00
1 Anchor	1450 lbs. @	.06½		94.25
25 Barrels Flour		6.25		156.25
1 Whaling Gun				45.00
50 Bomb Lances		3.50		175.00
Specie				97.60
				2689.33

We wish things sold at cost rather than have them brought home.
To Capt. George White. New Bedford, Nov. 30, 1858.

The figures given in the above list disclose that, on the average, the commodities were sold for about twice as much as they cost. This rate of return, though departed from in many individual instances, was perhaps representative of most mid-century whalers. The more responsible agents and owners gave their captains price lists which were calculated to yield a profit of about 100% per voyage on the slop-chest transactions. But these set prices applied only to sales to the crew. In dealing with outsiders the masters were commonly instructed to secure the best terms possible. And these "best terms possible" often pushed the profits well above 100%.

A case in point was the voyage of the ship *Montreal* which closed in 1862. For this cruise the slop-chest alone showed a net profit of $3149.56. During the four-year period 1856–1860, however, the ship *Adeline* made total sales aggregating only $2663.36 — an amount appreciably less than the net profits of the *Montreal.* A somewhat smaller amount, $2-458.29, was sold by the ship *Fabius* on her seventh and last voyage (she was wrecked on Solidad Reef, off the coast of California, in 1865). But even these sums were princely when contrasted with the $400.32 slop-chest profit made by the ship *James Maury* on her first voyage, 1845–1848. A normal and representative cruise, on the other hand, was that of the bark *Mars,* 1857–1859. Her figures were: [3]

[3] All figures given in this paragraph were taken from the original account-books of the vessels mentioned. These volumes are now in the stacks of the New Bedford Public Library.

Net Amount of Slops and Goods Sold to Crew	$1273.30	
Net Amount of Slops and Goods Sold for Recruiting	1869.12	
Total Sales		3142.42
Invoice (Cost Price) of Slops and Goods Sold		1559.59
Net Profit (slightly over 100%)		1582.83
Slops Not Sold and Appropriated to Ship's Use	487.61	
Invoice of Slops Remaining on Board	70.34	

The owners considered average slop-chest profits of 100% by no means extortionate, nor even unreasonable. The return was calculated for the entire voyage of two to four years, and not annually. The risks were undoubtedly great, and were further increased by the common practice of desertion, which made it impossible to secure payment for the supplies which had been advanced. Another source of loss existed in the form of unsold goods; for articles which had reposed in the hold of a whaler for many months could be disposed of only at a heavy discount, and in many instances constituted a total loss. In fact, most agents instructed their captains to sell surplus slop-chest supplies at cost, or even at a net loss, rather than to bring them back to the home port. And finally, the position of the master as seller and disciplinarian formed such a complete monopoly that the temptation to sell above the competitive prices prevailing on shore was well-nigh irresistible.

The resulting high prices, often combined with grossly inferior quality, made the average foremast hand regard the slop-chest as a necessary but iniquitous institution, through which the owners and master profiteered outrageously at his expense. But in addition to this double indictment, the forecastle also nourished more specific slop-chest grievances, generally incited by the less scrupulous masters. Noteworthy among the causes of such grievances were the refusal to advance cash when in port, thus compelling the men to buy articles on credit from the slop-chest and to sell them ashore at a loss in order to secure a small amount of spending-money; arbitrary raising of prices; failure or refusal to quote exact prices at the time of sale, so that a man would not know how much he had been charged until the final settlement was made, months or even years afterward; entering a charge in the buyer's account for a larger amount than the price quoted; and general arbitrary and autocratic treatment from one who was both captain and monopolist.

These sharp practices, where they existed, were much encouraged by the prevailing custom of giving the master a share in the proceeds of the slop-chest. Virtually every captain was allowed a percentage of the total sales or of the net profits; and under either method he frequently secured an appreciable sum. In the case of the four voyages cited in a preceding paragraph, for instance, the master of the *Montreal* received one-half of the profits, or $1574.78; the captain of the *Adeline* was given 10% of the total sales, or $266.33; the owners of the *Fabius* allowed 7½% of the total sales, or $184.37; and the master of the *James Maury*, entitled to one-half of the profits, received $200.16.

But even this mid-century system of slop-chest management, however inadequate, was a real advance over the conditions of the preceding decades, when it was customary for the captain to conduct the slop-chest as his own private perquisite, and to pocket the entire sum extracted from it. With no restraints except those imposed by his own judgment, the grasping and unscrupulous captain resorted to extortionate abuses which filled his pocketbook at the expense of his crews.

Lieutenant Charles Wilkes, U.S.N., who commanded an exploring expedition which brought him into intimate contact with whaling conditions in the Pacific, was unsparing in his criticisms of this early system. He cited an instance in which the master of a crew of thirty men made $1800 during a single voyage, and asserted that such personal profits sometimes ran up to $2000 or $3000. He was so impressed by the gravity of the abuses that he recommended that all slops be supplied to whaling crews at an advance over cost just sufficient to cover the expenses of handling and of storage, and that the captains be deprived of all interest in the greatly reduced receipts.[4] Unfortunately, however, these admirable recommendations were supplanted by milder reforms. In the end the captains

[4] See Wilkes, Charles, "Narrative of the United States Exploring Expedition," V, pp. 496–502.

1855 **Cruising in the Ochotsk Sea**

Ther. 6 A.M. 55

close in shore, lowered the three boats but it being so near dark, they had to give it up and come on board. Middle part calm and light airs from the Southward heard whales around the ship all night At ½ past 4 A.M. had breakfast, and seen whales, lowered the three boats, the Captain taking the waist boat, the 2nd mate being sick, it also breezed up fresh from N.N.W. a few minutes before meridian the bow boat struck a large whale and lost their line, before they could shift ends in the boat

Lat. by Obs. 59..09 N
Long. .. Acct 143..20 E

Barometer 6 A.M. 29.30
Ther. 6 A.M. 46

Tuesday Sept. 18th Begins with fine weather and moderate winds from S.W. plenty of whales in sight, also the City of Ochotsk, during the afternoon the larboard boat struck another 200 whales, and the irons drew, at 6 the boats came on board middle part, a gale of wind from N.W. and clear weather, hove the ship to under double reefed maintopsail and foretopmast staysail, head to the Westward, latter part the gale moderated a little, set double reefed topsails, and wore ship heading N E.

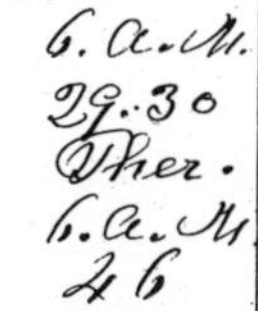

Lat. by Acct. 58..55 N.
Long. " Do 143..30 E.

Barometer 6 A.M. 29.70
Ther. 6 A.M. 43

Wednesday Sept. 19th Begins with strong winds from N.W. and clear weather, the ship under double-reefed topsails and courses, heading N.N.E. seen one whale, during the afternoon seen plenty of whales, but too rugged to lower, middle part strong winds from N.N.E. heading N.W. latter part fresh breezes from the same quarter with clear weather saw several whales, lowered all the boats

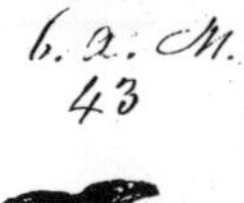

Lat. by Acct 59..10 N
Long. " Do 143..25 E

CRUISING IN THE OCHOTSK SEA
A Page From the Log of the Whaler *Adeline*

were provided with definite price-lists and their gains restricted to a given percentage of the income accruing from the operation of the slop-chest.

Individual slop-chest accounts ranged from one dollar to well over $125, with a tendency to congregate in the region between $50 and $75. An analysis of the figures of two voyages will present a picture typical of hundreds of other cruises. Of twenty-six men shipped by the *William C. Nye* on her first whaling voyage, 1851–1854, twenty-two made purchases of more than $25, leaving only four who succeeded in remaining below that figure. Of those owing more than $25, twenty accounts were for more than $50; seventeen, for more than $75; eleven, for more than $100; and seven, for more than $125. On the other hand, the accounts of the last voyage of the ship *Fabius*, 1862–1865, showed that twenty-two men bought less than $25 worth of slops, while twenty-six hands exceeded that sum. Of the latter group, twenty-one owed more than $50; fourteen, more than $75; thirteen, more than $100; and five, more than $125. The individual debts ranged from $156.54 down to only $1.00. On the *William C. Nye*, however, the smallest bill called for $3.12 and the greatest one for $171.12, with one abnormally heavy debt of $310.46.[5]

But the articles purchased were surprisingly lacking in variety. The range of choice of the average slop-chest was limited, and even within this restricted sphere certain articles ruled as prime favorites. Tobacco and various articles of clothing formed the staple commodities of slop-chest commerce; and, except for occasional investments in knives, cloth, thread, or eating utensils, the average foremast hand bought largely of these staples. This fact appears clearly in two itemized accounts, both taken from a voyage of the bark *Marcella*, 1854–1856. The first is characteristic of the larger accounts which were made up of a rather long list of purchases; while the second represents the smaller bills belonging to men who were unusually thrifty, who had been on board for only a short time, or who were fortunate enough to begin the voyage

[5] These figures were secured by analyzing the original account-books, now in the New Bedford Public Library.

with a well-stocked chest. In both instances certain charges which did not represent actual purchases have been omitted.[6]

Antone de Castra dr. to Bark *Marcella*

1854				
June 26	2 Jack Knives	@	.40	.80
	2 Frock Shirts		1.20	2.40
	2. pr. Thin pants		1.20	2.40
	2 pr. Thick pants		2.25	4.50
	2 pr. Drawers		1.30	2.60
	2 Thin shirts		1.00	2.00
	1 Guernsey frock		2.00	2.00
	2 pr. stockings		.75	1.50
	1 Reefing jacket		5.00	5.00
	1 Tar hat		.50	.50
	1 Neck Comforter		.25	.25
	1 pot, pan, & spoon		.60	.60
	1 Sheath knife		.25	.25
June 29	2 pr. Shoes		1.50	3.00
	2 lbs. tobacco		.30	.60
	1 doz. pipes		.16	.16
Sept. 6	5 lbs. tobacco		.30	1.50
	1 Jack knife		.40	.40
Dec. 22	2 lbs. tobacco		.30	.60
1855				
Jan. 21	Cloth at Augustine Bay 21 yds.		.20	4.20
	3 lbs. tobacco		.30	.90
Feb. 13	Thread		.12	.12
Aug. 7	2 lbs. tobacco		.30	.60
Sept. 22	1 Thick shirt		2.00	2.00
Oct. 10	1 pr. Thick pants		2.25	2.25
	1 pr. blankets		3.00	3.00
Oct. 29	3 lbs. tobacco		.30	.90
Nov. 24	2 yds. Cloth		.17	.34
Dec. 13	2 yds. Cloth		.17	.34
1856				
Jan. 23	2 lbs. tobacco		.30	.60
Feb. 22	2 lbs. tobacco		.30	.60
March 28	2 lbs. tobacco		.30	.60
May 22	2 lbs. tobacco		.30	.60
Total Slop-chest Bill				48.11

[6] *Ibid.*

Antone Belveira dr. to Bark *Marcella*

1854			
Sept. 4	5 lbs. Tobacco	.30	1.50
	2 Frock Shirts	1.20	2.40
Nov. 18	2 Thin shirts	1.00	2.00
Dec. 22	4 lbs. tobacco	.30	1.20
1855			
Jan. 31	Cloth at Augustine Bay 16 yds.	.18	2.88
Feb. 13	Thread	.12	.12
Aug. 7	2 lbs. tobacco	.30	.60
Sept. 22	1 lb. tobacco	.30	.30
Dec. 17	2 yds. Cloth	.17	.34
1856			
Jan. 27	½ lb. tobacco	.30	.15
March 28	1 lb. tobacco	.30	.30
May 22	3 lbs. tobacco	.30	.90
Total Slop-chest Bill			12.69

At certain rare intervals, however, the whaleman demanded cash, instead of goods advanced on credit. Chiefly this was just before he sallied forth for a few precious hours of shore leave. The commercialized dissipation and recreation open to sailors in most ports required at least a modicum of money; and since the average foremast hand was chronically penniless, he was forced to ask the master to advance a few dollars and to charge the amount against his prospective earnings. If, for any or no reason, the captain refused, there was no recourse. But if all the circumstances were propitious, the applicant would be set ashore with a few dollars (seldom more than $5 or $10) in his pocket. Whatever the exact sum might be, it was promptly added to the other debit items in the man's account.

The total amount of cash so drawn during the course of a single voyage ranged from zero to $100 per man. Seldom (and then usually in the case of a mate or boatsteerer) did it exceed the latter figure. Thus during the seventh voyage of the ship *Fabius*, 1862–1865, 53% of the men received advances ranging from $25 to $45, inclusive; 36% contrived to exist for three years with less than $25 in cash; and 11% drew more than $45. Only one individual was able to secure more than $75 from the captain.

Whaling accounting, however, did not always distinguish carefully between cash advances and other debit items. At times the cash accounts and the slop-chest accounts were merged, and the two sums would appear together in the final settlement as "Slops and Cash." Upon other occasions the amounts loaned and the interest charges were combined in a single entry, without distinguishing between principal and interest and without giving the rate of interest used in the calculation.

But whether or not the rate was given specifically, virtually all cash advances were accompanied by charges known as interest, exchange, or insurance. There was a singular lack of uniformity in the use of these terms; but the first two seem to have been employed interchangeably to denote ordinary interest, while insurance referred to an additional and separable charge, presumably made in order to cover the heavy element of risk involved in such transactions. In actual practice, however, interest and insurance were frequently merged.

Some concrete instances will illustrate both the varying terminology and the average or customary rates. During the years 1860–1861 the master of the ship *Canton* charged 20% exchange on cash advances. The ship *William C. Nye*, on her third whaling voyage, 1858–1861, also charged "Exchange on Cash," but at the rate of 25%. During the years 1843–1845 the bark *Canton* exacted a similar percentage, but termed it "Interest on Cash." Still earlier, during the first voyage of the bark *Minerva,* 1836–1839, both interest and insurance were combined under a single item of 25%. From 1859 to 1862, however, the ship *James Maury* separated these two charges by exacting 12% insurance in addition to 18% interest.[7] And finally, during an earlier cruise of this same vessel, each of the three terms is used in the account of one George Shaw. Here, under date of January 27, 1848, the following entries occur:

[7] This list might be multiplied indefinitely by other instances taken from the original account-books in the New Bedford Public Library and elsewhere.

20% Exchange on Cash $28.38	5.68
25% Int. & Ins. Charged at Prime Cost per Agreement on $53.50 is	13.37

As this list of typical entries indicates, the average whaleman was called upon to pay a charge of about 25% on all cash advances made to him.

It should be noted, however, that this was the rate per voyage, not per annum, and that the same rate applied uniformly to all advances, whether made at the beginning or near the end of a cruise. No attempt was made to proportion the charges to the length of time during which the respective sums remained unpaid. Thus if a foremast hand secured an advance of $10 soon after sailing, so that the debit remained active for as long as forty months, his account would be charged with an additional $2.50 (assuming the rate to be 25%); while if he contracted another $10 debt three years later, and only four months before the final settlement, he would likewise be given a debit of $2.50. Witness the typical case of John Rodney, steward of the ship *Canton* during the years 1852–1855. From April 27 to May 8, 1853, inclusive, Rodney obtained four advances of cash, aggregating $16.50; between April 27 and May 10, 1854, he drew $14.50; on November 28, 1854, the captain allowed him to have $24.50; and on March 24, 1855, he was given a final $20.50. But on each of these entries, scattered over a period of almost two years, he paid interest at the uniform rate of 25%.

Criticism of the heavy interest charges seems justified in large part by this practice of levying uniform rates per voyage. A charge of 25% to cover interest and insurance for a period of three years was certainly not unreasonable. But the same reasoning which justified a rate of 25% for three years necessarily condemned it when exacted for a period of only one year. When calculated on an annual basis, a percentage which was moderate for a long period became extortionate and usurious for a short period. A whaleman who paid a charge of 25% for an advance extending over three full years was paying interest at the rate of 8⅓% per year. But if he obtained another loan a year later, the annual rate of interest would be 12½%; if he waited another year, the rate per annum would be 25%; and if he borrowed again only six months before the end of the voyage, he would be paying interest at the rate of 50% annually. And since the rates of 8% to 15% per annum secured by the owners on the longer advances were certainly adequate, justification for the usurious percentages on the shorter loans seems signally lacking.

Such justification, even if attempted, would have been all the more difficult because the crew members were allowed no interest whatever upon their accumulating earnings, which, under the lay system, remained undistributed until the close of the completed voyage. If annual or semi-annual settlements had allowed the whalemen to receive a part of the earnings actually due them, it would have been unnecessary, in many instances, for them to ask for cash advances. In principle, at least, there would seem to have been no reason for charging interest on cash advances which would not have applied equally to the allowance of interest on accumulated but unpaid earnings. If the owners were without their funds until the close of the voyage, at least some members of the crews were likewise without their earnings for a similar length of time. And since the owners were the self-appointed custodians of those earnings, the undistributed lays served to fill up the financial gaps caused by the cash advances. From the standpoint of the employers, the result was largely an exchange of like values. To the foremast hands, however, there was no quid pro quo—except at a 25% premium! Noblesse oblige was hardly applicable to the financial relationships existing between the forecastle and the counting-room.

But exploitation, inherent in the lay system even at its best, was often intensified, in the hands of over-thrifty agents or masters, into usury and extortion. The rate of 14% per annum, paid by the men on the ship *Magnolia* during the years 1854–1858, was easily outdone by other vessels. During her eleventh whaling voyage, 1867–1869, the bark *Marcella* exacted a rate of 50% per voyage (though during this same cruise a total of $490.95 was advanced to some thirty persons without entering any interest charges whatever). Throughout these same years the captain of the ship *Adeline* was also charging 50% per voyage, though in one instance he outdid himself by increasing the rate to 60%. Nor did these vessels monopolize the excessive interest charges. Small wonder that interest and insurance payments were significant items on the debit side of crew account-books!

In addition to slop-chest purchases, cash advances, and interest charges, certain minor but interesting items often found their way into the debit columns. Among such entries were orders assigning given sums to third parties. Petty financial obligations, arising in various ways during the course of a long voyage, could not always be settled to the mutual satisfaction of the bargainers. In such cases, if the creditor was insistent, it was common to make out an order directing the agents, at the time of the final settlement, to pay the required sum to the person specified and to subtract a like amount from the lay of the debtor. These orders often found their way into the hands of the captain, who incorporated their terms into the accounts of the two men concerned. Because of the extra bookkeeping and the danger of numerous complications, however, the agents sought to restrict this practice to the officers and the more responsible foremast hands.

Desertion, too, caused extra charges. If a whaleman escaped from his vessel, his account was usually debited with the expenses involved in hiring another man to take his place, with the wages of any shore labor necessitated by his absence, and with the fees connected with his apprehension and arrest (if, as often occurred, he was discovered and returned to his captain). In the crew accounts of the ship *William C. Nye*, for instance, is to be found this receipt, strikingly suggestive of a large number of similar transactions:

Ship *Wm. C. Nye* To Port of Honolulu Dr.
For Arrest of Manuel Laurence $20
Honolulu Novr. 29, 1853.

Received Payment,
W. C. Parke.

For those deserters who were able to escape recapture, however, debit entries were of no practical effect. For on the one hand the owners perforce bore any net losses shown by the accounts; and on the other they were entitled to confiscate not only the deserters' shipboard effects, but also their net earnings, if any.

Now and again a foremast hand would be debited with the sum for which he had purchased certain articles at a ship's auction. On whalers, as on merchant vessels, the captain or mate customarily sold at auction the effects of men who died at sea. If the deceased member of the crew had been popular, or if there were a desire to help his family, there was often a deal of generous rivalry in bidding for his effects. Usually, however, the successful bidders had to postpone payment until the end of the voyage. At that time the various amounts realized at auction would be collected and given to the dead man's family, if known. If no members of his family were forthcoming (and in the polyglot whaling crews of the later days this was often the case) the appropriate sum, according to law, was to be delivered to a consul or to the courts.

Whenever the responsibility for stolen property could be fixed, the amount of such loss was also charged against the account of the offending seaman. The captain arbitrarily determined the value of the lost or damaged goods; and, as was expected, his estimates were generous. A striking illustration of this occurred in January, 1854, while the ship *James Maury* was anchored in Margaretta Bay. Strangely enough, four members of the crew succeeded not only in deserting, but also in stealing a whaleboat. The master must have been sorely vexed; for the accounts show that, instead of proportioning the loss amongst the four deserters, each one of them was charged with the full price of a new whaleboat.

Still other debit items, occasionally appearing in individual accounts, included the cost of medical attention received in foreign ports; stray board bills which had been paid by the captain for some man fortunate enough to spend several consecutive days or weeks ashore; various consular fees; payments involved in the process of hiring new men and of discharging old ones; and a series of fines, police fees, jail costs, and other expenses incurred on behalf of men who were arrested by police or by port authorities. The following entries, taken from the accounts of two separate voyages, illustrate the nature and variety of these incidental items. Some of the expenditures were borne by the owners; but a good percentage of them represented charges against individual members of the crew.

Miscellaneous Debit Entries From the Crew-Accounts of the Ship *Alice Mandell,* 1851–1855

Gov. Fees Native Seamen	$3.50	Shipping 2 men	$6.00
Shipping 3 Men	3.00	Shipping fees	3.00
Arrest of 3 Men	45.00	Fort & Police Fees *	19.00
Sending them on board	10.00	Shipping fees 4 men	12.00
Men Shipped	25.00	Charges 4 Seamen	37.00
Fort Bill Wm. Thompson *	6.00	Catching Deserter	17.00
Medicine for L. Potter (not chgd)	10.00	Catching 3 Deserters @ 15-E	45.00
Fort bill *	47.00	Jail fees 3 Kanakas	6.00
Paid for 2 Deserters	5.00	Advances 5 Chinese $20 each	100.00
Paid for John Clay	2.00	Hong Kong-Shipping 6 Men	6.00
4 Men Shipped	8.00		
for Kannaccka	1.00		
Fort fees *	13.00	Fort fees *	15.00

* These items refer to charges in connection with a military prison, instead of an ordinary jail.

Miscellaneous Debit Entries From the Crew-Accounts of the Ship *Wm. C. Nye,* 1851–1853

Medicine & Doctors fee	$7.84	Shipping 14 men	$18.00
Board bill	8.40	Certificate of Desertion	2.00
Visiting Ship	5.00	Discharging 17 men	12.50
For Securing 2 Natives	3.00	For Articles 2 Natives	2.00
Consul fees for shipping 1 man	4.50	For Shipping 2 Natives Governor's fees	7.00
Shipping 12 Men & 2 Certificate	16.00	Discharging 6 Men & 2 Certificate	7.00

In addition to these seagoing debits, which originated with the master during the course of a voyage, there were also landlubber debits growing out of the activities of the agents in the home port. Towering above all other items in this second series was the outfitter's bill. To the whaleman the outfitter was a parasitical nemesis. Existing primarily to provide necessary equipment which the penniless and inexperienced hands were unable to furnish for themselves, the outfitter gradually added other functions which placed him at the very center of an entangling web of commercialized exploitation. The hapless occupant of the forecastle was deceived, cheated, and robbed at every turn. And nowhere more than in paying for the outfit which was placed in his hands just before sailing.

Soon after the vessel bearing any given seaman and his outfit had put to sea, the outfitter presented his bill to the ship's agent and received payment in full, usually in the form of a note payable in thirty days to six months. The agent, in turn, entered the amount of the bill on the debit side of the man's account, where it remained until the end of the voyage. During the many intervening months, however, the original sum (seldom less than $60 and commonly not exceeding $100) grew materially through interest charges. As in the case of cash advances by the master, the rate was often as high as 25% per voyage. But even a lower rate was quite sufficient to swell the amount to be eventually collected. A seaman who shipped on the *Samuel Robertson* for a voyage during the late fifties, for instance, paid $75.82 in settling for an outfitter's bill of $65.93. The difference, or $9.89, represented 15% interest for a cruise of fifteen months.

An outfitting bill might be supplemented, too, by an infitting bill, often presented by the same firm. Returning whalers were boarded by competing agents, or runners, seeking to entice the forecastle-weary hands into their respective establishments for a rough-and-ready cleansing and rejuvenation. After securing a bath, a shave, and a haircut, donning an outfit of cheap, but clean, "shore togs," and accepting a small sum of cash for immediate spending-money, the transformed whaleman was pronounced ready to sally forth to seek the delights of port society. But these attentions were carefully confined to those men whose accounts indicated a clear credit balance for the settlement which was to come several days later. If a seaman were so unfortunate as to possess only a debit balance, it was often impossible for him to secure these infitting services without agreeing beforehand to ship for another voyage at the first opportunity. Such an "opportunity" usually appeared from three days to three weeks later; and it meant, for the foremast hand, a fleeting glimpse of the home port between two seemingly endless voyages, each occupying from thirty to forty-five slow-moving months.[8]

The task of fitting a whaler for sea also explained a charge in the agents' accounts; and if and when she returned, laden with oil and bone, the laborious discharging of the cargo added still another debit. Both fitting and discharging might have been done, properly enough, by the regular crews. But tradition decreed that prospective hands be dumped aboard their vessels only a few hours before sailing, and that the hardened veterans of a long voyage were free to go ashore as soon as they saw the anchor down and the sails furled. And so the beginning and the ending of a whaling cruise required shore labor in considerable quantities. While the owners and agents acquiesced in this system, they insisted that the members of each crew must reimburse them for the additional wages bill. Hence the presence in virtually every man's account of two more debit items — one for fitting ship and the other for discharging the cargo.

Theoretically the exact amount of expense entailed by these operations was to be prorated among the men aboard each vessel. But in practice the agents adopted a simpler expedient: they collected two arbitrary sums from each whaleman. These sums, deemed sufficient to cover the extra outlay, ranged from \$5 to \$15 for fitting the vessel for sea, and from \$2 to \$10 for discharging and cleaning her. The amounts appearing most frequently were \$10 for fitting and \$5 or \$6 for discharging. For five successive voyages, covering the period 1848 to 1868, the ship *James Maury* was fitted for sea at a cost of \$10 per head; but larger cargoes or a greater attraction for barnacles caused her discharging and cleaning bills to advance from \$6 to \$9 per man. The *Samuel Robertson*, sailing in 1857, exacted \$15 at the beginning of her cruise and \$10 at its conclusion; while the bark *Minerva*, returning in 1839, cost her crew only \$5 apiece for fitting and \$2 each for discharging.

The owners exacted still another tribute from their crews

[8] For a more detailed description of the activities of the outfitters, see Chapter VI.

by charging for a medicine-chest. Such a necessity might well have been considered a vital part of the equipment, to be provided by the agents just as harpoons and whale-boats and foodstuffs were furnished. In the early years of the industry, however, the treatment of illness at sea depended upon the master and the rough-and-ready administration of such simple remedies and common medicines as he may have had the foresight to take aboard. Later a standardized medicine-chest, including a supply of the more familiar drugs and directions for their use, was developed. But the owners felt that the provision of such a chest, however simple and inexpensive, was an additional financial burden which their crews must assist in bearing. Consequently a medicine-chest charge of $1 to $2 per head began to appear in the crew-accounts; and this custom, once started, continued throughout the remaining history of American whaling. The amount assessed (seldom more than $2) was relatively unimportant in itself; but it added another link to the long chain of whalemen's debits.

The final type of debit entry was due to sums advanced by the agents to third parties, practically always kinsmen of men who were away at sea. Such advances of money, while carefully restricted to the families of officers, boatsteerers, and the most responsible foremast hands, were sometimes necessitated by the exigencies of whaling life. In exceptional circumstances the agents might consent to assist in meeting an emergency even without express authorization from the man whose account would be affected. Thus some family crisis no doubt explained the item of $100.63 paid to the master's wife during a voyage of the ship *Brighton* which ended in 1847. More commonly, however, a departing whaleman would request his agents to advance reasonable sums to specified dependents whenever they might apply for them. During the sixth whaling voyage of the ship *James Maury*, for instance, the wives of the master, second mate, cooper, and a boatsteerer received numerous advances. Or a man at sea might send back a written request to advance definite amounts to dependents or creditors.

To avoid inconvenience and risk, the agents were loath to pay out money in this manner unless they felt assured of the existence of serious need and of the dependability of the whaleman whose account was to be debited. At times, however, it was impossible to escape some degree of obligation; and in such rare instances the transaction gave rise to the last of the long list of debits charged against the crews while away at sea.

Credits, by contrast, were few in number. In many instances one lone entry, consisting of the amount of the lay, was left in single-handed opposition to its smaller, but numerous, antagonists. Normally the strength of the lay was sufficient to prevail against its crowding adversaries; but frequently it was dragged down to defeat. And every defeat meant that some whaleman actually found himself in debt to the owners for whom he had been working throughout many months.[9]

Occasionally, however, the lonesome lay was joined by two small and not-too-valiant brothers-in-arms. One of these was the entry for bounties and rewards. In order to stimulate alertness on the part of the look-outs, balanced precariously on the cross-trees high above the deck, whaling captains sometimes offered rewards to those hands who "sang out" long and often. The terms of these awards, varying widely, were determined at the pleasure of the respective masters. The winner might be the man who sighted the first whale of the voyage, or the hand who announced the greatest number of "Blo-o-o-ws" during the entire cruise, or the seaman whose victims yielded the largest quantity of oil. Actual capture and stowing away of the oil was usually insisted upon: the man who sighted a whale which escaped was seldom honored.

The prizes were limited both in amount and in variety. The more penurious captains sought to induce watchfulness by offering a few pounds of tobacco, a handful of cigars, or an article of clothing from the slop-chest. More commonly, however, the rewards consisted of cash; and occasionally these cash payments were placed upon a steady and business-like basis. During a voyage extending from 1842 to 1845, for instance, the bark *Canton* had a standing offer of four dollars

[9] See Chapter X for a detailed account of the lay system.

for every one hundred barrels of oil secured from whales reported by any member of the crew; and the accounts show that thirteen men shared in the total bounty, securing sums ranging from fifty cents to $15.38 each.

In some instances the owners authorized their captains to offer bounties, and regarded the slight additional outlay as a part of the operating expense of their vessels. At other times masters and mates subscribed to purses later presented to those who had qualified under the terms of the awards. An unusually generous purse of fifty dollars was so raised by the four officers of the ship *William C. Nye* during her first whaling voyage, 1851–1854. Evidently the officers were eager to make a striking success of the vessel's maiden cruise. One hopes, therefore, that they proved to be more successful with lance and harpoon than they were with the pen! The agreement is reproduced in full:

Dec. 2nd 1853

the undersigned do agree to pay to any one man that raises the most oil taken the remainder of the voyage the some of Fifty Dollars on the arival of the ship which the agent of the *Wm. C. Nye* will please pay when these wrigtings ar presented and Charge each of the undersigned with his respective share as folows

Charles H. Adams	$20
James M. Riley	15
Charles A. Evarts	10
Wm. H. Wilson	5
	$50

Manuel Laurence is entitled to this bounty.

C. H. Adams

But whether the rewards were offered by the owners or by the officers, payment was seldom made until the end of the cruise; and even so, the slowly cumulative credits of any given whaleman were always small. The individual who earned fifteen to twenty dollars as bounty money during a single voyage was fortunate above the average. Many cruises, in fact, were no more lucrative than that of the bark *Mars* from 1857 to 1859, when the bounties earned by the entire crew amounted to only $24.25.

A third credit entry, appearing timidly in a few accounts,

consisted of sums due from third parties. At times the petty obligations between various members of a crew found their way into the accounts of the men concerned. Among the figures pertaining to the eleventh voyage of the bark *Marcella,* for instance, is a notation to the effect that a certain seaman had been doing the mate's washing for an entire year, and that he asked to have his wages as a laundryman added to his earnings as a whaleman. Other men, on other vessels, became creditors through the performance of personal services, through the sale of belongings, or through the production of elaborate scrimshaw work. Whatever the occasion for the debt, the deferred receipt of payment sometimes gave rise to this last and least important of the credits.

But there were also several "credits in kind," consisting of goods or utilities for which no monetary equivalent was exacted by the owners. Witness the opportunities for visiting strange lands, gratification of a love of the sea, and a plentiful supply of whalebone for the monotony-routing scrimshaw work. Such matters, comprising inevitable and incidental by-products, and involving no additional outlay, would scarcely deserve even passing mention were it not that they did add to the scanty list of satisfactions offered by a whaling cruise, and as such were not to be entirely overlooked.

A similar item was that of "free room," including a bunk in the forecastle or steerage. Obviously living quarters for the crew constituted a *sine qua non* of any whaling voyage, to be provided by the owners as a matter of course. It would have been both absurd and impossible to attempt to determine the value of the forecastle to the various foremast hands, and to collect appropriate sums in rent. But the agents might have levied purely arbitrary charges for the use of the bunks, just as they did for providing a medicine-chest, for advances of cash, and for fitting and discharging their vessels. Their failure to take advantage of such a conceivable rent-charge left the crews with the maritime equivalent of "free rent."

By far the most important of the "credits in kind," however, was food. It was customary on board all American whalers for the owners to provide all meals throughout the duration of a voyage. Due in part to the limitations of shipboard space and to the lack of refrigeration, and in part to the thrift and parsimony of agents and owners, the coarse, staple foods taken to sea were surprisingly inexpensive. But even after allowing for the execrable fare that was commonly furnished, "free board" for a period of two to four years was an important matter to the men in the forecastles.[10]

Just how important it was, however, in accurate terms of dollars and cents, no one seems to have taken the trouble to ascertain. Figures showing the food cost per whaleman, either by the voyage or by the year, are signally lacking; but a helpful approximation may be reached by determining the cost of the food consumed during the course of a voyage and then dividing by the number of men in the crew.

If, then, the sperm whaler and the combined sperm and right whalers are taken as the most significant types, and if thirty is regarded as the average number of men per vessel, it would seem that during the middle decades of the nineteenth century the food cost per individual ranged from a minimum of about $150 to a maximum of about $300 per voyage. If three years is used as a convenient figure for the average length of voyage, the annual food bill per whaleman varied from $50 to $100; while the same individual cost, calculated on a daily basis, reached the surprisingly low figure of fourteen cents to twenty-eight cents. Three meals provided at a total outlay of fourteen to twenty-eight cents — from a trifle less than five cents to somewhat more than nine cents per meal! And if (as was actually true during many seasons) the average time at sea of all returning sperm whalers was forty months and more, instead of three years, these annual and daily figures would be still further reduced.[11]

The maximum figure of $300 per man per voyage was de-

[10] See Chapter VII for a description of whaling food and dining etiquette.

[11] These figures make no distinction between officers and men. To attempt to do so would necessitate raising the food cost per officer and lowering it per foremast hand; for all of the choicest and most expensive articles of food were reserved for the officers or for the entertainment of guests by the captain and mates. It must be borne in mind, too, that these sums are mere approximations, used because of the lack of more precise information; and that in every phase of whaling the variations from normal were legion. Nevertheless, there is valid ground for believing that the majority of large, long-cruising vessels fell between these minimum and maximum figures.

rived from a list of foodstuffs taken aboard by the ship *Three Brothers* before sailing from New Bedford in 1865. This list is contained in one of several "Memorandum Booklets, Containing Lists of Outfits for Whaling Voyages." [12] Such booklets, partly printed, partly in long-hand, with many of the items checked and rechecked, were used as a means of verifying the cargoes before putting to sea. The list of foodstuffs carried by the *Three Brothers*, here reproduced in full, suggests similar items checked for the other voyages, made by the bark *Mars*, the bark *Marcella*, and the ship *Adeline* between the years 1852 and 1865. Many entries are accompanied by prices: others stand alone.

List of Foodstuffs Taken Aboard by the Ship *Three Brothers*, 1865

Flour, Baked	120 bbls. @	$ 7 & 7/8	Vinegar	2 Casks — 206 gals. @	16¢
Flour, Packed	70 bbls. @	9½	Pepper Sauce	½ doz.	
Flour, Packed	30 bbls. @	8½	Souchong Tea	3 Chests 110 lbs.	
Mess Beef	50 bbls. @	13½	Hyson Tea	12 lbs.	
Mess Beef	125 bbls. @	10.75	Coffee	17 Boxes 1020 lbs. @	12½¢
Prime Pork	75 bbls. @	25.50	Chocolate	1 Box 25 lbs @	36¢
Mackerel	2 half bbls. @	9.50	Hops	2 Boxes	
Salt Fish	500 half bbls. @	5.00	Essence Spruce	½ doz.	
Pork Hams	1 bbl. 195 lb.	22¢	Ginger	6 lbs.	
Molasses	1007 gals. @	37½¢	Pepper	20 lbs.	
Sugar	2 Boxes		Allspice	4 lbs.	
Sugar, Crushed	30 lbs.		Cassia	1 Mat	
Butter	967 lbs.		Ground Mustard	2 lbs.	
Cheese	254 lbs.		Nutmegs	1 lb.	
Indian Meal	4 bbls.		Saleratus	4 Galls.	
Rice	100 lbs.		Cloves	1 lb.	
Corn	31 bu.		Fine Salt	½ bu.	
Beans	14½ Bu. @	1.75	Coarse Salt	1 Cask	
Dried Apples	479 lbs. @	11¢	Lemon Syrup	½ doz. Bottles	
Cucumber Pickles	1 Bbl.		Sage	4 lbs.	
Onions	40 bu.		Summer Savory	4 lbs.	
Potatoes	120 bu.		Gherkin Pickles	½ doz. ½ gal. jars	
Turnips	15 bu.		Mustard Seed	2 lbs.	

Preserved Meats

4 doz. Cans Roast Beef	@	$7.50	$30.00
1 doz. Cans Roast Mutton	@	7.50	7.50
8 doz. Cans Soups and Bouilli	@	7.00	56.00
2 doz. Cans Lobster	@	5.50	11.00
4 doz. Cans Clams	@	6.00	24.00
1 doz. Cans Oysters	@	6.50	6.50
1 doz. Cans Turkey	@	9.00	9.00
4 doz. Cans Green Corn	@	5.50	22.00
			$166.00

[12] Seven of these booklets are in the New Bedford Public Library.

Obviously many of these items were intended only for the use of the officers upon special occasions. But without attempting to distinguish between the ultimate uses of the various articles, and after allowing for a generous margin of error in estimating the cost of the entries for which no prices are given, it is apparent that this original cargo represented an outlay of $7,000 to $8,000. The amount expended for fresh provisions and for other foodstuffs during the remainder of the cruise could not have raised the total above $9,000, or about $300 per man for a crew of thirty hands. And this figure, for several reasons, is an unusually high one. The main cargo was purchased at the high prices which were prevalent in 1865; the cost estimates, where they have been necessitated by the absence of definite price quotations, have been made upon a generous basis; and the agents of the *Three Brothers*, since she was making her first whaling voyage, were naturally anxious to encourage success by providing her with a good outfit.

The lower figure of $150 per man per voyage was obtained by combining the items of the *Three Brothers* with an earlier list compiled by the Honorable Joseph Grinnell, Member of Congress from New Bedford, in 1844. Mr. Grinnell drew up a list of the foodstuffs carried by the whalers of his day, and accompanied it with a record of the prices paid; but instead of giving the quantity of each article required by a single vessel, he showed the amount consumed annually by the entire whaling fleet. The amounts consumed per vessel, however, were secured by employing the figures given for the *Three Brothers*. In view of the negligible changes in whaling shipboard life between 1844 and 1865, and because of the virtual identity of the average whaleman's food at the beginning and at the close of this period, it seemed entirely reasonable to assume that a crew of thirty men would require approximately the same amounts of the staple foodstuffs in 1844 as in 1865.

By using the articles and prices given by Grinnell in connection with the quantities consumed by the later vessel, then, it was found that the original cost of a cargo of foodstuffs in 1844 was $4160.28. The outlay for fresh provisions during the remainder of the cruise would raise the total to about $4500, or $150 per man per voyage for a crew of thirty men.

The articles and prices used in this compilation, together with the quantities consumed annually by the entire fleet at sea on January 1, 1844, were as follows:[13]

43,868	bbls. beef and pork	@	$8.50
40,692	bbls. flour	@	5.25
14,014	bushels Indian corn	@	55¢
263,306	lbs. rice	@	3½¢
169,734	lbs. cheese	@	7¢
205,578	lbs. butter	@	13¢
134,831	lbs. dried apples	@	4¢
3,381	bushels beans & peas	@	1.25
5,754	bbls. vinegar	@	3.50
1,237	bushels corn meal	@	3.50
32,726	bushels potatoes	@	35¢
193,136	lbs. dried codfish	@	3¢
320,933	gals. molasses	@	27¢
52,192	lbs. black tea	@	35¢
3,650	lbs. green tea	@	60¢
33,661	lbs. raisins	@	5¢
206,336	lbs. sugar	@	7½¢
206,336	lbs. coffee	@	8¢

But whether the amount were $150 or $300 per voyage, the food consumed by the whaleman constituted his only significant credit in kind. Although it never appeared in his accounts, this "free board" constituted far and away his largest source of income, with the exception of the lay itself. And when added as a silent and unannounced companion to the long list of entries which did stride across the pages of his accounts, its very unobtrusiveness only emphasized the whaleman's preeminent position as a dealer in long-run debits and credits, a juggler of financial pros and cons.

[13] Together with other figures relating to the whaling industry, this list was given by Joseph Grinnell, in a pamphlet entitled, "Speech on the Tariff, With Statistical Tables of the Whale Fishery." The pamphlet, published in 1844, is now in the New Bedford Public Library.

CHAPTER XII

PROFITS AND THE COUNTING-ROOM

THE most significant and certainly the most troublesome feature of whaling, as viewed by an entrepreneur, was its element of risk. No other well-established and legitimate industry was subject to such wild and unpredictable fluctuations of fortune. Financial returns ranged from ruinous losses to fabulous gains; and the whaling merchant became not only a vender of oil and bone, but also, and preëminently, a dealer in risks.

These risks were both varied and severe. Three major types, however, may be distinguished. Labor risks involved the possibility that illness, desertion, or death might bring about the loss of sums which had been advanced to certain members of the crews. Physical risks were focussed upon the loss of equipment, cargo, or vessel through storm, fire, mutiny, or the misfortunes of the chase. And business risks resulted largely from the extreme irregularity of the financial returns attendant upon the cargoes obtained. For whaling cargoes fluctuated violently with regard to quantity, quality, selling price, and length of the accompanying voyage.

The labor risks of the entrepreneur require only a brief additional mention. It is obvious that if, under the lay system, a given whaleman died, deserted, or was incapacitated before the amount of his lay had come to equal the sum of the various advances which had been made to him, the difference constituted a net loss to the owners. Such losses were partially offset, it is true, by the agents' confiscation of the effects and unpaid lays of those deserters whose accumulated earnings chanced to be in excess of their total debits. But in the end the balance was heavily against the owners, and added appreciably to their financial uncertainties.

The physical losses involved in the destruction of vessels, cargoes, and equipment were indeed formidable when considered in the absolute; but when compared with the size of the entire fleet at sea during any given year they formed a smaller percentage of the whole than the nature of the industry would seem to have justified. One contemporary writer stated that the number of vessels which were totally wrecked each year seldom exceeded one per cent of the entire fleet; while the losses resulting from partial wrecks and all other major forms of accident added only a half of one per cent. Insurance underwriters rarely charged a premium of more than two and one-half per cent per annum, since such a figure enabled them to pay good dividends.[1] Account-books and policy forms of whaling insurance companies provide further corroborative evidence by showing charges varying from one per cent to nine per cent for an entire voyage. This rather wide range in rates was accounted for by differences in the size and condition of vessels, the type of whaling to be carried on, the probable length of a proposed cruise, and the nature of the regions to be visited.

The heaviest financial burdens, however, were those connected with business risk. It was estimated that during the middle years of the nineteenth century approximately ten per cent of all American whaling vessels made voyages which resulted in a net loss to their owners.[2] Even after allowing for a reasonable margin of error in the figures underlying this estimate, and after allocating one and one-half per cent of all losses to physical factors and not more than a like percentage to labor risks, it is evident that deficits arising out of the quantity, quality, or price of the cargoes comprised more than half of the total losses. It is probable, too, that this estimate of ten per cent was not unduly liberal, since many years witnessed a far higher proportion of losing voyages.

In 1837, for instance, 53 vessels paid profits; eight made "saving voyages" (in whaling parlance this term described a venture which escaped an actual loss, but in which the gain was negligible); and 20 incurred deficits, including nine severe

[1] Wilkes, Charles, "Narrative of the U. S. Exploring Expedition," V, pp. 494 ff.
[2] *Ibid.*

losses. In other words, virtually 25% of the 81 voyages were losing ones.[3] Of 68 whalers which were expected to arrive in New Bedford and Fairhaven during the year 1858, no less than 44, or 64.7%, were placed in the loss column for an aggregate amount of not less than $1,000,000.[4] In 1871 a still greater catastrophe stunned the industry; for during that season the entire Arctic fleet, with the exception of only five whalers, was caught in the ice and smashed into kindling-wood. Thirty-three vessels, together with all equipment, carried down with them an investment of more than $1,500,000.[5] And again in 1876 the same episode was repeated, though fortunately on a smaller scale.

But the wide fluctuations of whaling cargoes and earnings appear unmistakably even in the relatively successful operations which seldom, if ever, chanced to show actual losses. Jonathan Bourne, during a career of 53 years as an unusually prosperous New Bedford whaling merchant, acted as agent for 24 vessels, aggregating 7461 tons. This fleet, which made a total of 148 voyages during a period of 4421 months, secured an average catch per voyage of 487 barrels of sperm oil, 1136 barrels of whale oil, and 12,504 pounds of whalebone. Clearly a record which permitted justifiable pride! But behind such an eminently satisfactory average were countless variations of vessels and cargoes. Witness even the earnings of the *Bartholomew Gosnold,* one of the most consistently successful members of the fleet. During thirteen consecutive cruises, ending in the respective years given, this whaler yielded cargoes which sold, approximately, for the following amounts: [6]

1836	$33,000	1858	$ 57,000
1839	30,000	1862	$ 43,000
1843	33,000	1866	105,000
1847	39,000	1870	83,000
1851	21,000	1876	81,000
1854	63,000	1880	27,000
1885	48,000		

[3] Starbuck, A., "History of the American Whale Fishery," p. 149.
[4] *Ibid.*
[5] Old Dartmouth Historical Society Sketches, No. 45, p. 43.
[6] *Ibid.,* No. 44, pp. 11 and 21.

The full extent of entrepreneurial risk, however, was reflected most clearly in the extraordinary variations in the size and value of individual cargoes. The records of the industry are replete with instances in which vessels returned to port with catches which were either ruinously small or astoundingly profitable. Among the unlucky voyages which must have tried the souls of owners and crews, for example, were those of the brig *Emeline*, of New Bedford, which returned after 26 months at sea with only ten barrels of oil; of the *Clifford Wayne*, of Fairhaven, which was forced to put back into port at a loss of $10,000 after a mutiny; and of the *Benjamin Rush*, of Warren, which lost an entire boat's crew in an encounter with an "ugly" whale off the coast of Japan, circumnavigated the globe, and reported only 90 barrels of oil after a full year's cruising.[7]

For consistent and long-continued ill fortune, however, few records equal that of the bark *Tempest*, of New London. The *Tempest* put to sea on May 21, 1857, under command of Captain G. L. Allyn, a veteran whaleman who had made several successful voyages. On September 8, after having visited Spitzbergen and East Greenland without success, Captain Allyn found himself near the Azores without having sighted a single whale. Continuing southward he rounded the Cape of Good Hope and entered the Indian Ocean, where on the last day of the year he finally filled his try-pots for the first time during the voyage. Thence the vessel was steered over the regular cruising grounds of the South Pacific, the North Pacific, and the Sea of Okhotsk without a break in the continuous round of poor luck. After three years of cruising, so little oil had been obtained that the voyage was practically a total loss; and at Honolulu the *Tempest* was finally turned over to another master.[8]

But the accounts of surprisingly successful voyages afford an even more striking illustration of the capricious fortune which often attended a whaling cruise. Preëminent in this respect was the colorful history of the ship *Envoy*, of New Bedford. During a period of fourteen years, 1833–1847,

[7] See Starbuck, A., "History of the American Whale Fishery," p. 149.
[8] Brown, J. T., in "Fisheries and Fishery Industries," VII, p. 293.

her cargoes had yielded $200,000 to the owners. When she returned to port in 1847 the underwriters refused to insure the weather-scarred "blubber-hunter" for another cruise, and she was sold as an old hulk. In the end, however, she was purchased by a certain Captain W. T. Walker and again fitted out for sea. Upon leaving port, still entirely uninsured, both vessel and equipment represented a total investment of only $8,000 — and showed it only too plainly! From the first, nevertheless, the venture prospered amazingly. Two surprisingly short cruises in the North Pacific yielded bountiful returns; the cargo of a wrecked whaler, discovered *in extremis*, was purchased at a heavy discount; several shipments of oil were sent home as freight, and the ship rapidly refilled after each emptying of the casks; and finally, in 1852, Captain Walker arrived at San Francisco with his vessel again filled to overflowing. Here he disposed of both the *Envoy* and her last cargo; and a final accounting showed that the four years' cruise in a condemned hulk had yielded a grand total of $138,-450 — an astounding sum when compared with the original investment of $8,000! [9]

Such superlatively good fortune, however, was not confined to the *Envoy*. The *Corinthian* came into port in 1862, after a four years' cruise, with a cargo valued at the extraordinary figure of $275,000; the *Alaska* returned in 1880 with 3,255 barrels of sperm oil; the *Loper*, operating during the late twenties and early thirties, completed three Pacific voyages, yielding 6581 barrels of sperm oil, within the short space of sixty-two months; and the *Adeline Gibbs* secured, during a single cruise, 133 pounds of ambergris — a remarkable amount of this precious substance which sold, in 1878, for $23,231.25. Still other voyages, too, were exceptionally profitable; and among them were to be found the following: [10]

[9] The facts concerning the *Envoy* are common knowledge in the literature of whaling. The best and most accessible accounts are in Starbuck, A., "History," p. 147, and in Brown, J. T., *loc. cit.*

[10] The records of the *Corinthian, Alaska,* and *Adeline Gibbs* are given by Brown, J. T., writing in "Fisheries and Fishery Industries," VII, p. 293. The material pertaining to the remaining voyages listed was taken from Alexander Starbuck's painstaking "History of the American Whale Fishery," pp. 145–148. Mr. Starbuck, writing at a time when men who had sailed on many of these voyages were still alive, and when countless documents were to be had for the

Vessel	Port	Sailed	Returned	Cargo	Value
Pioneer	New London	1864	1865	1,391 bbls. whale oil 22,650 lbs. whalebone	$150,060
Montreal	New Bedford	1850	1853	195 bbls. sperm oil 3,823 bbls. whale oil 31,700 lbs. whalebone	136,023
Coral	New Bedford	1847	1850	3,350 bbls. sperm oil	126,630
Sheffield	New Bedford	1849	1853	7,000 bbls. whale oil 115,000 lbs. whalebone	124,000
Favorite	Fairhaven	1850	1853	300 bbls. sperm oil 4,300 bbls. whale oil 72,000 lbs. whalebone	116,000
Wm. Hamilton	New Bedford	....	1838	4,181 bbls. sperm oil	109,269
Russell	New Bedford	1846	1849	2,650 bbls. sperm oil	92,000
Sarah	Nantucket	1827	1830	3,497 bbls. sperm oil	89,000
South America	Providence	1847	1849	200 bbls. sperm oil 5,300 bbls. whale oil 50,000 lbs. whalebone	89,000
Uncas	Falmouth	1828	1831	3,468 bbls. sperm oil	88,000

Truly an imposing array of outstanding successes! But unfortunately every item might be matched with the record of some other vessel which put to sea bravely, and never came back. And it might be matched many times over by other voyages which resulted in net losses or in negligible gains. Business risks, like labor risks and physical risks, gripped the industry; and this grip was not to be broken by those strokes of fortune which, through their very scarcity, only served to emphasize the uncertainties and irregularities of the occupation.

Such a formidable group of varied and heavy risks obviously necessitated measures designed to offset them. In the case of the labor losses these alleviating factors took the form of counter-gains rather than of prevention. Thus the "long lay" resulted not only in relatively low wages but also in wages which were proportioned to profits. An appreciable saving was secured by the whaling owners through having their crews assembled and equipped for them by the special class of merchants known as outfitters, who were eager to perform this

asking, performed an invaluable service by incorporating in his volume, in tabular form, great quantities of documentary material.

service without fee in return for the privilege of selling the men their outfits at exorbitant prices. Still other counterbalancing gains were the high interest rates which were charged upon all cash advances; the profits which accrued from the sale of goods from the slop-chest; and the confiscation of the abandoned effects and accumulated but unpaid earning of those deserters whose accounts showed a credit balance.[11]

Against the physical risks of the industry the whaling owners used a triple shield, made up of the lay system, the diversified ownership of single vessels and of fleets through the holding of fractional shares, and a well-established scheme of marine insurance. The device of the lay not only automatically decreased the wages bill in proportion to any losses which might be sustained, but also obviated the necessity of paying any wages whatever in those cases in which a voyage proved to be a total failure. As compared to common business practice, under which an employer paid the going rate of wages with small regard to the financial success or failure of his own particular enterprise, such a plan evidently lightened appreciably the financial burdens of shipwreck and of other forms of physical disaster.

The possibility of achieving some dispersion of risk through the diversified ownership of single vessels was at once too desirable and too obvious to be ignored in an occupation like whaling. Consequently it became a common practice to divide the ownership of both vessels and outfits into fractional shares of ½, ¼, ⅛, 1/16, or 1/32, and to dispose of these shares among a relatively small group of investors, seldom more than ten or twelve per vessel. The case of the bark *Lagoda,* of New Bedford, which made twelve voyages during the years 1841–1886, and was owned throughout this period by a slowly-changing group of four to eight individuals, was typical.[12] Outfits, as distinct from the vessels themselves, were also financed on the same fractional basis. Thus during one voyage of the bark *Minerva,* her outfit, which cost $11,458.92, was subscribed by nine different persons in the following proportions: one in-

[11] See the two preceding chapters for a more detailed discussion of the matters referred to in this paragraph.

[12] An account of the voyages of the *Lagoda* is given in the Old Dartmouth Historical Sketches, No. 45, pp. 33–42.

Sperm Whaling—"The Capture"

From a lithograph by Endicott & Co., New York, 1862, after drawings by A. Van Best and R. S. Gifford, corrected by Benjamin Russell

vestor, ¼ share @ \$2864.73; four investors, ⅛ share @ \$1432.37 each; and four other investors, 1/16 share @ \$716.18 each.[13]

Such a system of fractional ownership possessed the double advantage of allowing persons of moderate means to invest in the industry, and of making it possible for the wealthy merchants to scatter their holdings over a large number of whalers. The individual possession of an entire whaling vessel and its outfit was quite beyond the reach of the ordinary investor. As early as 1844 the average value of all vessels in the sperm fishery, including their outfits, was about \$38,000 each. Whalers and outfits in the right whale fishery cost, on the average, about \$28,000 each; and even the 73 smaller craft in the Atlantic sperm fishery represented an average expenditure of about \$14,000 each.[14] Modest savings were quite helpless in the face of such demands; but they might very well suffice to purchase a share entitling the holder to 1/32, 1/16, or ⅛ of the proceeds of a given voyage.

But even the whaling capitalists, who were well able to finance entire voyages, deliberately scattered their holdings over a number of vessels. Men of wealth commonly bought shares in many different whalers and outfits. Thus Jonathan Bourne, a well-known merchant of New Bedford, amassed a fortune which would have enabled him to purchase an entire fleet of seventeen ships and barks. Significantly and characteristically, however, he chose to invest his wealth in fractional shares of forty-six different whalers rather than to own and operate seventeen vessels outright.[15]

The greatest safeguard against the inevitable hazards of whaling, however, lay in the system of marine insurance which was readily available to owners throughout the period after 1820. The main features of this early scheme of insurance,

[13] These figures were taken from an original manuscript Day-Book of Charles R. Tucker, now in the New Bedford Public Library. Mr. Tucker was the agent and part owner of the *Minerva,* and one of the prominent whaling merchants of New Bedford.

[14] These figures were compiled by Joseph Grinnell, and published in 1844 in a booklet called "Speech on the Tariff, with Statistical Tables of the Whale Fishery." See Appendix B for a more detailed account of both the figures and the reference.

[15] Old Dartmouth Historical Sketches, No. 44, p. 31.

as used by whaling owners, will be suggested by the consideration of a representative policy. On December 4, 1830, Charles W. Morgan and Samuel Rodman, Jr., joint owners of the ship *Magnolia,* took out insurance on both vessel and outfit in the amount of $11,000 for a whaling voyage "from New Bedford to the Pacific Ocean and elsewhere." This sum covered less than one-third of the entire investment, for the ship was valued at $20,000 and the outfit at $17,000 additional. The premium charged for the entire voyage was 7%, or $770. The policy stipulated that no responsibility would be assumed for any partial loss to outfit, cargo, or vessel unless it amounted to at least 5% of the value thereof; but in other respects the extent of the protection granted was sweeping. "Touching the adventures and perils which the said undersigned are contented to bear, and take upon them in this voyage, they are of the Seas, Fire, Enemies, Pirates, assailing Thieves, Restraints, and Detainments of all Kings, Princes, or People — and all other losses, and misfortunes, which have or shall come to the damage of the said ship and outfits or any part thereof, to which Assurers are liable by the rules and customs of Assurance in Boston." [16]

The practice of spreading risks was also carried into the field of insurance; for in this case the policy was issued in the name of, and the payment of any proper liability guaranteed by, fourteen different persons. Each man's name was signed to the document by attorney; and the extent of his liability, as well as the amount of the premium to which he was eligible, was shown specifically. Thus two men were liable for $1375 each, five for $917 each, five for $458 each, one for $688, and one for $687; while in apportioning the premium $96.25 was to go to each of the first two men, $64.19 to each of the next five, $32.06 to each of the five in the third group, $48.16 to the following man, and $48.09 to the last one.

But business risk, the last of the three great groups of whaling hazards, could not be covered by insurance. Neither could it be eliminated. Instead, as in other industries, it was

[16] This policy was found in a set of original whaling insurance policy forms now in the New Bedford Public Library. The manuscript does not give the name of the company; but from internal evidence it seems certain that the policy was written by a local mutual organization at New Bedford.

borne by the entrepreneur as one of his characteristic and most onerous functions. The whaling merchant, however, did enjoy two alleviating circumstances in the lay system, which proportioned wages to profits, and in the possibility and not too infrequent occurrence of magnificent profits.

Whaling profits were perhaps the least stable of all the varying elements in this uncertain industry. Consequently contemporary estimates lacked both definiteness and agreement. Lieutenant Wilkes, writing about 1843, declared that the profits in the whale fishery had been great. Something more than a decade later Lewis Holmes stated rather vaguely, in "The Arctic Whaleman," that the industry as a whole was reasonably profitable for the higher officers, and as tempting as any other to the investor. He added that the ports which were seriously engaged in whaling compared favorably with inland manufacturing and farming communities "in enterprise, wealth, educational appliances, and in all the comforts, and even the luxuries, of life." Alexander Starbuck, after an exhaustive search amongst whaling records which ended in 1876, was able to report only that profits were subject to great fluctuations, and that large gains were balanced by many accidents and losses. This commonplace was repeated in the mid-eighties by J. T. Brown, who simply remarked that profits fluctuated widely and were very uncertain.[17]

Such emphasis upon the variations and uncertainties of entrepreneurs' gains was fully justified by the facts. For success, when it came, was sometimes as enriching as failure was ruinous. Unusual good fortune might attend not only the conduct of a single voyage, as already shown, but also the operations of an entire season and the management of a given vessel over a long period of years. The year 1849 affords an excellent illustration of a most gratifying season; for the 154 whalers then in the Arctic returned with a huge catch which sold for $3,419,622, whereas the total value of both ships and outfits was only $4,650,000.[18]

[17] See Wilkes, C., "Narrative of the U. S. Exploring Expedition," V, pp. 493 ff.; Holmes, Lewis, "The Arctic Whaleman," p. 286; Starbuck, A., "History," p. 145; and Brown, J. T., in "Fisheries and Fishery Industries," VII, p. 293.

[18] Starbuck, A., "History," p. 148.

Among the outstanding cases of long-continued success was that of the bark *Lagoda,* of New Bedford. Throughout nine consecutive voyages, occupying the years 1841–1873, the cargoes of the *Lagoda* yielded profits ranging from 29.6% to 363.5% per voyage. Excluding the first cruise, the lowest rate of profit secured on any one of the following eight ventures was 66.9%. Exact dates and profits were as follows:

Sailed	Returned	Absent			Profits %
		Years	*Months*	*Days*	
Oct. 9, 1841	Sept. 15, 1843	1	11	6	29.60%
Nov. 8, 1843	May 26, 1846	2	6	18	120.57%
Aug. 25, 1846	June 13, 1849	2	9	18	66.96%
July 1, 1850	Apr. 23, 1853	2	9	22	177.19%
Nov. 3, 1853	May 26, 1856	2	6	23	100.00%
July 19, 1856	June 28, 1860	3	11	9	96.89%
Aug. 27, 1860	Apr. 18, 1864	3	7	21	363.50%
July 25, 1864	May 26, 1868	3	10	1	219.00%
July 25, 1868	June 5, 1873	4	10	10	115.25%

Even this sequence of triumphs was finally broken, however, by the tenth, eleventh, and twelfth voyages, which yielded, respectively, a loss of $14,460.47, a small gain of $6,414.44, and another loss of $10,253.55.[19]

In striking contrast to such generous profits were other figures given out by a prominent whaling merchant in 1846. These calculations comprised an estimate, based on long personal and family experience and wide acquaintance with the conditions of the industry, of the average profits which might reasonably be expected by sperm and right whalemen in either the American or British fleets. It was held that a sperm whaler of 350 tons, fitted for a four-years' cruise at a total outlay of 10,920 pounds sterling for vessel, outfit, and interest, would yield, on the average, a profit of about 580 pounds ster-

[19] These figures were taken from the Old Dartmouth Historical Sketches, No. 45, pp. 33–38. They form part of an extended analysis of the accounts of the *Lagoda* made by Benjamin Baker, a long-time confidential employee of the firm which operated this vessel throughout her whaling career. Mr. Baker thus had complete access to all the private accounts and papers pertaining to these cruises. The figures in the "Absent" column furnish an admirable illustration of the increasing length of voyage which was common to the entire industry during this period; and the dates of returning and of the next sailing show the brief spaces of time which a whaler was allowed to spend in her home port.

ling, or 5.3% for the entire period. On the other hand, a right whaler of 250 tons, fitted out for a voyage of two years at a cost of 5,500 pounds sterling, similarly calculated, might be expected to show an average profit of about 720 pounds sterling, or thirteen per cent for the cruise.[20] In other words,

[20] These figures form part of an analysis of the whaling industry contained in a letter written by Charles Enderby to T. R. Preston, under date of October 31, 1846. This letter, revised and augmented, was published during the following year as a "Proposal for Re-Establishing the British Southern Whale Fishery." Mr. Enderby was the head of the largest whaling firm in Great Britain, and had succeeded his father in the business. The Enderby house-flag was known the world over; and the son was eminently qualified, as a result of lifelong experience, thorough acquaintance with all phases of the industry, and a wide knowledge of whalemen and their problems, to speak with authority concerning the financial aspects of whaling. Although the bulk of the pamphlet was given over to British conditions, he referred frequently to American whaling as well; and in the analysis of profits from which these figures were taken he stated specifically that reference was had to both American and British whalers. The detailed figures were as follows:

SPERM WHALING

Ship of 350 tons, fitted for 4-years' cruise @ £26 per ton	£ 9,100
Interest for 4 years @ 5%	1,820
Total cost of equipment	£10,920
Returns of 150 tons sperm oil @ £80 per ton	12,000
Less shares of the crew	3,500
Net Returns	8,500
Add value of the ship (original value minus outfit and depreciation)	3,000
Net returns plus value of the ship	11,500
Deduct original cost of ship and equipment with interest	10,920
Leaves as profit	£ 580

RIGHT WHALING

Ship of 250 tons, fitted for 2-years' voyage @ £20 per ton	£ 5,000
Interest for 2 years @ 5%	500
Total cost of equipment	5,500
Returns of 170 tons Whale Oil, @ £ 18 per ton	3,060
Returns of 15 tons Sperm Oil, @ £ 60 per ton	900
Returns of 7 tons Whalebone, @ £180 per ton	1,260
Total Returns	5,220
Less shares of the crew	1,500
Net Returns	£ 3,720
Add value of the vessel (original investment of £5,000 minus outfit and depreciation)	2,500

the average rate of profit obtained in right whaling was about 6.5% per annum, while that secured in sperm whaling was only about 1.3% yearly.[21]

Quite different, these average rates of profit, from those secured by such rarely fortunate vessels as the *Lagoda!* But if figures of 1.3% to 6.5% per annum are to be taken as a rough average, it is evident that there must have been a large number of moderate losses and at least some heavy ones in order to offset so effectively the soaring rates of the exceptionally successful whalers. And in truth, the records of whaling afford numerous instances of precisely such heavy losses. The disastrous seasons of 1837 and 1858; the Arctic destruction of 1871, when virtually the entire fleet was swept away; the heavy toll exacted by the extreme northern latitudes again in 1876; the depredations of the Confederate cruisers *Shenandoah* and *Alabama* during the Civil War; and the lists of vessels which were wrecked or reported missing with all hands — these were only the outstanding cases among a long line of catastrophes which outweighed many bonanza voyages.[22]

The financial results of American whaling, then, covered the whole range between ruinous losses and magnificent profits. But, though the available figures do not warrant precise and conclusive assertion, it is evident that the cases at each extreme offset each other so effectually that the long-run, normal rate of profit for the industry as a whole was an essentially modest one. And this was true in spite of the fact that many considerable fortunes were begotten of the union of forecastle and counting-room.

Net Returns plus value of the vessel	6,220
Deduct cost of original equipment, including vessel and interest	5,500
Leaves as profit	£ 720

[21] This discrepancy between the average rates of profit in the two branches of the industry was largely responsible for the drift away from sperm and into right whaling — a movement which became marked during the decade of the forties and continued throughout the ensuing years. The greater profitableness of right whaling was due in part to the increasing demand for whalebone and for heavy lubricating oils, and in part to the progressive penetration of the Arctic whaling grounds. The whales found in the Arctic were not only more numerous than those frequenting warmer waters, but they also yielded more oil per animal of a given size, because of their heavier coats of blubber.

[22] See the following chapter for a more detailed account of the severe tonnage losses which occurred during and after the Civil War.

In fact, the American whale fishery admirably illustrated what has been termed, aptly enough, the "prizes and blanks" theory of business profits. This concept holds that in any field of business activity, but particularly in the more hazardous ones, the individual instances of exceptionally high profits serve merely to offset other individual cases of disastrous reverses and of appreciable losses. Specific illustrations of great profits do not necessarily betoken an unusually prosperous industry: they may be only a counterpoise which is essential in order to bring the returns for the industry as a whole up to an ordinary level.[23] Such an analysis, whatever may be its claim to general acceptance, seems peculiarly applicable to the conditions of the whaling industry. For certainly the lottery of individual whaling voyages yielded both prizes and blanks aplenty.

Thus the normal rate of profit for the industry as a whole resulted, in large part at least, from the cancellation of extremes. But in spite of the staggering variations in its constituent elements, this normal rate of profit continued to absorb a surprisingly steady percentage of the total net proceeds of the industry. Several independent sets of figures agree in showing that approximately seventy per cent of the net proceeds of American whaling went to the entrepreneurs, leaving the remaining thirty per cent for officers and men. As early as 1834 a writer in the *North American Review* stated that, on the average, whaling owners secured sixty-nine per cent of the net income of the whaling industry, and that officers and men were rewarded with the balance of thirty-one per cent.[24] Similar estimates of seventy per cent and thirty per cent, respectively, were repeated in 1844 by an early whaling statistician.[25] And finally, a chance sampling of seven voyages, for which suitable and accurate accounts were available, yielded average figures of 69.7% and 30.3%. The sums and percentages pertaining to each voyage are given below.[26]

[23] See especially Marshall, Alfred, "Principles of Economics," 7th Edition, p. 621.

[24] Williams, J. R., writing in the *North American Review* for January, 1834, Vol. XXXVIII, p. 105.

[25] Grinnell, Joseph, "Speech on the Tariff, With Statistical Tables of the Whale Fishery," p. 9.

[26] The figures for the six voyages of the *Minerva* and the *Marcella* were

It should be remarked, however, that the sums allocated to the owners did not comprise business profits proper; for the latter appeared only after certain further deductions, such as depreciation charges and interest payments, had been made. Similarly the thirty per cent of the net proceeds which went to officers and men by no means represented actual cash received; for this gross amount was whittled away in order to pay for the numerous charges against the lay which accumulated during the course of a cruise. Thus one voyage of the bark *Minerva*, 1836–1839, yielded gross lays of $2,768.21, or 24.75% of the net proceeds. But the amount of cash actually paid out to officers and men at the time of final settlement was only $1546.58, or 13.8% of the net proceeds. The difference of $1,221.63 represented the cumulative debit entries for the voyage. Or, stating the matter in other terms, charges and advances made during the cruise absorbed 44.1% of the total wages bill and 10.9% of the entire net proceeds.

But what determined whether the earnings of a given firm's voyages, however divided between owners and crews, were to be great or small? The facts of the industry made it painfully and unmistakably evident that luck was by no means a despicable factor in bringing about whaling success or failure. Was it, however, more important than the possession of capital, connections, and business ability?

The possession of adequate capital was particularly advantageous in whaling because it was often necessary to tide over long periods of severe losses. Good connections, too, could not fail to prove helpful in an industry which was so strongly

taken directly from the original account-books, now in the New Bedford Public Library. Those for the *Lion* were found in an early article entitled "Notes on Nantucket," dated August 1, 1807, and published in the Massachusetts Historical Society Collections, Series 2, Vol. III, p. 30.

			Crew's Share (Gross Lays)		Owner's Share	
Vessel	Voyage Made	Net Proceeds	Amount	Per Cent	Amount	Per Cent
Lion	1805–1807	$37,298.27	$13,045.53	34.98	$24,252.74	65.02
Minerva	1836–1839	11,182.75	2,768.21	24.75	8,414.54	75.25
Minerva	1839–1840	18,688.00	6,113.17	32.70	12,574.83	67.30
Minerva	1840–1841	11,041.00	3,387.88	30.70	7,653.12	69.30
Marcella	1840–1841	18,646.00	5,583.59	29.95	13,062.41	70.05
Marcella	1844–1847	22,202.00	6,619.90	29.80	15,582.10	70.20
Marcella	1847–1850	21,594.00	6,286.59	29.20	15,307.41	70.80

localized in a few small New England ports. Paradoxically enough, American whaling combined world-wide voyages at sea with an intimate, closely-twined provincialism on land. The leading families of New Bedford, Fairhaven, Nantucket, New London, and Sag Harbor constituted an aristocracy of oil and bone in which membership was a real business asset. Still, success might be attained, and was attained, with little capital, with ordinary conections, and with only indifferent luck.

Business ability, however, was a *sine qua non* for conspicuous long-run success. In spite of a prominent element of chance, which at times seemed quite overpowering, whaling was not a mere lottery. Over a long period large earnings were due primarily to sound judgment, good management, and that real, if indefinable, combination of qualities called business ability.

Management and judgment were required in many respects. A "full ship" in itself did not necessarily guarantee large profits. Fully as important in determining the financial outcome of the enterprise were the economy, efficiency, and attention to detail with which the vessel had been repaired and equipped; the knowledge of market prices and conditions which made possible the most advantageous sale of a cargo; the choice of supplies which had been put on board; and the care with which the master had been instructed regarding the policies to be followed during the voyage.

Nowhere was sound judgment more important than in this matter of selecting and instructing the firm's masters and mates. Detailed orders covering all the exigencies of a three-years' cruise were obviously impossible; and since communications were often delayed for periods of many months, it was essential to give the captain wide latitude in making decisions, as well as in changing plans and policies. In such circumstances a resolute, skillful officer, calm in emergencies and sound in judgment, was a prize indeed; whereas a master who was incapable, lazy, cowardly, or unscrupulous might cause greater losses than storm or stranding.

During the course of one voyage of three years, one month, and twenty-seven days, for instance, only a small cargo had been obtained. And yet the log-book showed that 320 days had been spent at anchor without a reasonable excuse on the part of the master. In this case the owners had been guilty of a double mistake: they had given the captain permission to engage in trading for his personal gain, thus tempting him to linger in port, to waste the ship's stores, and to neglect opportunities for the pursuit of whales; and they had chosen a man who allowed himself to fall an easy prey to such temptations.[27]

And even the apparently fortuitous process of filling a vessel with oil and bone was by no means removed from the influence of business ability. A single "full ship" might well represent the outcome of mere good fortune. But a succession of "full ships," returning season after season to the same firm, indicated a keen business sense in the selection of masters and mates, in the formulation of shrewd and far-sighted policies, and in the successful adaptation of means to ends.

[27] Enderby, Charles, "Proposal for Re-Establishing the British Southern Whale Fishery," p. 53. See also Scammon, C. M., "Marine Mammals of the Northwestern Coast of North America," pp. 216 ff., for a good discussion of the relative importance of business ability and luck in making for long-time whaling success.

PART III

DERELICT

CHAPTER XIII

CIVIL WAR, PETROLEUM, AND THE ARCTIC

THE apogee of American whaling was clearly marked by the decade 1850–1860. With the Civil War came a group of forces which relentlessly tore down the structure of prosperity erected after 1830. Some of these forces struck sudden blows at the fishery; others worked slowly and cumulatively. But all were destructive; and following their ravages, whaling activities entered upon a period of decline which passed gradually into irretrievable ruin and ultimately into virtual extinction.

Foremost among the proximate causes of decay was the loss of tonnage due to the Civil War and to the Arctic. The years of hostilities brought to the whalemen captures and burnings by the Confederate cruisers *Alabama* and *Shenandoah;* transfers to foreign flags and to the merchant service; vessels which rotted at their wharves instead of sailing the high seas; and a heavy sacrifice of tonnage, if not of money, in the form of the "Great Stone Fleet" of forty whalers which was sunk in an effort to blockade Charleston and Savannah harbors. And soon after the wreckage of war came the exactions of the Arctic. In 1871 thirty-three vessels were crushed in the ice off Point Belcher; while in 1876 twelve more whalers were destroyed by the ice-locked waters of the far north.

But physical losses, no matter how heavy, cannot bring about the permanent ruin of an entire industry without simultaneous sapping of recuperative power. American whaling, in particular, had several times suffered and quickly repaired disasters as crushing even as these. This time, however, the forces opposed to recovery were too fundamental. The increasing use of petroleum after the Civil War provided both illuminants and lubricants of such a nature that they rapidly undermined the markets for whale oil and sperm oil. Due in part to this progressively victorious competition of petroleum, and in part to the more profitable employment of labor and capital on shore, the whaling industry was drained of much of its capital at the very time when funds were sorely needed for replacements. An old and waning fishery could attract but little of the flood of new capital which was pouring into the feverish exploitation of the natural resources of the nation. New Bedford herself, queen of whaling ports, devoted more and more of her savings to the construction of cotton mills rather than to the equipment of whaling vessels.

At the same time the diminution and shyness of the game required longer and more uncertain voyages, thus increasing financial risk and necessitating the employment of capital throughout a greater period of time in order to reach the same goal — a "full ship." The character and efficiency of the crews, too, had so deteriorated as to add heavily to the burdens of the whaling merchants. And those burdens were not reduced, assuredly, by such factors as the general business depression of 1873 and the increased costs which were traceable to post-war high prices. Heightened costs constituted a peculiarly trying problem to the whaleman, because a falling demand prevented his prices from rising in the same proportion as his expenses.

It was the Civil War, however, which delivered the first and most obvious attack upon the industry's long period of prosperity. During the space of five years the fleet was cut in half. Between January 1, 1861, and January 1, 1866, the whaling tonnage fell from 158,746 to 68,536 — a loss of 57%; and the same half-decade witnessed a 49% shrinkage in the number of vessels — from 514 to 263.[1]

Much of this destruction was due to the Confederate cruisers *Alabama and Shenandoah.* The former, operating in the

[1] Figures taken from the annual statistics of the *Whalemen's Shipping List.*

Atlantic, pounced upon the slow-sailing and unarmed Yankee whaling vessels whenever an opportunity presented itself. Once sighted, the sluggish, deep-laden whalers could not fail to be overhauled and to fall an easy prey to this lithe, graceful panther of the seas. The *Shenandoah,* on the other hand, struck but a single belated and deadly blow at the whalemen. Appearing suddenly in Behring Straits in 1865, she surprised and captured virtually the entire Arctic fleet, supposedly safe because of its vast distance from the seat of hostilities. Within three or four days twenty-five whalers were burned and four others were converted into Confederate transports.

Many Northern whalemen insisted that the enemy cruisers set fire to their prizes in order to lure other craft into the vicinity. Passing vessels, seeing the flames and leaving their courses in an attempt to succor some other vessel naturally assumed to be in distress, would also fall prey to the marauding craft which still hovered in the vicinity of the first victim. As a matter of fact, several captures were made in precisely this manner by the *Alabama.* But it is probable that her commander, instead of deliberately setting a bait for other whalers, was only destroying prizes which at the time could neither be manned nor taken into port.

The sinking of the "Great Stone Fleet" in 1861 involved another heavy sacrifice of whaling tonnage. Some forty of the older whalers were purchased by the Federal Government, laden with stones, and deliberately sunk off Charleston and Savannah harbors in an effort to make the navigable channels unsafe for blockade runners. The cargoes of stones were intended to prevent the hulks from being raised or washed away after settling into place in the channels. As an experiment in blockading tactics the enterprise was dubious; but as a drain upon the resources of the whaling industry it was eminently successful. For although the vessels were purchased by the Federal Government at a fair price, their destruction left a gap in the ranks of the whaling fleet which, as it proved, was never to be filled.

Still other vessels were lost to whaling through sale, transfer, or decay. Owners who were unwilling to run the heavy risks of whaling in war time sold their craft to foreign capitalists or transferred them to the merchant service under foreign registry. Some whaling merchants who refused to suspend operations put their vessels under the Hawaiian flag and kept them in the Pacific. Other operators, still more cautious, kept their whalers tied up to the home wharves, where they rotted away to such an extent that when the war closed they were in need of extensive repairs, if not entirely unseaworthy.

For several years after 1865 it seemed that the fortunes of the industry might be revived, in spite of the ravages of war. Prices deserted their war-time heights, it is true; but they fell away gradually, rather than precipitously, and continued to remain above the level of 1860. The average prices of oil and bone for the decade beginning with 1860 present a clear picture of the course of the market: [2]

Year	Sperm Oil (per gallon)	Whale Oil (per gallon)	Whalebone (per pound)
1860	$1.31½	$ .44¼	$ .66
1861	1.42½	.59⅓	.88
1862	1.61	.95¼	1.62
1863	1.78	1.28	1.80
1864	2.25	1.45	1.71
1865	2.55	1.21	1.37
1866	2.23½	.73¼	1.17½
1867	1.92	.82	1.02½
1868	1.78	1.01¾	1.24
1869	1.35½	.67¼	.85

Such relatively favorable prices, together with the removal of war risks, served to release the potential energy which had been held in leash. As a result the annual figures of the

[2] These average prices were taken from the annual figures published by the *Whalemen's Shipping List.* It should be noted, however, that such averages might be the result of wide fluctuations during the course of any given year. Thus the ship *James Maury,* arriving at New Bedford during the fifth month, 1868, sold her best whale oil for 78¢ per gallon, despite the fact that the average price for the year was $1.01¾. Similarly, her whalebone brought only 86¢ per pound, although the average price of all bone sold during that year was $1.24 per pound. Her sperm oil, however, sold for $1.80 per gallon, whereas the average price was $1.78. Prices varied, too, with quality. The accounts of this same voyage, for instance, showed that the Grade A sperm oil was disposed of for $1.80 per gallon; the Grade B sperm oil ("dark and sour") for $1.50; and the Grade C sperm oil ("black and stinking") for $1.25. The differences in the quality of the whale oil on board were less striking. The best quality brought 78¢ per gallon, whereas only three cents less was secured for both brown and black whale oil. Whalebone prices varied with the length and the

Whalemen's Shipping List showed that the whaling fleet had increased from 263 vessels, of 68,536 tons, on January 1, 1866, to 336 vessels, of 74,519 tons, on the same day three years later.

The hopes accompanying this brief revival, however, were rudely shattered by the abandonment, in 1871, of virtually the entire Arctic fleet of that season. This catastrophe was clearly traceable to the hazardous strategy to which the growing scarcity of whales had forced the industry. In the never-ending search for new and more fertile whaling grounds it had become customary for the right whalers operating in the North Pacific to penetrate farther and farther into the recesses of the Arctic Ocean. Entrance was made through Behring Straits during mid-summer and the hunt continued until late September or early October, just before the ice began to form for the long winter. As the catches of succeeding seasons grew smaller and smaller, the temptation to prolong operations beyond the period of safety became irresistible. Each year the fleet escaped the ice by a narrowing margin; and in 1871, when the sea froze over at an unusually early date, the margin was wiped out. Thirty-three vessels, with their cargoes and crews, suddenly found themselves fast within the grip of solid field-ice — a grip more tenacious than steel, which would not be relaxed until the following summer. Long before then even these heavily-built whalers would have been crushed like eggshells in the furious winter storms; and it would have been madness to attempt to spend the winter on the nearby coast, barren of both food and fuel.

The only alternative was to transfer the combined crews of some 1200 men to the five barks which, several miles to the south and around a long point, were still in clear water. By a rare combination of skill and good fortune this was completed without the loss of a single life; and when the tragically over-manned little flotilla reached Honolulu, several weeks later, death was still without a victim.

The property loss, however, was enormous. Twenty-two of the abandoned vessels were sailing from New Bedford; and

place of origin. Long bone was more valuable than short; and Arctic or Northwest bone brought a higher price than that from the South Seas.

the loss to that port alone, including vessels, outfits, and cargoes, was placed at $1,090,000.[3] The total fleet represented a sacrifice of something more than $1,600,000. In addition, too, heavy damages were suffered by the owners of the five barks which chanced to be available for transport purposes. These vessels not only furnished supplies for 1200 men en route to Honolulu, but they lost a large proportion of an entire season's whaling. So unusual was the resulting situation that the owners of the five whaler-transports considered it proper to petition Congress for reparations in the sum of $275,-042.95. After being buffeted about amidst the shoals and storms of committee hearings for almost twenty years, the petition was at length allowed in part. On February 21, 1891, long after many of the actors in the drama had passed away, Congress awarded damages to the owners on the basis of $138.89 for each passenger carried from the scene of the disaster to Honolulu. These awards, made as follows, marked the final outcome of this tragic episode:[4]

Vessel's Name	Damages Claimed	Damages Awarded	Number of Men Carried and Compensation per Man		
Bark *Progress*	$51,094.30	$26,111.32	188	@	$138.89
Bark *Midas*	51,052.90	19,861.27	143	@	138.89
Bark *Europa*	71,100.00	33,889.16	244	@	138.89
Bark *Lagoda*	51,033.25	23,611.30	170	@	138.89
Bark *Daniel Webster*	50,762.50	21,527.95	155	@	138.89

The annihilation of the Arctic fleet of 1871 dealt the fishery a body blow at the very time when it was struggling to stand up in the face of numerous adverse forces. Something less than $2,000,000 worth of capital was snatched away at a moment when it was not only inadvisable, but practically impossible to replace it. Within the twelve months of the year 1871

[3] This was the figure given by Pease, Z. W., and Hough, G. A., in their booklet entitled "New Bedford, Massachusetts: Its History, Industry, Institutions, and Attractions," p. 31.

[4] Figures taken from Old Dartmouth Historical Sketches, No. 45, p. 43. With the exception of Starbuck, A., "History of the American Whale Fishery," pp. 103–108, this is the best account of the loss of 1871.

no less than sixty-eight vessels, of 16,671 tons, passed out of whaling, leaving the industry with only 220 whalers and a tonnage of 52,701.

In spite of this dwindling strength, however, the pitiless Arctic had not yet secured full revenge for the daring intrusions into the solemn stillness of northern waters. In 1876 twelve more vessels were frozen in for the winter and had to be abandoned, this time with a loss of fifty lives and of $660,000 worth of property. Again in 1888 the unleashing of a heavy gale sent down a fleet of five craft off Point Barrow.[5] And over and above these large-scale calamities was the constant warfare of wind, wave, and cold against individual whalers, — a state of unrelenting hostilities which yielded a formidable list of whaling casualties. The Arctic was not to be invaded with impunity!

But war and weather were not alone in attacking the whalemen. They were strongly abetted by the slow-moving, fundamental forces of demand and supply. In the growing use of petroleum and its accompanying products, in fact, was to be found the most potent single cause of whaling decay. The superiority of natural gas for illuminating purposes was obvious and unquestioned; and every gas jet took the place of several spermaceti candles or of the coarser whale oil illuminants. Throughout the sixties, seventies, and eighties the market for whaling illuminants was slowly but surely crumbling before the inroads of natural gas.

At the same time a similar struggle was taking place in the field of lubrication. The heavier grades of whale oil, which had constituted the standard lubricants before the Civil War, were rapidly supplanted by various petroleum products; and as a result the market for whale oil went still further into eclipse. Since whale oil was used for both illumination and lubrication, however, its price sank somewhat less rapidly than that of sperm oil, which was restricted more closely to the one field of illumination.

Whalebone alone remained without a satisfactory substitute,

[5] An account of the catastrophes of 1876 and of 1888 is given in Pease, Z. W., and Hough, G. A., "New Bedford, Massachusetts: Its History, Industry, Institutions, and Attractions," p. 31.

and increased in price even more rapidly than the other two products declined. Unfortunately, however, the demand for it did not carry the price to the point where it became profitable to hunt bone alone; and consequently this one favoring influence could not overcome the powerful combination of factors which were making for decay.

The post-war records of production and of prices for the three major whaling products told an interesting story of joint cost. Since whale oil and whalebone were secured from the same animal, the right whale, their production tended to change in about the same proportion. In 1880 whale oil imports had fallen to one-fourth of their 1860 figure, and whalebone had dropped to one-third of the earlier amount. (This discrepancy in percentages was probably occasioned by the fact that more care was exercised in saving the higher-priced bone, while less attention was paid to the handling of the lower-priced oil.) In the case of whale oil, the demand fell off in about the same degree as the production, so that the price in 1880 was roughly the same as that prevailing in 1860. During the same period the demand for whalebone increased so heavily that the later price was two and one-half times as great as the earlier. Sperm oil, on the other hand, coming from a separate species of whale but often sought by right whalers as well as by sperm whalers, suffered a drop in demand which was even greater than the fifty per cent decrease in supply; and consequently its price fell materially. The following figures, taken at four-year intervals from the annual summaries of the *Whalemen's Shipping List*, indicate the relation between supply, price, and (by implication) demand during the two decades 1860–1880:

	SPERM OIL		WHALE OIL		WHALEBONE	
Year	Production (barrels)	Average Price (per gallon)	Production (barrels)	Average Price (per gallon)	Production (pounds)	Average Price (per pound)
1860	73,708	$1.41½	140,005	$ .49½	1,337,650	$ .80½
1864	64,372	1.78	71,863	1.28	760,450	1.80
1868	47,174	1.92	65,575	.82	900,850	1.02½
1872	45,201	1.45½	31,075	.65½	193,793	1.28½
1876	39,811	1.40½	33,010	.56	150,628	1.96
1880	37,614	.99	34,776	.51	464,028	2.00

But if demand was increasingly disappointing, supply was no less so. Indications that the whales were either more shy, or more scarce, or both, multiplied everywhere. Even the old reliable whaling grounds were showing unmistakable signs of exhaustion. More and more frequently individual vessels would spend weeks on the best whaling grounds, in the very midst of the season, without a single capture. The constant, desperate searches for new and fertile grounds found added perils and hardships rather than increased production. A "full ship" became a goal so difficult, so elusive and fickle, and so long postponed, that it seemed a mirage, leading its followers through month after month and across mile after mile of fruitless effort. The average length of voyage increased alarmingly, and seemed destined to grow until it had converted the whalemen into modern Rip Van Winkles who worked and searched, instead of sleeping, while time passed them by, and finally brought them back, near strangers, to a changing world. Only the Arctic continued to afford a reasonable opportunity for finding whales in large numbers; and every season the masters of the right whalers were tempted to challenge a gray death by entering the frigid zone earlier, by penetrating farther and farther north, and by remaining later.

This growing difficulty in finding whales, rather than in capturing them after they had been sighted, naturally led to much speculation and to many surmises. Had the whaling population of the world's waters been decimated by the long years of slaughter? Or had even these greatest of all living creatures condescended to become shy and wary? Unfortunately there was (and is) no conclusive answer. Certain calculations concerning the mortality rate among several species of whales, however, were too interesting and too significant to be ignored. They were mere estimates, it is true; but they were erected upon a basis of known facts, and had some claim to validity.

For the period 1835 to 1872, inclusive, the average number of vessels employed annually in American whaling was 524. This fleet captured annually an average of 96,625 barrels of sperm oil and 172,448 barrels of whale oil. Taking 25 barrels as the average yield of a sperm whale, the number required to produce 96,625 barrels could be ascertained by a simple process of division. But there were certain forms of slaughter which brought no oil into the try-pots of the whalers. Because of stress of weather or other exigency, some carcasses were lost or had to be cut adrift before they had been relieved of their blubber. Other animals escaped after having been harpooned, but with mortal wounds which soon led to death. Still others, killed after a long, hard chase, cheated their captors by sinking before their very eyes. Allowing an additional ten per cent for such unproductive deaths, it developed that some 4,253 sperm whales had been killed annually for 38 years, making a grand total of 161,614 for the entire period.

The calculations for whale oil were somewhat more complicated. This product was derived from four main kinds of whales, the right whale, bowhead, humpback, and California gray whale. The first three, too, were guilty of a higher percentage of sinkings after death than was the cachalot. This was particularly marked in the case of the humpback. But by allowing twenty per cent for unproductive killings in this branch of the industry, and by taking 60 barrels as the average yield per animal contributing to the supply, it was found that the American whale oil harvest caused the death of some 3,450 whales per annum. This average, recurring annually for thirty-eight years, accounted for a total of 131,100 animals.

By combining the figures for sperm oil and whale oil, the grand total of all whales killed by American vessels from 1835 to 1872, inclusive, was placed at 292,714. In addition there was the slaughter caused by the whalemen of all other nations and by American whalers both before and after these two dates. By recalling, in connection with such figures, that the whale possesses very slow reproductive powers, giving birth to young only after a period of gestation which is appreciably longer than that of human beings, it becomes evident that the theory of an actual scarcity of game was by no means based upon pure fancy.[6]

But whether due to an excess of deaths over births or caused by an increased wariness on the part of the whales, it was undeniable that production was falling off rapidly and that a "full

[6] These interesting calculations were made by Scammon, C. M., "Marine Mammals of the Northwestern Coast of North America," pp. 243 f.

ship" necessitated ever longer voyages. Such cruises meant, for the crews, an absolute decrease in earnings per month, since it now required more months in which to earn the same fractional share of a given cargo. And they brought to the owners the unwelcome necessity of employing a larger amount of capital throughout a longer stretch of time in order to attain the same net gain.

A whaler which was compelled to spend forty-eight to fifty months in filling up with oil and bone obviously incurred heavier bills for refitting and had to replenish her stores oftener than if the same cargo had been secured in the thirty-three to forty-two months which sufficed before the Civil War. Expenditures were further increased, too, by the growing scarcity of many articles which entered into the building and equipment of a whaling vessel, and by a high general price level combined with the falling prices of sperm oil and of whale oil. Even the striking rise in the price of whalebone was offset by a drop in production which resulted in a smaller gross income from this source. The 1,337,650 pounds of bone imported in 1860, selling at an average price of 80½¢ per pound, brought approximately $1,077,000. But the 150,628 pounds of 1876, at an average price of $1.96 per pound, sold for only $295,000; and even the larger crop of 1880, amounting to 464,028 pounds and selling, on the average, for $2.00 per pound, yielded a gross sum which was almost $150,000 less than the receipts of 1860.[7]

Nor was there any effective means of shifting the financial burdens which followed in the wake of increased expenses and lessened income. For the time being, at least, whaling was conforming with vexatious strictness to the trying conditions of an industry of diminishing returns.

This accelerated decline in the profits of the whale fishery came at the very time, too, when the opportunities for amassing wealth on land were unsurpassed. The hectic exploitation of the natural resources of the United States after the Civil War presented chances for gain on a gigantic scale which could not

[7] These figures were taken from the summaries of production and the calculations of average prices which were published annually by the *Whalemen's Shipping List*.

have been matched by the whaling industry even in its palmiest days. Neither new capital nor lusty young manhood could be induced to go into whaling when one of the richest land areas of the world lay at their doors and promised untold wealth to those who could harness it. Even New England, far from the tempting prairies and mines of the West, found in the cotton mill a substitute for harpoon and whaleboat. To many an old whaleman it must have seemed little short of an outrage when New Bedford herself, suckled and grown strong on oil and bone, denied further capital to her battered old whalers and poured her savings into cotton mills. Great, tall smokestacks, symbols of a newer industry and of greater profits, came to dwarf and replace the sturdy masts and yards which had seen Cape Horn and Java Head times without number and which had weathered the shrieking gales of both the North Atlantic and the South Pacific. New capital ignored the demands and needs of whaling, now obviously senescent, and flowed into manufacturing, mining, farming, and commerce; and even old capital deserted the rapidly thinning ranks of the whaleships whenever opportunity offered. Consequently the victims of decay, shipwreck, fire, reef, shoal, and Arctic ice, once gone, could never be replaced.

But capital was not the only truant; for men, especially those of character and of ability, were also deserting the industry in large numbers. More and more the intelligent and ambitious young American refused to go to sea, even in New England, and least of all on a whaler. This drift away from the sea began as early as the thirties, was greatly accelerated during the fifties by the lure of California gold, and was completed during the halcyon days of internal development which followed the Civil War. And as the better types of Americans forsook the forecastles, their bunks were filled by criminal or lascivious adventurers, by a motley collection of South Sea Islanders known as Kanakas, by cross-breed negroes and Portuguese from the Azores and the Cape Verdes, and by the outcasts and renegades from all the merchant services of both the Old World and the New. This extraordinary mixture of races, nationalities, and types had also characterized the fishery, it is true, throughout the forties and fifties. But after 1865 the dilution of the labor force by an ever-mounting percentage of ignorance, incompetence, and general inefficiency proceeded at a still swifter pace. By 1880 the dregs of American-born men comprised only one-third of the 3,896 hands who manned the New Bedford whaling fleet. Another third was made up of Portuguese; and the remainder included negroes, Kanakas, and scattered individuals from most of the great ports of Europe and of Asia.[8]

This progressive deterioration in the character, skill, and efficiency of the crews lay like a rock in the path of whaling success. It would be too much to say that it was the main cause, or even a major cause, of the decline of the fishery. For in the pre-war decades the industry had prospered mightily with crews not markedly superior to these later ones. But certainly such hands added nothing to the effectiveness of an occupation which was already struggling in its death-throes; and this remained true despite the ridiculously high lays, translatable into low earnings, which were paid. Sweated labor, combining low efficiency with low wages, has never led to industrial dominance. Any revival of whaling prosperity would have had to come about in spite of, rather than with the assistance of, the men in the forecastles.

[8] An account of these later whaling crews is given by Brown, J. T., in "Fisheries and Fishery Industries of the United States," VII, pp. 218 ff.

CHAPTER XIV

DISINTEGRATION AND DECAY

THROUGHOUT the last quarter of the nineteenth century it was apparent that American whaling was doomed. The forces of destruction were doing their work so well as to permit no doubt of that. Or if doubt there were, it must have wilted under the prime solvent of disheartening facts. Even false and lingering hopes, strongly fortified by tradition and desire, could hardly withstand the broadsides of statistical evidence which were published by the *Whalemen's Shipping List.* The annual summaries of this authoritative organ of the industry were more eloquent of whaling decadence than many reams of discussion. Year after year the post-war decades witnessed the same process of disappearing vessels, of shrinking tonnage, and of fading production. The figures for the period 1846 to 1886, taken at five-year intervals, presented a striking contrast between pre-war activity and post-war decline. 1846 marked the apex of the fishery: 1886 saw but the ghost of the earlier fleet.

Year	Number of Vessels	Tonnage	Sperm Oil (barrels)	Whale Oil (barrels)	Whalebone (pounds)
		Pre-War Period of Activity			
1846	735	233,189	95,217	207,493	2,276,939
1851	553	171,971	99,591	328,483	2,869,200
1856	635	199,141	80,941	197,890	2,592,700
1861	514	158,746	68,932	133,717	1,038,450
		Post-War Period of Decline			
1866	263	68,536	36,663	74,302	920,875
1871	288	69,372	41,534	75,152	600,655
1876	169	38,883	39,811	33,010	150,628
1881	177	39,426	30,598	31,677	368,322
1886	124	29,118	———	———	———

This picture of progressing decay was completed by the further losses which came with the turn of the century. In 1896 there were still 77 whalers, of 16,358 tons; but by 1906 even this small fleet had been cut in half, leaving only 42 vessels, of 9,878 tons. Most of these, too, were small craft. The average whaler was never a large vessel, even when measured according to contemporary sailing-ship standards; but during the years she grew still smaller, instead of larger. In 1846 the average number of tons per vessel was 317: in 1906 it had fallen to 235. And, as was to be expected, this downward trend of tonnage, vessels, and products was closely followed by the number of men and the amount of capital. At the height of the fishery 20,000 men and more than $20,000,000 worth of capital were afloat in all latitudes and longitudes; whereas in 1880 the $2,891,650 worth of invested capital was amply manned by only 4,198 seamen.[1]

During the period of sixty years bounded by 1846 and 1906 almost 700 American whaling vessels disappeared from the seas. What became of them? To those who knew the mind of the seaman, who endowed every sailing vessel with a distinct personality, the fate of these whaleships was comparable in interest to the fate of men. Many left readily accessible records. Thus it was known that fifty were destroyed by Confederate cruisers, that forty were sacrificed in the sinking of the "Great Stone Fleet," and that fifty more were lost in the Arctic disasters of 1871, 1876, and 1888. As for the others, their dead hulks lie strewn about the reefs and ocean floors of the seven seas. A few were broken up at home; but most of them found their last resting-places under water, as a good ship should. Some were "never heard from again," which is a way of saying that they are guarding the bones of their crews anywhere from the Cape of Good Hope to the Sea of Okhotsk; some were sold into the merchant service, and were finally scuttled after a shamed old age spent in carrying cheap

[1] The best sources of information for the later years of the industry are to be found in the files of the *Whalemen's Shipping List* and in the monumental work, "Fisheries and Fishery Industries of the United States." In the latter, see especially the writings of Brown, J. T., in Vol. VII, and those of Clark, A. H., in Section V, Vol. II. The most readable secondary account, as well as the most recent one, is that by Jenkins, J. T., "History of the Whale Fisheries."

and heavy commodities; some were pounded to pieces on the uncharted reefs and shoals of the South Pacific; a few were sacrificed to the fury of mutineers or the treachery of South Sea natives; others were burned to the water's edge; some met a turbulent, racking fate in the whirl of an Indian Ocean typhoon; and some were crushed by ice-floes or field-ice, and perhaps, for a season or two, haunted the frigid waters as rudderless and dismasted derelicts.[2]

The long and picturesque career of the ship *Maria* was reminiscent of many others. This vessel, intended for use as a privateer, was built in 1782. After the Revolution she was bought by William Rotch, well-known among the earlier whaling merchants, and fitted out as a whaler. In this capacity she made no less than twenty-seven voyages, occupying a period of seventy years, for her purchaser and his descendants. Soon after the close of the Revolutionary War, while lying in the River Thames, the *Maria* displayed the new American flag; and by so doing she earned the reputation of having been one of the first vessels to introduce John Bull to the Stars and Stripes. Finally, in 1863, having wrested cargo after cargo of oil from cachalot and bowhead, right whale and humpback, she was sold into Chilean hands. There the veteran whaler was put into use as a common carrier; but only three years later she proudly elected to sink rather than to continue in such ignominious drudgery. Carrying dirty, miscellaneous cargoes in the muddy waters of a foreign coasting trade was no fit occupation for a vessel which had spent her life in hunting the mightiest of big game throughout all the deepest seas. And the *Maria* knew it!

Together with the destruction of the old whalers went a decided change in the fortunes of the ports from which they had sailed. One after another the smaller whaling ports of New England and of Long Island were forced out of the fishery. Even the leaders found it impossible to keep up their earlier pace. New Bedford, it is true, still possessed 123 vessels in

[2] See Scammon, C. M., "Marine Mammals of the Northwestern Coast of North America," pp. 244 ff., for an interesting discussion of the careers and fates of many of the old American whalers.

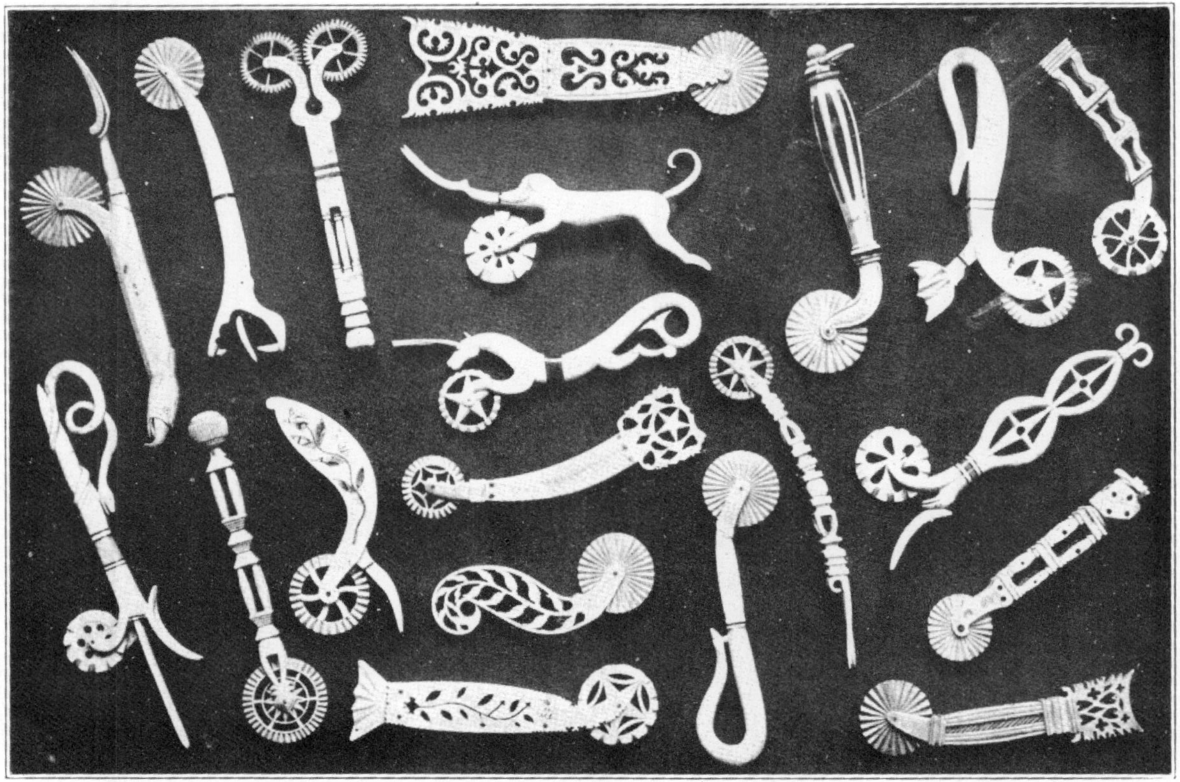

"JAGGING WHEELS"

Carved by whalemen out of the jawbone and teeth of the sperm whale, and used to crimp pastry around the edge of a pie

1880 (though this was only half of the fleet which she had owned thirty years before). But in the same year New London retained only five whaling craft; and the fleets of Sag Harbor and of Fairhaven had been blotted out completely. Provincetown alone, still clinging to small, short-voyage sloops and schooners, was able to show an apparent increase by advancing from sixteen vessels in 1850 to nineteen in 1873. But even this slight gain was deceptive; for the larger fleet of 1873 displaced a total tonnage which was 310 tons less than that of 1850. By 1880 only nine ports — New Bedford, Provincetown, Edgartown, Boston, New London, San Francisco, Westport, Marion, and Dartmouth — were still sending out whaling vessels. New Bedford alone, with 123 craft, boasted of three-fourths of the total; Provincetown was an insignificant second with twenty small "plum-pud'ners"; and the seven remaining communities had but midget fleets of one to seven vessels each. The juxtaposition of the figures for three widely separated years presents a striking picture of the decline of the leading ports: [8]

	Whaling Vessels 1850	Whaling Vessels 1873	Whaling Vessels 1880
Provincetown	16	19	20
Sag Harbor	23	2	0
Fairhaven	46	5	0
New London	48	20	5
New Bedford	237	129	123

The one port which suffered the most complete and ruinous reversal of fortune, however, was Nantucket. At one time the largest and proudest whaling port in the New World, Nantucket fell upon evil days even before many of her smaller rivals. Throughout the fifties and sixties a decreasing number of captains and merchants struggled on valiantly in the face of a growing maze of handicaps; but in 1870 the battle ceased. On June 14 of that year, when the brig *Eunice H. Adams* let

[8] The figures for 1850 and 1873 were taken from Scammon, C. M., "Marine Mammals," pp. 241 f.; while those for 1880 were taken from Clark, A. H., in "Fisheries and Fishery Industries of the U. S.," Section V, Vol. II, p. 3.

go her anchor in the harbor, the final cargo of oil had arrived. Yet less than thirty years before, in 1842, Nantucket had been at the zenith of her career with a fleet of 36,000 tons!

This relatively early demise of whaling on the island was due both to the misfortunes which were then afflicting the entire industry and to certain factors peculiar to Nantucket. One self-imposed handicap lay in the traditional preference for sperm oil and the consequent refusal to take up Arctic right and bowhead whaling, though the latter branches of the fishery had proved more lucrative. The drain of manpower to the California gold mines which took place during the fifties was much more difficult to replace in a small island community like Nantucket than on the mainland. And a great fire which swept the town in 1846 may have had some effect in lessening its recuperative powers.

But the most obvious difficulty, and the one to which the inhabitants attributed many of their woes, was a sand-bar which prevented deep-draught vessels from coming into port. Relatively small vessels could be floated over the bar without difficulty; but entering the harbor was a precarious matter for larger craft. Attempts to secure the opening of an adequate channel were made repeatedly, but in vain. Congress, as usual in such matters, was apathetic; and the task was too great to be undertaken by the town itself. Consequently the bar remained, and kept away (in the minds of the islanders, at least) not only larger craft, but also much potential prosperity.[4]

But the whole New England seaboard, for almost a century the whaling headquarters of the world, was likewise doomed. During the post-war years the whale fishery's center of gravity was shifting slowly but surely from Atlantic to Pacific. For a time Honolulu was able to retain her time-honored supremacy as a port for recruiting and refitting; but after 1880 the palm passed to San Francisco. For some time the Golden Gate had been a convenient base for operations in the North Pacific and in the Arctic waters north of Behring Straits — the only re-

[4] For further details regarding the dramatic decline of whaling at Nantucket see Hussey and Robinson, "Catalogue of Nantucket Whalers, and Their Voyages From 1815 to 1870"; and Douglas-Lithgow, R. A., "Nantucket: A History," pp. 374 ff.

gions in which whales were still being captured in any numbers. Its advantages expanded tremendously, however, with the coming of steam whaling. The original cost of a steam whaler was about three times that of a sailing vessel of equal capacity, and the expenses of operation were greatly increased by a heavy coal bill and a crew which necessarily included several skilled mechanics. Such vessels, in order to be profitable, had to spend a larger proportion of the year actually on the whaling grounds, and a smaller percentage in going to and from port. Carrying a cargo to the Atlantic coast, via Cape Horn, was out of the question; and consequently many vessels which were still owned in New Bedford transferred their registry and base of operations to San Francisco. Catches of oil and bone were shipped to the Eastern market over the new transcontinental railway; and these railway connections, plus her facilities for mechanical repairs and for general refitting, soon enabled the Golden Gate to supplant Honolulu as the chief whaling base of the Pacific.[5]

But even San Francisco whaling was but a shadow of the past. During the nineties the whaling merchants were forced to derive their income more and more from bone, rather than oil; and whalebone alone, in its growing scarcity, was too narrow a base for a great industry. In spite of the high price of bone and the meticulous care taken to save it, the Arctic fleet sailing from San Francisco grew smaller and smaller. Bone might be wanted badly; but oil, the mainstay of the fishery, was no longer desired. In 1897 the average price of whalebone at San Francisco was $4.00 per pound, but the average price of whale oil was only thirty cents per gallon. And even $4.00 bone could not make amends for thirty cent oil, especially when both were being obtained in quantities which diminished alarmingly.

The Atlantic, however, fared even worse than the Pacific. In 1892 there were only thirty-two vessels still pursuing whales on both sides of the equator; and when the turn of the cen-

[5] The comparative advantages and disadvantages of steam and sail for whaling vessels are discussed in Jenkins, J. T., "History of the Whale Fisheries," p. 248. For a description of the extent and causes of the rise of San Francisco as a whaling port see Goode, G. B., in "Fisheries and Fishery Industries of the United States," Section IV, p. 42.

tury loomed, in 1898, Atlantic whaling had been further reduced to fourteen craft and a catch of 12,520 barrels of sperm oil.

In truth, the entire American whaling fleet was being ground into bits between the upper and nether millstones of Pacific decline and Atlantic decay. For the decade ending with 1905 the whole industry was able to employ, on the average, only fifty-one vessels per year. And in 1906, just sixty years after New World whaling had reached its zenith with a fleet of 735 vessels and 233,000 tons, only a beggarly flotilla of forty-two whalers remained afloat. New Bedford, still in the lead, contributed 24 vessels; San Francisco, 14; Provincetown, three (small "plum-pud'ners," as always); Norwich, Connecticut, one; and all the other erstwhile whaling ports, from Nantucket to Mattapoisett — not a single sail!

But even this was not the end. Disintegration and decay continued to work their will upon the industry for something more than another decade, until only the bark *Wanderer* and the ship *Charles W. Morgan* remained. And when the *Wanderer*, tired of dragging out her days in the uncongenial atmosphere of the twentieth century, at length piled up on the rocks at Cuttyhunk in 1924, the *Morgan* was left as the sole (and inactive) survivor of a fleet which once whitened every sea. Venerable but pathetic, never venturing from port, her deserted mast-heads and silent forecastle stand in tragic contrast to the era of world-girdling voyages and oil-soaked decks of which she is the last mute reminder. Perhaps, after all, it would have been better had she, too, yielded to Davy Jones's ardent wooing, as did so many of her sisters, while still in the full flush of youth!

APPENDIX A

A Consular Report on Whalers and the Whaling System [1]

Consulate of the United States
Paita, September 1st, 1858.

Honorable John Appleton,
Assistant Secretary of State,
Washington.

Sir,

I have the honor to submit to the Department a report on Whalers and the Whaling System as pursued and carried on from several ports in the United States — and I hope, although I shall be compelled to go somewhat into details, the tediousness therefrom will be compensated for by the information contained.

I feel confident that there is no branch of trade or enterprise entered into in the United States in which so large a capital is invested, that is so

[1] This interesting and authoritative report was found amongst the original manuscript Consular Letters of the Department of State. Reference to these documents, which are preserved in the State Department Library at Washington, is by port and year; and in this case both means of identification are given in the heading of the report itself. In spite of its length the letter, including two appended tables, is reproduced in full. Paita was an admirable spot in which to observe the conditions of whaling life; for it was the favorite recruiting port for the hundreds of whalers which visited the great Off-Shore Grounds. A later consul, writing on December 8, 1863, said of it: "It has only been of special value, heretofore, to the whaling fleet of New England, as a periodical resort for giving liberty to seamen after long and tedious cruises at sea, for seeking medical and surgical relief, and for refreshing with vegetables, especially the onion, their most valuable antiscorbutic, which is only grown in the interior of this part of the coast." And when he wrote this report Consul Ringgold, characterized as "an educated and a kind hearted gentleman" even by his rather critical and suspicious successor, had had five years of intimate contact with the situation which he described. Unfortunately the proposed report on the "ill treatment of Seamen and its causes," mentioned in the last paragraph, was never called for; but in a later despatch of June 16, 1860, the same consul described a "frequent practice" which was causing him much official annoyance. In order to escape the payment of three months' wages to discharged seamen, as required by law, many whaling masters formed the habit of lying off and on instead of coming to anchor in the harbor. In this way they avoided the necessity of submitting the ship's papers to the consul, and were free to discharge or to leave men on their own terms. Then, upon touching at another port some time later, these hands would be reported as deserters, and certificates of desertion taken out for them. In this manner the whaling captains not only saved the advance wages, but also secured legal permission to confiscate the earned but unpaid lays of their victims! And these victims, left ashore in a penniless condition, were usually forced to apply to the consul for assistance.

little understood except by the immediate persons concerned. Even intelligent merchants in other ports, who speculate in the results of a whaling voyage, know little and care less from whence and by what means they are taken. I shall therefore not confine myself to merely giving the number of vessels and men employed the amount of tonnage — average quantity of oil taken etc. etc. but do my utmost to lay before you an exact account of the Whale System — the advancement and improvements introduced into modern whaling in the one hand — the abuses and acts of injustice on the other.

From after the War of 1812 we may date the commencement of American Whaling in the Pacific. Commodore Porter in that erratic, yet bold expedition in the Essex, by destroying the English Whale fleet on this side of land, gave the impetus and laid the foundations to a vast trade which the Eastern States with their characteristic energy, and especially Massachusetts, eagerly grasped and from which they have reaped millions of profit.

At that period I doubt whether there would have numbered more than ten or twelve American Whalers in this Ocean. But at the Close of the war, English Capitalists on this branch, having either been ruined or fearing to again entrust their vessels so far from home and for so long a time, left the Pacific free — and consequently the American Trader found this immense space entirely at his disposal.

For many years so successful were the voyages and in comparatively so short a space of time were they performed, that nothing but old condemned merchant vessels, patched up, sometimes newly rigged with the necessary boats and whaling gear were sent to this Coast — but latterly, that is within the last twelve years a great change has taken place and a much superior class of vessels has been constructed. This is owing to the circumstance that whales have either become much more scarce, which some of the most experienced masters affirm — or they have become more knowing, more cautious and wilder — as others equally experienced assert. In all probability a combination of these causes would be nearer the truth. The sperm whale is endowed with an extraordinary acuteness of hearing, and the smallest splash of a paddle or an order given in too loud a tone, will be the signal for instant disappearance — and as they are either taught by nature or experience that they can, and vessels cannot go "dead in the wind's eye" — they dart off in that direction and are soon lost to sight. It has been found, therefore that a fast sailing clipper, although she cannot compete with the whale when thoroughly alarmed, yet when he is seen from aloft she can by quick sailing and proper manœuvering get more rapidly within lowering distance and thereby have a great advantage over the old tubs that formerly went to sea. Many full and numbers of half clippers have been of late years added to the whaling fleet and others are annually being built. As a matter of course new tight and commodious vessels render the labors of the Seamen much less and their comforts much greater.

There is no more of that everlasting pumping, patching, and caulking — these have been more or less done away with — the men have dry bunks to sleep in — and when not cutting in whales or trying out oil, their work is comparatively light.

It was also found that the System of bad beef, bad pork and worse biscuit turned out in the end an unprofitable economy. For good men, accustomed to wholesome food at home and shipped at small lays or shares, would desert with the hope of bettering themselves and the master would, consequently, be compelled to take any class whether good or bad to make up his compliment giving at the same time better lays and a large advance. I am satisfied that there are no vessels afloat as a general rule that have better provisions and in greater abundance than whalers.

Beside the salt provisions and small stores, these vessels touch at some port at least every six months and lay in large supplies of fresh meat, vegetables and fruits. On one occasion I knew a master to buy some sixty barrels of sweet potatoes, paying a high price for them although told they were too old to keep. His reply was that there were no others to be had (which was a fact) and if he took them on board the men would be satisfied even if he had to throw them into the sea in two days. It will be perceived from the foregoing that there has been considerable improvement in the building of vessels and the care for the comforts of the men and I wish sincerely my report could cease here. I should be only too glad if over the picture which I shall be now compelled to sketch, I could draw a curtain. I would be glad for the sake and name and reputation of our countrymen engaged in whaling but I having imposed upon myself this (which I conceive to be not a task but a sacred duty) I will not shrink from doing justice to the mariner, after having bestowed that praise which was due to the owner and Master.

In order to make myself thoroughly understood, it will be necessary to explain that generally speaking when a vessel is being fitted out for a whaling voyage to the Pacific Ocean from the U. States, a shipping master is applied to, who if he has not on hand a sufficient number of men immediately sends his *runner* through the interior of Massachusetts, N. Hampshire, Vermont, N. York and even as far as Ohio to pick up what are termed green hands. As these Shipping Masters make large profits upon each and every man whom they procure, they use every means in their power to excite the fancy and stimulate the ambition of those whom chance may throw in their way. They return to the port with ten or fifteen fine able bodied young men who are shipped by the proper authority and are then told they can amuse themselves at their boarding house until the day of sailing — and that the shipping master has orders to supply them with clothing adapted to and necessary for the voyage they are about to undertake.

The time of sailing arrives. All hands are huddled on board — their chests into which they have had no chance of looking are put into the fore-castle when the vessel is about getting under way. They have already signed a receipt for it. The order is given to make sail and off they speed on a voyage of four years.

Each sailor is charged in the Owner's books with an average outfit of seventy (70$) dollars. But this sum is not paid the Shipping Master until the vessel has been six months at sea. By many Owners interest is charged on this sum for outfit, from the day of sailing until the return of the vessel.

When at sea the sailor opens his chest for the first time, when it is discovered alas too late that he has received including the value of the chest and all in it, property to the amount of twenty to twenty five dollars. The result is that in the beginning of the voyage the men are not only dissatisfied but they are compelled upon the first appearance of cold or rugged weather to seek warm clothing from the Slop Chest which in many instances is placed on board by the owners as a profitable speculation, they paying the Master a small Commission for his trouble.

The lay or share of a greenhand is from the one-one-hundred and eightieth (1/180) to the one-two-hundredth 1/200 — that is one barrel of oil, for every hundred and eighty or two hundred that are taken. And I will now show what profit accrues to a green hand.

A Sperm Whale Ship of 350 tons will take on a voyage of four years 1200 barrels of Sperm oil. This is a liberal average. The share or part which the Sailor who has the 1/180 lay will have coming to him will be two hundred and ten (210) gallons or equal at the present home prices to two hundred and sixty two (262 25/100) dollars and twenty five cents but from this sum there are to be made sundry deductions. Ten per cent discount on the amount of oil taken for leaking and shrinkage is always made and very frequently three per cent for insurance, although if the vessel is lost with five hundred barrels of oil on board and it is fully covered by insurance the owners recover all and the men get nothing — because the charge is not made on the men until the vessel is safe home — and the policy of insurance is of course made out in the name of the owner. Or as an old sailor once expressed himself to me when a vessel was burnt in this bay — 'it's no use Sir to give me a certificate the Owners play an open and shut game — if the vessel gets home I pay for insurance but if she is lost they pay the insurance and pocket the profits.' And it is so.

Then the accumulated interest (not always charged) on the original seventy dollars outfit and twelve per cent per annum on any money given as liberty money during the voyage. Finally ten (10$) dollars is charged every man whether discharged on the coast or at home for "fitting Shipping and Medicine Chest" — a phrase the meaning of which I have never been able to have satisfactorily explained to me.

The following is the result of the Seaman's voyage of four years:

Sailors share reduced to money		$262.25
Less fitting shipping and Medicine Chest	$10.00	
10 pr. ct. discount on $262 25/100	26.22	
3 pr. ct. insurance on $262 25/100	7.86	
Money originally advanced	70.00	
Interest on same	16.80	
Cash advanced during voyage	30.00	
Interest on same 1 pr. ct. pr. month	7.20	
Clothing which he was compelled to draw owing to his bad outfit	40.00	
		208.08
Amount to be received at the end of Voyage		$ 54.17

But allowing that every dollar of the $262 25/100 should be handed to the Seaman at the end of the voyage — that the Seventy dollars first advanced should be admitted as a gratuity — that the outfit should have been properly and honestly put up, that even the money advanced and the clothes given on the voyage and all the interest upon these various sums should be thrown in as a gift to encourage the young man to continue a profession for which he may by this time have conceived some liking — what would be his monthly wages? A sum so preposterously small that I feel almost ashamed to mention it. It seems incredible that an intelligent active young American should pass through four years of labour (not to mention dangers from both sea and monster) separated from family and country at the rate of five dollars and twenty-two cents ($$5\frac{22}{100}$) per month. Yet such is the case.

I grant as the Owners and Masters contend that this is his first voyage, that he had no experience — that if he conducted himself properly and felt disposed he can take advantage of knowledge he has now gained and demand from the Owners a much better lay and perhaps if recommended by the Master, a better position such as boatsteerer — but before he can reap the benefit of these advantages we must be assured that he has really obtained them, we must know whether he is not entirely destitute or even in debt to the ship and thereby compelled to reship almost immediately; we must know what class of Masters and Mates he has had over him — whether they have treated him kindly, whether they have been willing to advise or instruct or whether they have (which I am sorry to say is generally the case) left him to himself and that sink of immorality the forecastle.

It is very clear no matter what may be the amount of his instruction or what the sum of knowledge gained on the voyage the pecupiary benefit is merged into the hands of the owners.

But there are other and more serious results springing from this System of small pay (which by the way is almost too mild a term) to the country at large.

The vast amount of uncultivated and unpopulated territory at home would naturally lead us to discourage in every possible way the emigration of our native born citizens. Every arm is useful and when we find that three or four thousand young men yearly sail from the United States in whalers and becoming disgusted desert and either from shame or moral corruption never return or if they do return in after years are no credit to their country, we are naturally led into the reflection that the cause which produced this must be of the most serious nature. And I am satisfied on my own mind that the small profit as the result of labour is more the cause than the bad treatment which they sometimes receive at the hands of the Masters and Mates.

It is almost impossible to expect that twenty five or thirty men can exist at sea for the length of six months without some slight disagreement and when we take into consideration on the one hand that men may be obstinate self-willed or perverse or that on the other the Master or Mate may be too exacting or prone to tyrannize it is not at all astonishing that there should be complaints of minor or greater character frequently entered at the different Consulates.

But I must defer to a future occasion observations upon the personal ill treatment of Seamen and its causes and should the Department desire and approve of it I will with pleasure place before it several cases of an exaggerated character that have come under my official and personal knowledge. And in conclusion I will now place in tabular form the number of whale vessels sailing out of the United States with their tonnage average number of men, average amount of oil, etc. Also a table showing the amount of profit derived by the owners from these voyages from which last it will be seen that it would be very easy for them to largely increase the pay of their men and still make an enormous interest on their money.

I have the honor to be, Sir

with great respect, Your obedient Servant

FAYETTE M. RINGGOLD

TABLE A

Showing the Number of Vessels, Tonnage, Number of Seamen and Average Catch and Value of Oil and Bone Taken in Whaling Vessels Sailing from the United States

To What Ports Belonging	Number of Tons	Total No. of Vessels	No. of Sperm Vessels	No. of Whale Vessels	No. of Vess. Rep. or Fittg	No. of Seamen Employed	Average Time of Voyage	Average Quantity of Sperm Oil Taken Annually Barrels	Average Quantity of Whale Oil Taken Annually Barrels	Average Quantity of Whale Bone Taken Annually Pounds	Value of Sperm Oil Taken Annually Dolls. Cts.	Value of Whale Oil Taken Annually Dolls. Cts.	Value of Whale Bone Taken Annually Dolls. Cts.	Total Value Dolls. Cts.
New Bedford	109,845	320	217	103	50	8,000	4 yrs.	86,300	87,500	875,000	3,417,750.00	1,930,477.50	612,500.00	5,960,727.50
New London	18,733	65	47	19	15	1,625	" "	18,400	16,150	161,500	724,500.00	356,107.50	113,050.00	1,193,657.50
Fairhaven	16,530	47	35	12	10	1,175	" "	14,000	10,200	102,000	551,250.00	224,910.00	71,400.00	847,560.00
Nantucket	11,829	36	34	2	9	900	" "	13,500	1,700	17,000	531,562.50	37,485.00	11,900.00	580,947.50
Provincetown	3,314	31	15	16	2	620	" "	6,000	13,600	136,000	236,250.00	299,880.00	95,200.00	131,330.00
Westport	4,252	20	20	..	4	500	" "	8,000	......	......	315,000.00	....	....	315,000.00
Sag Harbor	5,929	20	15	5	1	500	" "	6,000	4,250	42,500	23[illegible],250.00	93,712.50	29,750.00	359,712.50
Matapoiset	3,791	19	18	1	5	475	" "	7,600	850	8,500	299,250.00	18,742.50	5,950.00	323,942.50
Edgartown	5,757	18	11	7	3	450	" "	4,400	5,950	59,500	173,250.00	131,197.50	41,650.00	346,097.50
Warren	5,512	16	12	4	4	400	" "	4,800	3,400	34,000	189,000.00	74,970.00	23,800.00	287,770.00
Portsmouth	2,805	10	10	..	..	250	" "	4,000	......	......	157,500.00	....	....	157,500.00
Sundry Small Ports	14,757	59	47	12	10	1,475	" "	18,800	10,200	102,000	740,250.00	224,910.00	71,400.00	1,036,560.00
	203,062	661	480	181	113	16,370	4 yrs.	193,300	153,850	1,538,000	7,571,812.50	3,392,392.50	1,076,600.00	12,040,805.00

TABLE B

Showing the value of the Whaling Vessels, interest on same, annual expenses, etc., etc.

Estimated value of the 661 Whale Vessels Sailing from the United States including their outfits, provisions and the advances made to Seamen on the day of Sailing, at the rate of 25,000$ each	$16,525,000.00
Six pr. ct. pr. annum interest on Same	991,500.00
Ten pr. ct. pr. annum for ware and tare	1,600,000.00
Two and a half pr. Cent insurance	413,125.00
Fresh Supplies purchased by the Masters equal to about 1200$ pr. annum each	793,000.00
Amount of Money paid to Masters Officers and Crew being their Shares of the Oil taken — equal to one third of the gross value of the products	4,013,601.33
Total Amount of Money invested including interest, etc., etc.	$24,336,226.33
Value of the Annual Amount of Oil taken Showing a clear yearly profit of 46 p. ct.	12,040,805.00
Difference between the whole Capital invested and the yearly profit	$12,295,421.33

APPENDIX B

Desertions, Discharges, and Deaths in Relation to the Size of Whaling Crews [1]

Table I

Desertions, Discharges, and Deaths, In Comparison With the Size of the Original Crews, for Fifteen Whaling Voyages Made by Eight Different Vessels During the Years 1843–1862

Name of Vessel	Number of Whaling Voyage	Years Absent	Number in Original Crew	Number of Desertions	Number of Discharges	Number of Deaths
Endeavor	First	1843–45	25	5	8	2
Endeavor	Second	1845–47	28	2	4	..
Endeavor	Third	1847–51	25	14	7	..
Fabius	First	1844–46	33	4	..	..
Fabius	Fourth	1851–54	33	5	5	..
Fabius	Fifth	1854–57	31	3	13	..
Montreal	First	1850–53	37	4	3	..
Montreal	Second	1853–57	36	12	6	2
Montreal	Third	1857–62	39	30	79	7
Wm. C. Nye	First	1851–54	37	16	5	1
Wm. C. Nye	Second	1855–57	36	4	3	1
Alice Mandell	(?)	1851–55	31	9	11	..
Triton	Fourth	1854–58	31	11	6	1
James Maury	Fourth	1855–59	36	14	10	2
Adeline	Second	1856–60	31	10	6	..
TOTALS			489	143 (29.2%)	166 (33.9%)	16 (3.3%)

[1] The figures for both tables included in this appendix were obtained through an examination of the original manuscript account-books for the voyages listed. It will be noted that the third voyage of the *Montreal* and the fourth voyage of the *James Maury* are given in both tables. But since one table is concerned with the original crews, and the other with the total crews, or all men shipped during a voyage, no duplication is involved. The two tables are entirely independent of each other. See Chapter V, on Forecastle and Cabin, for an interpretation of the significance of these figures with regard to the problem of labor turnover in the whaling industry. All of the account-books used were found in the New Bedford Public Library.

Table II

Desertions, Discharges, and Deaths, In Comparison With the Total Number of Men Shipped at Any Time During the Course of a Given Voyage, For Twenty-three Whaling Voyages Made by Ten Different Vessels During the Years 1839–1879

Name of Vessel	Number of Whaling Voyage	Years Absent	Total Number of Men Shipped	Number of Desertions	Number of Discharges	Number of Deaths
Benjamin Tucker	First	1839–43	39	10	1	1
Benjamin Tucker	Second	1843–46	35	2	3	..
Benjamin Tucker	Third	1846–49	32	..	4	..
Benjamin Tucker	Fourth	1849–51	33	5	3	..
Benjamin Tucker	Fifth	1851–55	33	7	6	2
Brighton	Second	1844–47	48	5	8	1
Brighton	Third	1847–50	31	10	1	3
James Maury	First	1845–48	34	5	3	1
James Maury	Second	1848–51	30	7	2	..
James Maury	Third	1851–55	32	8	6	2
James Maury	Fourth	1855–59	53	14	10	2
James Maury	Fifth	1859–62	38	13	4	..
James Maury	Sixth	1864–68	39	11	6	3
Isaac Howland	(?)	1854–59	88	19	18	..
Magnolia	(?)	1855–58	83	15	44	1
Montreal	Third	1857–62	158	30	79	7
Wm. C. Nye	Third	1858–61	71	6	11	1
Fabius	Seventh	1862–65	44	5	4	3
Adeline	(?)	1866–69	100	22	46	1
Marcella	Eleventh	1867–69	36	9	2	..
Marcella	Twelfth	1870–73	27	16	3	..
Marcella	Thirteenth	1873–76	23	5	..	1
Marcella	Fourteenth	1876–79	34	3	3	1
TOTALS			1141	227 (19.9%)	267 (23.4%)	30 (2.6%)

APPENDIX C

Average Crew Earnings for Thirty Whaling Voyages Made by Four Representative Vessels During the Years 1836–1879

Table I

Cash Balances of Twenty-Three Crews Shipped During the Years 1836–1879[1]

	Marcella	*James Maury*	*Minerva*
FOREMAST HANDS			
Number of Accounts	175	85	63
Lowest Balance	$226.95 Minus	$184.17 Minus	$ 83.37 Minus
Highest Balance	278.51	461.35	149.43
Median Balance	11.79 Minus	103.20	000.00 (Sic)
BOATSTEERERS AND COOPERS			
Number of Accounts	46	24	18
Lowest Balance	$271.82 Minus	$ 51.85 Minus	$ 54.86 Minus
Highest Balance	539.66	1554.85	359.41
Median Balance	73.15	283.41	96.01
MATES			
Number of Accounts	32	18	14
Lowest Balance	$323.77 Minus	$183.69 Minus	$ 37.19
Highest Balance	2531.97	2042.19	1121.56
Median Balance	342.07	491.43	273.93
MASTERS			
Number of Accounts	6	1	2
Lowest Balance	$ 276.39	$2016.41	$ 734.29
Highest Balance	1450.13	....	1846.74
Median Balance	994.03	....	......

[1] Negative balances, signifying that certain whalemen found themselves in debt to the owners at the time of settlement, are followed by the word "Minus." Other figures signify that the balance was due to the whaleman. The figures for the bark *Marcella* were taken from the accounts of twelve consecutive voyages, 1845–1879; those for the ship *James Maury,* from six consecutive voyages, 1845–1868; and those for the bark *Minerva,* from five consecutive voyages, 1836–1846.

Note on Appendix C.

The figures included in the five tables of this appendix were taken directly from the original manuscript crew-accounts of the four vessels concerned. All voyages were made from New Bedford, and the crew-accounts are now in the New Bedford Public Library. The members of this particular quartet were chosen because they were well-known and representative whalers, because their accounts were available for several consecutive voyages, and because, taken together, their voyages covered the decades of the industry's greatest prosperity and of its early decline. Six hundred and eighty-one individual accounts yielded figures indicating the cash balances received or due; and three hundred and two individual accounts showed the amounts of the gross lays which were payable. Only accounts belonging to the more stable whalemen, who completed an entire voyage or a substantial portion of one, were included. Certain others were omitted because they involved factors which, though by no means uncommon in the industry, were hardly characteristic of the majority of whalemen. Such excluded accounts, numbering well over one hundred in the aggregate, belonged to men who had deserted, who had died during the early stages of a cruise, who had been on board for only a few weeks, or who had received monthly wages in place of, or in addition to, the more familiar device of the lay. These figures, if used, would have lowered materially the average earnings secured. But since the object of the analysis was to arrive at certain suggestive averages arising from the earnings of the ordinary whaleman, who was neither a mere transient nor unusually unfortunate or unscrupulous, the lowest rungs of the whaling account ladder were omitted. Only in the accounts of the *Fabius* (Table V) were the figures for all hands, regardless of circumstances, brought into the computations.

The proper interpretation of any *average* figures for the whaling industry, however, demands that the averages be accepted as suggestive rather than definitive. In an occupation which was subject to so many wide fluctuations it was practically impossible to secure figures which would yield an average of meticulous accuracy. And averages dealing with earnings were particularly treacherous because of the numerous and complicated ramifications of the lay system. (See the chapters on Earnings and the Lay, and Debits and Credits, for a detailed discussion of the many qualifications and abuses of this peculiar scheme of wage payment). Nevertheless the results yielded by hundreds of individual accounts, spread over thirty separate voyages, and extending over a period of forty-three years, are not without some significance. Particularly interesting are the median figures for lays and balances. It will be noted that the median cash balance for 175 foremast hands, shipped during twelve voyages of the *Marcella,* was actually a minus amount! And sixty-three hands on the *Minerva,* by a curious stroke of fortune, yielded a median figure which was exactly zero; while the median balance received by 85 hands on the *James Maury* proved to be the relatively munificent sum of $103.20.

Table II

Gross Lays Earned by the Members of Thirteen Whaling Crews Shipped Between 1836 and 1868[1]

	Marcella	*James Maury*	*Minerva*
FOREMAST HANDS			
Number of Accounts	70	83	39
Lowest Lay	$ 10.70	$ 77.49	$ 22.46
Highest Lay	217.05	630.39	155.73
Median Lay	99.82	323.09	88.41
BOATSTEERERS AND COOPERS			
Number of Accounts	15	27	12
Lowest Lay	$ 58.79	$ 103.91	$ 77.20
Highest Lay	496.17	1801.95	328.90
Median Lay	287.92	643.35	153.48
MATES			
Number of Accounts	9	21	9
Lowest Lay	$ 35.66	$ 215.99	$ 170.02
Highest Lay	932.56	3932.73	934.40
Median Lay	528.62	1169.72	315.74
MASTERS			
Number of Accounts	9	5	3
Lowest Lay	$ 984.47	$2144.50	$ 650.06
Highest Lay	4978.39	7244.83	1334.86
Median Lay	1692.98	4473.63	698.92

[1] These accounts cover some of the same voyages which were employed in Table I. The crew-accounts of the bark *Marcella* offered figures for four consecutive voyages, 1845–1856; those of the ship *James Maury,* for six consecutive voyages, 1845–1868; and those of the bark *Minerva,* for three consecutive voyages, 1836–1841.

Table III

Cash Balances of the Foremast Hands Shipped by Three Whalers for Twenty-Three Voyages Made Between 1836 and 1879 [1]

Cash Balances		*Marcella*	*James Maury*	*Minerva*
Over $200	Minus	2	..	..
$200 to $151	Minus	5	2	..
$150 to $101	Minus	12	..	..
$100 to $ 51	Minus	40	3	16
$ 50 to $ 1	Minus	38	7	15
$ 0 to $ 49	Plus	40	21	12
$ 50 to $ 99	Plus	14	9	8
$100 to $149	Plus	12	15	12
$150 to $199	Plus	7	8	..
$200 to $299	Plus	5	13	..
$300 to $399	Plus	..	4	..
Over $400	Plus	..	3	..
Total Number of Accounts		175	85	63

Table IV

Gross Lays of the Foremast Hands Shipped by Three Whaling Vessels for Thirteen Voyages Made Between 1836 and 1868 [2]

Lays	*Marcella*	*James Maury*	*Minerva*
$ 0 to $ 49	24	..	6
$ 50 to $ 99	14	2	18
$100 to $149	16	4	10
$150 to $199	13	8	5
$200 to $299	3	27	..
$300 to $399	..	16	..
$400 to $499	..	19	..
$500 and over	..	7	..

[1] The Minus Accounts signify that the balances were due from whalemen to owners — or, otherwise stated, that these hands returned from voyages of two to four years actually in debt to the owners for the amounts indicated. The Plus Accounts, on the other hand, indicate sums which were due from owners to whalemen. See the chapter on Debits and Credits for a detailed discussion of the numerous items which resulted in these final cash balances. These figures constitute simply another arrangement of the accounts of foremast hands which were employed in Table I. The same accounts have been used for both tables.

[2] These figures were secured from the same foremast hand accounts which formed part of Table II. They were taken from four consecutive voyages of the bark *Marcella*, 1845–1856; from six consecutivc voyages of the ship *James Maury*, 1845–1868; and from three consecutive voyages of the bark *Minerva*, 1836–1841.

Table V

Crew-Accounts of the Ship *Fabius,* of New Bedford, for Seven Consecutive Voyages, 1844–1864 [1]

Voyages	Net Proceeds	Number of Men With Balances Due Them	Lowest Balance Due	Highest Balance Due	Number of Men in Debt to Ship	Lowest Debt	Highest Debt	Number of Men With an Even Balance
1st Voyage Arrived Feb. 14, '46	$ 33,390	21	$ 4.87	$ 292.12	6	$ .61	$117.61	2
2nd Voyage Arrived Jan. 8, '49	43,002	25	25.54	397.74	10	52.04	100.04	0
3rd Voyage Arrived Feb. 14, '51	46,210	26	39.12	366.42	0	....	.....	1
4th Voyage Arrived March 13, '54	31,770	23	40	412.70	5	2.89	59.54	0
5th Voyage Arrived March 17, '57	100,829	29	45.21	1586.95	2	28.67	25.34	0
6th Voyage Arrived April 18, '62	65,219	15	24.19	789.90	8	9.69	100.41	0
7th Voyage Sailed Oct. 16, '62; Lost Jan. 27, '65, on Solidad Reef, off California Coast [2]	81,383	14	93.90	1206.32	11	68.41	152.42	0

[1] These figures include the accounts of all hands except masters and mates. As is evidenced by the sums secured as net proceeds, the *Fabius* was successful in all of her voyages; but the last three cruises were profitable enough to excite special comment. It was a rare occasion indeed when whalemen who were not officers received cash balances of $1200 or $1500! True, such sums, even on the bonanza voyages, were restricted to coopers or boatsteerers, who commonly shipped at the best lays; but on the last three cruises of the *Fabius* even the foremast hands obtained respectable balances. Such lucky voyages as these constituted the bait which kept merchants and whalemen in the fishery in spite of counteracting losses and disappointments.

[2] Although the *Fabius* was wrecked, she did not go to pieces until after her cargo had been removed and sent home in other vessels.

APPENDIX D

A Cross-Section View of the American Whale Fishery on January 1, 1844[1]

Table I

Figures Descriptive of the Industry as a Whole

	Number of Vessels	Number of Men	Average Number of Men per Vessel	Average Length of Voyage, Including Time in Port
Sperm Whaling	242	6,776	28	44 months
Right Whaling	329	9,212	28	27 "
Atlantic Sperm Whaling	73	1,606	22	15 "
Totals	644 *	17,594		

* Made up of 595 ships and barks, 41 brigs, and 8 schooners, displacing 200,484 tons.

Table II

Whaling Vessels, Tonnage, and Seamen Sailing From the Ports of the New Bedford District

Ports	Number of Vessels (Ships, Barks, and Brigs)	Tonnage	Number of Men
New Bedford	219	69,803	5,888
Fairhaven	45	14,350	1,170
Dartmouth	1	387	30
Westport	11	1,982	253
Mattapoiset	10	1,938	230
Sippican	7	1,335	161
Wareham	6	1,366	138
Total	299	91,161	7,870

[1] The statistical material included in the seven tables of Appendix D was taken from compilations made by Joseph Grinnell, Member of Congress from New Bedford. Mr. Grinnell made use of these figures in preparing a speech on whaling which he delivered in the House of Representatives on May 1, 1844; and during the same year he published his remarks in the form of a sixteen-page pamphlet entitled, "Speech on the Tariff, With Statistical Tables of the Whale Fishery." While many of the calculations are admittedly estimates, and are so labelled, Mr. Grinnell's official position and the ready access to reliable sources of information which it gave him, plus his own familiarity with the conditions of the whale fishery, confer a strong degree of presumptive accuracy upon his compilations.

Table III

Cost of a Typical Whaling Vessel, Exclusive of Outfit, as Shown by Figures Arising From the Construction of a Live Oak Ship of 351 Tons at New Bedford in 1841

Yellow pine plank		39,520 feet	$1,691.09
Yellow pine timber		18,915 feet	431.50
Live oak timber		3,557 feet	3,521.60
White oak plank		2,561 feet	399.64
Knees		249 feet	436.22
White oak timber		2,921 feet	756.32
Locust timber		151 feet	213.80
Mahogany timber		255 feet	37.25
Sundry pieces timber bought by the stick		....	36.25
Cordage		21,577 pounds	2,524.59
Duck		135 pieces	2,167.75
Sheathing copper	8,796 pounds		
Copper bolts	4,800 pounds		
		13,596 pounds	3,311.16
Spikes	4,409 pounds		
Forelocks	150 pounds		
Rudder joints	654 pounds		
Sheathing nails	350 pounds		
Coppering nails	565 pounds		
		6,128 pounds	1,409.44
Sundry articles — augers, plates, bobstay work, etc.			350.00
Copper cooler and other copper work			166.00
Chain cables		15,840 pounds	1,188.00
Chain sheets and ties		950 pounds	104.50
Fluke chains, etc.		1,663 pounds	133.04
Anchors		3,800 pounds	342.00
Blacksmith's bill			1,970.56
Painter's bill			697.00
Rigger's bill			390.00
Sailmaker's bill			561.50
Blockmaker's bill			550.00
Amount paid for labor, carpentering			7,008.41
Amount paid for labor, caulking			528.69
Amount paid for labor, joiner work			298.41
Total Outlay			$31,224.72

Table IV

List of the Principal Articles Required to Outfit a Vessel for a Voyage in Sperm Whaling or in Right Whaling, Together With the Amount of Each Article and the Cost According to the Prices Which Prevailed on January 1, 1844

Article	Sperm Whaler					Right Whaler				
	Amount		Price		Cost	Amount		Price		Cost
Oil Casks	2,800 bbls.	@	$ 1.25	=	$3,500.00	2,800 bbls.	@	$ 1.25	=	$3,500.00
Beef & Pork	240 bbls.	@	8.50	=	2,040.00	163 bbls.	@	8.50	=	1,385.00
Flour	220 bbls.	@	5.25	=	1,155.00	155 bbls.	@	5.25	=	813.75
Corn	75 bus.	@	.55	=	41.25	59 bu.	@	.55	=	27.50
Beans & Peas	14 bus.	@	1.25	=	17.50	14 bu.	@	1.25	=	17.50
Corn Meal	5 bbls.	@	3.50	=	17.50	5 bbls.	@	3.50	=	17.50
Tobacco	2,500 lbs.	@	.11	=	275.00	2,000 lbs.	@	.11	=	220.00
Rice	1,200 lbs.	@	.035	=	42.00	1,200 lbs.	@	.035	=	42.00
Potatoes	150 bu.	@	.35	=	52.50	130 bu.	@	.35	=	45.50
Cheese	800 lbs.	@	.07	=	56.00	600 lbs.	@	.07	=	42.00
Butter	900 lbs.	@	.13	=	117.00	800 lbs.	@	.13	=	104.00
Dried Apples	600 lbs.	@	.04	=	24.00	600 lbs.	@	.04	=	24.00
Vinegar	10 bbls.	@	3.50	=	35.00	7 bbls.	@	3.50	=	24.50
Cod Fish	800 lbs.	@	.03	=	24.00	800 lbs.	@	.03	=	24.00
Molasses	1,600 gals.	@	.27	=	432.00	1,200 gals.	@	.27	=	324.00
Tea, Black	250 lbs.	@	.35	=	87.50	200 lbs.	@	.35	=	70.00
Tea, Hyson	20 lbs.	@	.60	=	12.00	12 lbs.	@	.60	=	7.20
Raisins	200 lbs.	@	.05	=	10.00	100 lbs.	@	.05	=	5.00
Sugar	1,000 lbs.	@	.075	=	75.00	800 lbs.	@	.075	=	60.00
Coffee	1,000 lbs.	@	.08	=	80.00	800 lbs.	@	.08	=	64.00
Duck, Heavy	60 pieces	@	18.00	=	1,080.00	40 pieces	@	18.00	=	720.00
Duck, Light	36 pieces	@	8.00	=	288.00	25 pieces	@	8.00	=	200.00
Tar	20 bbls.	@	2.25	=	45.00	10 bbls.	@	2.25	=	22.50
Whale Boats	6 boats	@	60.00	=	360.00	6 boats	@	60.00	=	360.00
Oars	7 sets	@	8.50	=	59.50	7 sets	@	8.50	=	59.50
Boards	4,000 ft.	@	20.00M	=	80.00	4,000 ft.	@	20.00M	=	80.00
Nails, Composition	700 lbs.	@	.22	=	154.00	500 lbs.	@	.22	=	110.00
Copper, Sheathing	8,500 lbs.	@	.21	=	1,785.00	7,000 lbs.	@	.21	=	1,470.00
Cordage	8,500 lbs.	@	.10	=	850.00	7,000 lbs.	@	.10	=	700.00
Tow Lines	3,000 lbs.	@	.12	=	360.00	3,000 lbs.	@	.12	=	360.00
Try Pots	3 pots	@	60.00	=	180.00	3 pots	@	60.00	=	180.00
Cloth	6,000 yds.	@	.09	=	540.00	5,000 yds.	@	.09	=	450.00
Iron Whaling Crafts	4,000 lbs.	@	.15	=	600.00	4,000 lbs.	@	.15	=	600.00
Clothing, Ready-made	....			=	2,800.00	....			=	2,700.00
Labor in Port	....			=	2,500.00	....			=	2,300.00
Total Cost of Outfits					$19,774.75					$17,129.45
Cost of a Typical Vessel					31,224.72					31,224.72
Combined Cost of Vessel & Outfit					$50,999.47					$48,354.17

Table V

Estimated Cost of Outfitting the American Whaling Fleet of 644 Vessels Which Was at Sea on January 1, 1844

Estimated Cost of All Articles Consumed Annually	$3,132,199.48
Estimated Wages Paid Annually for Labor Employed in Outfitting	= $ 548,060.00
Estimated Wages Paid Annually for Labor Employed in Coopering	= $ 165,238.98
Estimated Annual Cost of Articles and Labor	$3,845,498.46

Table VI

Estimated Value, at the Time of Sailing, of the 644 Vessels Which Comprised the American Whaling Fleet on January 1, 1844
(These figures include both the vessels and their outfit)

242	Ships, Barks, and Brigs in the Sperm Fishery, @ an Average Value of $38,000 each	$9,196,000
329	Ships, Barks, and Brigs in the Right Whale Fishery, @ an Average Value of $28,000 each	$9,212,000
73	Ships, Barks, and Brigs in the Atlantic Sperm Fishery, @ an Average Value of $14,000 each	$1,022,000
644	Total	$19,430,000

Table VII

Estimated Value, Including Vessels, Outfits, *and Catchings at Sea,* of the American Whaling Fleet of 644 Vessels on January 1, 1844

242	Ships, Barks, and Brigs in the Sperm Fishery, @ an Average Value of $55,000 each	$13,310,000
329	Ships, Barks, and Brigs in the Right Whale Fishery, @ an Average Value of $40,000 each	$13,160,000
73	Ships, Barks, and Brigs in the Atlantic Sperm Fishery, @ an Average Value of $18,000 each	$1,314,000
644	Total	$27,784,000

Table VIII

Estimated Value of the Catchings of the American Whaling Fleet Which Was at Sea on January 1, 1844 (The figures in this Table have been derived from those in Tables VI and VII)

Catchings of 242 Ships, Barks, and Brigs in the Sperm Fishery, @ an Average Value of $17,000 each	$4,114,000
Catchings of 329 Ships, Barks, and Brigs in the Right Whale Fishery, @ an Average Value of $12,000 each	$3,948,000
Catchings of 73 Ships, Barks, and Brigs in the Atlantic Sperm Fishery, @ an Average Value of $4,000 each	$292,000
Total	$8,354,000

APPENDIX E

Number of Arrivals, Average Cargo of Oil, and Average Length of Voyage of American Whaling Vessels During Each Calendar Year from 1842 to 1856[1]

Year of Arrival	Number of Vessels Arriving		Average Cargo of Oil: Sperm Oil	Average Cargo of Oil: Whale Oil	Average Length of Voyage	
			Bbls.	Bbls.	Months	Days
1842	55	Sperm	1,973	135	41	8
	65	Atlantic Sperm	280	12	13	28
	74	2-Season Right	422	1,722	24	15
	13	1- “ “	122	1,602	10	15
	207	Total Arrivals				
1843	70	Sperm	1,641	124	41	13
	55	Atlantic Sperm	285	25	14	20
	90	2-Season Right	311	1,937	25	10
	15	1- “ “	92	1,398	11	28
	230	Total Arrivals				
1844	69	Sperm	1,419	293	43	0
	42	Atlantic Sperm	248	38	12	0
	112	2-Season Right	248	2,059	25	9
	7	1- “ “	69	1,176	11	14
	230	Total Arrivals				
1845	91	Sperm	1,291	387	43	21
	43	Atlantic Sperm	238	76	13	7
	101	2-Season Right	196	2,180	24	0
	8	1- “ “	55	796	12	4
	243	Total Arrivals				

[1] The figures included in this appendix were collected from the statistical summaries of the whaling industry which were published in January of each year by the *Whalemen's Shipping List.* Throughout each year this accepted organ of the whale fishery gathered its statistical material with care and systematic precision; and its summaries were considered authoritative by those who were familiar with the industry.

Year of Arrival	Number of Vessels Arriving	Average Cargo of Oil Sperm Oil Bbls.	Whale Oil Bbls.	Average Length of Voyage Months	Days
1846	42 Sperm	1,350	280	41	6
	48 Atlantic Sperm	259	14	14	7
	94 2-Season Right	225	2,034	30	2
	1 1- " "	...	2,005	12	2
	185 Total Arrivals				
1847	52 Sperm	1,505	219	45	12
	34 Atlantic Sperm	228	42	15	9
	150 2-Season Right	195	1,978	31	7
	1 1- " "	...	940	7	0
	237 Total Arrivals				
1848	52 Sperm	1,292	192	41	19
	31 Atlantic Sperm	303	18	16	21
	122 2-Season Right	222	2,187	33	2
	1 1- " "	...	727	7	19
	206 Total Arrivals				
1849	54 Sperm	1,284	168	41	21
	28 Atlantic Sperm	216	25	6	16
	102 2-Season Right	209	2,271	32	9
	1 1- " "	...	728	7	14
	185 Total Arrivals				
1850	48 Sperm	1,189	115	41	22
	33 Atlantic Sperm	235	14	9	14
	83 2-Season Right	189	2,217	33	20
	1 1- " "	...	485	7	16
	165 Total Arrivals				
1851	53 Sperm	1,170	333	44	4
	46 Atlantic Sperm	201	9	10	24
	83 2-Season Right	130	2,619	35	16
	33 1- " "	111	2,431	20	10
	1 Greenland	...	258	8	20
	216 Total Arrivals				
1852	50 Sperm	1,054	151	44	21
	54 Atlantic Sperm	202	38	13	1
	16 2-Season Right	128	1,989	33	14
	7 1- " "	85	1,852	17	21
	127 Total Arrivals				

Year of Arrival	Number of Vessels Arriving		Average Cargo of Oil Sperm Oil Bbls.	Whale Oil Bbls.	Average Length of Voyage Months	Days
1853	64	Sperm	1,004	153	44	1
	66	Atlantic Sperm	?	32	14	5
	59	2-Season Right	134	1,938	33	24
	24	1- " "	69	1,980	18	27
	213	Total Arrivals				
1854	34	Sperm	896	187	34	1
	69	Atlantic Sperm	207	114	15	7
	107	2-Season Right	92	1,778	34	18
	4	1- " "	7	603	12	29
	214	Total Arrivals				
1855	37	Sperm	913	76	41	18
	43	Atlantic Sperm	158	110	12	17
	79	2-Season Right	68	1,512	41	26
	1	1- " "	511	1,841	21	11
	160	Total Arrivals				
1856	47	Sperm	928	149	42	1
	67	Atlantic Sperm	131	203	15	11
	71	2-Season Right	106	1,668	38	12
	3	1- " "	174	1,925	20	10
	188	Total Arrivals				

APPENDIX F

Cargoes and Prices Secured by the Ship *James Maury* for Six Successive Whaling Voyages, 1848–1868 [1]

Whaling Voyage	Date of Arrival	Whale Oil (gallons)	Price per Gallon	Sperm Oil (gallons)	Price per Gallon	Whalebone (pounds)	Price per Pound
First	5th Month 1848	106,262	$ .285	21	$ .95	36,095	$ .225
Second	7th Month 1851	60,562	$ .435	2,632	$1.20	11,952	$ .38
Third	9th Month 1855	30,635	$ .76	682	$1.70	4,960	$ .49
Fourth	5th Month 1859	31,071	$ .45	3,242	$1.15	.....	.....
Fifth	9th Month 1862	52,962	$ .65	24,972	$1.37	5,570	$ .72
Sixth	5th Month 1868	11,384	$ .78	2,081	$1.80	8,163	$ .86

[1] These figures were taken directly from the *James Maury's* original manuscript account-book, now in the New Bedford Public Library. The wide variations in both quantities and prices afford an admirable illustration of the fluctuating fortunes which characterized the whale fishery throughout its history.

APPENDIX G

List of Articles Required to Outfit a Vessel for a Whaling Voyage[1]

Division I. — Slop-Chest Supplies. 43 Separate Kinds of Articles, including: Twilled Kersey Shirts, Striped Cotton Shirts, Calico Full Bosom Shirts, Calico Plain Shirts, All Wool Drawers, Cotton & Wool Drawers, Guernsey Frocks, Duck Frocks, Denim Frocks, Thick Trousers, Duck Trousers, Denim Trousers, Short Jackets, Pea Jackets, Reefing Jackets, Monkey Jackets, Tarpaulin Hats, Palmleaf Hats, Scotch Caps, Russian Caps, Bed Comforters, Neck Comforters, Mittens, Stockings, Southwesters, Blankets, Linen Thread, Yarn, Braces, Sheath Knives, Jack Knives, Sheaths & Belts, Iron Spoons, Tin Pots, Tin Pans, Needles, Shoes, Pumps, Boots, Brogans, Vests, Mounted Palms.

Division II. — Recruiting Supplies. 27 Separate Kinds of Articles, including: Tobacco, Pipes, Cigars, Powder, Muskets, Flints, Narrow Axes, Hatchets, Axe Hatchets, Axe Handles, Hardware, Bleached Cottons, Unbleached Cottons, Blue Cottons, Blue Denims, Boots, Shoes, Pumps, Paints, Oil Soap, Bar Soap, Furniture Prints, Fancy Prints, White Lead, Linseed Oil, Sperm Candles, Naval Stores.

[1] This list comprises a synopsis of the contents of a 48-page pamphlet, entitled "Outfits for a Whaling Voyage," which was issued by H. S. Kirby, a ship-chandler situated at No. 8 Commercial Wharf, New Bedford. Such long pamphlet lists were checked over by owners or officers as a convenient means of assuring themselves that all of the multifarious articles required for a whaling voyage were actually on board before sailing. This particular pamphlet is included in that part of the Daniel B. Fearing Collection which is now in the Treasure Room of Widener Library, Harvard University; but similar lists may be found both in the same collection and in the New Bedford Public Library. The one here reproduced was not dated, but from the nature and variety of the articles mentioned it is evident that it was issued during or after the mid-century decades of greatest whaling activity. The thousands of individual articles carried by a whaler, comprising more than one thousand different *kinds* of goods, were classified under twenty-five major divisions, or groups. For purposes of concrete illustration, the lists included under the first three divisions are reproduced *in toto*. For the remaining divisions, however, only the number of separate kinds of articles is given. Not every whaler, of course, set out with all the supplies listed under these twenty-five headings. The "plum-pud'ners" and other small Atlantic craft managed to complete their cruises with a smaller number of commodities. But the larger sperm and right whalers, setting out for long Pacific voyages, could dispense with very few of them. Checking and stowing away supplies for a world-girdling whaling voyage of forty to fifty months was no sinecure!

Division III.—Nautical Supplies. 31 Separate Kinds of Articles, including: Chronometer, Thermometer, Barometer, Steering Compass, Boat Compass, Telltale Compass, Mast-head Compass, Time Piece, Spy Glasses, Opera Glasses, Half-Hour Glasses, Log Glasses, Log Book, Log Slates, Common Slates, Pacific Directory, Horsburgh's Directory, Ink, Quills, Steel Pens, Memoirs South Atlantic, Blanks, Wafers, Lead Pencils, Paper, Memorandum Books, Account Books, Charts, Nautical Almanac.

Division	IV.	— Provisions and Cabin Stores.	58 Separate Kinds of Articles.
Division	V.	— Coopers' Tools.	49 Separate Kinds of Articles.
Division	VI.	— Nails.	19 Different Kinds.
Division	VII.	— Sail Needles.	10 Different Kinds.
Division	VIII.	— Crockery.	27 Separate Kinds of Articles.
Division	IX.	— Tin Ware.	49 Separate Kinds of Articles.
Division	X.	— Copper Ware.	27 Separate Kinds of Articles.
Division	XI.	— Whaling Craft, etc.	81 Separate Kinds of Articles.
Division	XII.	— Blocks, etc.	19 Separate Kinds of Articles.
Division	XIII.	— Cordage.	40 Separate Kinds of Articles.
Division	XIV.	— Spare Sails.	26 Separate Kinds of Articles.
Division	XV.	— Spare Spars.	9 Separate Kinds of Articles.
Division	XVI.	— Spare Paints.	19 Separate Kinds of Articles.
Division	XVII.	— Boats, Lumber, Oars, etc.	20 Separate Kinds of Articles.
Division	XVIII.	— Carpenters' Tools, Hardware, etc.	300 Separate Kinds of Articles.
Division	XIX.	— Miscellaneous.	41 Separate Kinds of Articles.
Total Number (Including Div's. I–III)			895 Separate Kinds of Articles.

In addition to these 895 separate kinds of articles, all except the smallest short-voyage vessels carried some, though not all, of the goods listed under the following 6 divisions.

Additional Memoranda.

Division XX.—Casks. 10 different types, comprising 14 different sizes, ranging from a 20-inch, 60-gallon cask to one of 46 inches and 290 gallons.

Division XXI. — Anchors. 19 different weights, varying from a small anchor of 150 pounds for a 30-ton vessel to one of 2700 pounds for use on a vessel of 700 tons.

Division XXII. — Try-Pots. 5 different sizes, ranging from 140-gallon (3 ft. 9 in. by 3 ft. 4 in.) to 220-gallon (4 ft. 9 in. by 3 ft. 8 in.).

Division XXIII. — Sheathing Copper. 11 different weights, from 16-ounce to 36-ounce.

Division XXIV. — Medicinal Supplies. Complete medicine chest, epsom salts, New England rum, Holland gin, brandy, port wine.

Division XXV. — Documents, Papers, and Specie. Letters of instruction, bills of sale, protections, copy of shipping papers, invoice of slops, letter-bag, ship's papers, ship's library, and a small supply of specie.

APPENDIX H

Account of Stores for Sloop *Susannah*, Capt. Peter Pease, from Martha's Vineyard to the West Indies, Whaling, Nov., 1763.[1]

7 bbls. Beef	7 bushels Salt	1 Spade
4 bbls. Pork	16 bushels Corn	30 Irons
3 bbls. Flour	2 bushels Potatoes	1 Inch Auger
1 bbl. Molasses	1 Hatch Bar	1 Pitch Fork
1 bbl. Rum	1 Tea Kettle	3 Compasses
1100 lbs. Bread	1 Tea Pot	1 side Good Leather
56 lbs. Sugar	1 Scraper	2 Good Pots
4 lbs. Chocolate	12 Lances	12 Codfish Hooks
2 lbs. Tea	120 fathoms Cordage	1 pair Canthooks
1 lb. Pepper	250 Copper Nails	3 Padlocks
50 lbs. Cheese	1 Pewter Basin	1 Iron Ladle
56 lbs. Rice	16 Runner Hooks	1 Iron Shod Shovel
4 bushels Beans	1 barrel Tar	3 Files
3 bushels Meal	1 Coffee Pot	14 lbs. Candles
4 bushels Turnips	1 Crow Bar	130 ft. Pine Boards

[1] This relatively short and primitive list of the articles carried on an early whaling voyage presents a striking contrast to the elaborate equipment described in Appendix G. It is interesting, too, from an antiquarian point of view, since it affords a glimpse of the details of American whaling equipment at a time when the radius of activities was still a short one. This particular cruise to the West Indies kept the *Susannah* away from home for only six months — and no doubt her small crew contrived to exist for that length of time with the articles here listed. For voyages which took a vessel to the north of New England, however, a somewhat different proportion of stores was carried. Rum, for instance, was held to be even more essential in a cold climate than in a warm one. This "Account of Stores" may be found in Banks, C. E., "History of Martha's Vineyard," Vol. I, p. 436. In his researches the author chanced upon the original account-book of the *Susannah,* and fortunately considered this portion of the material to be worthy of reproduction.

APPENDIX J

The Manufacture of Sperm Oil and of Spermaceti Candles[1]

Crude oil, or oil in its natural state, is that which is obtained from the blubber of the whale in the process of "trying out" on shipboard. The oil, then, which is taken from whale ships and carried to the oil manufactory, is said to be in its *crude* state. We will speak first of the manufacture of *crude sperm oil.*

The first step in the process of manufacture, is to take the oil in its crude state, and put it into large kettles, or boilers, and subject it to a heat of one hundred and eighty to two hundred degrees, and then all the water which happened to become mixed with the oil, either on shipboard or since, will evaporate.

Winter Strained Sperm Oil. In the fall, or autumn, the oil is boiled for the purpose of granulation during the approaching cold weather. The oil thus passes from a purely liquid into a solid state, or one in which it is in grains, or masses.

When the temperature of the atmosphere rises, or the weather slackens during the winter, the oil which has been frozen, but is now somewhat softened, is shovelled out of the casks and put into strong bags that will hold half a bushel or more, in order to be pressed. . . . The oil which is now obtained from this first pressing is called *winter strained sperm oil.*

Spring Sperm Oil. What remains in the bags after the first pressing, is again heated by being put into boilers, after which it is baled into casks again, and upon cooling, it becomes more compact and solid than it was before.

During the month of April, when the temperature is about fifty degrees, the oil becomes softened; it is then put into bags, and goes through a second process of pressing similar to the first. The oil from this pressing is called *spring strained sperm oil.*

Tight Pressed Oil. That which is left in the bags after the second pressing, is again melted, and put into tin pans or tubs which will hold about forty pounds each. When this liquid is thoroughly cooled, as each pressing makes what is left harder, in consequence of extracting the oil, the cakes taken from the tubs are then carried into a room heated to about ninety

[1] This material has been taken verbatim from a description given by Holmes, Lewis, in his volume on "The Arctic Whaleman," pp. 288–296. The author obtained his facts directly from a certain Charles J. Barney, foreman of Dr. Daniel Fisher's oil factory, in Edgartown, Martha's Vineyard. As a detailed account of industrial processes which have long since been forgotten, this description of the manufacturing phase of the whale fishery is not without some historical interest.

degrees; and as they begin to yield to the influence of this high temperature, or the remaining oil begins to soften the cakes, they are taken and shaved into very fine pieces, or ground up as in some instances, deposited in bags as hitherto, and put into the hydraulic press.

The room being at the temperature indicated above, and the bags subjected to a powerful pressure of three hundred tons or more, all the oil is extracted from them, and what is left is perfectly dry, free from any oily matter, and brittle. The oil thus obtained by this last pressing is called *tight pressed, or summer oil.*

Spermaceti. What remains after the several pressings, and the removal of all the oil, is called *stearine, or spermaceti.*

Spermaceti is not confined to the head matter of the whale, as some suppose . . . But . . . the spermaceti from the head oil is quite different from that of the body oil; the former presents fine, bright, transparent scales like small particles of isinglass, while the latter is more compact, something like dough. In cooling, one exhibits a sparry, crystalline structure, the other that of clay.

Head oil or *matter* is usually manufactured with the body oil of the whale, and mixed in proportion to one-third of the former to two-thirds of the latter.

Spermaceti Candles. That which remains in the bags after the hydraulic pressure is both dry and brittle. The oil, it is supposed, is wholly extracted, and nothing now remains but the spermaceti. Its color, however, is not white, but interspersed with grayish streaks, bordering on the yellow.

The spermaceti is put into large boilers adapted for the purpose, and heated to the temperature of two hundred and ten degrees. It is refined and cleared of all foreign ingredients by the application of alkali. Afterwards water is added, which, with a temperature of two hundred and forty degrees, throws off the alkali in the form of vapor. The liquid which remains is as pure and clear as the crystal water, and ready to be made into the finest spermaceti candles.

BIBLIOGRAPHY

I — Works Dealing With Early European Whaling[1]

HAKLUYT, RICHARD, "The Principall Navigations, Voiages, Traffiques, and Discoveries of the English nation." London, 1599.

McCULLOCH, J. R., "A Dictionary, Practical, Theoretical, and Historical, of Commerce and Commercial Navigation." London, 1856.

ANDERSON, ADAM, "An Historical and Chronological Deduction of the Origin of Commerce, From the Earliest Accounts to the Present Time." Dublin, 1790.

——, "Collection of the Whalefishery Acts of Great Britain, 1672–1800." This collection, including only the texts, without notes or comments, was made by Daniel B. Fearing. It is now in the Fearing Collection at Widener Library, Harvard University.

BRINNER, LUDWIG, "Die Deutsche Grönlandfahrt." Berlin, 1913. Abhandlungen zur Verkehrs — und Seegeschichte, Vol. VII. Im Auftrage des Hansischen Geschichtsvereins.

FISCHER, P., "Cetacees de Sud-ouest de la France." — Actes de la Societe Linneenne de Bordeaux, XXXIV. Bordeaux, 1881.

DUCERE, EDOUARD, "Dictionnaire Historique de Bayonne." — Commission des Archives Municipales Ville de Bayonne. Bayonne, 1911.

FOTHERBY, ROBERT, "Narrative of a Voyage to Spitzbergen in the Year 1613. At the Charge of . . . the Muscovy Company: With a Description of the Country, and the Operations of the Whale-Fishery." American Antiquarian

[1] For the sake of convenience, the titles here presented have been arranged under some nine headings which were logically suggested by the nature of the subject-matter. Within each section, too, the works are listed in order of importance rather than alphabetically. Outstanding volumes are placed at the heads of their respective sections; while lesser works follow in general, though not meticulous, order of significance. Occasionally, where the nature of the subject-matter renders it advisable, the same work is given in more than one division. Brief comments, designed to serve as rough characterizations, are presented in connection with the more important or unusual titles. All sections are meant to be suggestive rather than exhaustive. With few if any exceptions, all writings of significance either to the student or to the general reader have been included; but the literature of American whaling has been robbed of its best, not stripped bare. Reference is made to those works which are necessary in securing an adequate understanding of New England whaling; but large quantities of unimportant, repetitive, or untrustworthy manuscripts and writings have been omitted. The student who wishes to penetrate into these recesses of whaling literature, often interesting even though not essential, will inevitably gravitate toward the storehouses to be found in the New Bedford Public Library; in the Old Dartmouth Historical Society Museum, New Bedford; in the Daniel B. Fearing Collection in Widener Library, Harvard University; and in the Nantucket Whaling Museum.

Society's Transactions, Vol. IV. (Introduction and notes by Samuel F. Haven). Boston, 1860.

BROWN, R. N. D., "Spitzbergen." (One chapter on the early whalemen). London, 1920.

ELKING, HENRY, "A View of the Greenland Trade and Whale-fishery. With the National and Private Advantages thereof." London, 1725. (Second Edition.) Brief but illuminating.

MARKHAM, C. R., "On the Whale Fishery of the Basque Provinces of Spain." In Living Age, 1882. Vol. 153, pp. 52–56.

II — ACCOUNTS OF EARLY AMERICAN WHALING

STARBUCK, ALEXANDER, "History of the American Whale Fishery from Its Earliest Inception to the Year 1876." Waltham, Mass., 1878. (Also appended, as Vol. IV, to the Report of the United States Commission of Fish and Fisheries for 1875–1876). Easily the most valuable single work on the early history of American whaling. The narrative is continued down to the middle decades of the nineteenth century, though this later period is not treated as fully as the earlier ones. Contains much tabular and statistical material of great value.

MACY, OBED, "History of Nantucket and of the Whale Fishery." Boston, 1835. The best account of early Nantucket whaling. The author possessed the advantage of having been in close contact with many of the men and events described.

CRÈVECOEUR, JEAN HECTOR ST. JOHN DE, "Letters From an American Farmer." London, 1782, and Philadelphia, 1793. Letters IV–VIII contain descriptions of Nantucket and of Martha's Vineyard in which due respect is paid to whaling.

——, Massachusetts Historical Society Collections. Material of particular value with regard to early whaling is contained in Series I, Vol. III, pp. 157 and 161, and Vol. VIII, pp. 202 f; and in Series II, Vol. III, pp. 19 *et passim.*

——, "Discovery and Adventure in the Polar Seas and Regions, With an Account of the Whale Fishery." New York, 1833. This volume formed No. 14 of Harper's Family Library. The author's name did not appear.

WEEDEN, W. B., "Economic and Social History of New England, 1620–1789." Two Vols. Boston, 1894. Contains various references to whaling.

DOW, G. F., "Whale Ships and Whaling, a Pictorial History of Whaling During Three Centuries, With an Account of the Whale Fishery in Colonial New England." Salem, Massachusetts, Marine Research Society, 1925. An admirably complete pictorial record of whaling. Only 40 pp. of text.

——, American Antiquarian Society Proceedings, No. 57, pp. 28 *et passim.* Remarks on early whaling in the Pacific.

——, "The Whale Fishery." Boston, 1833. Numbers 18 and 24 of the Scientific Tracts. Author's name not given.

SPEARS, J. R., "The Story of the New England Whalers." New York, 1908. A secondary but readable account. The chapters on Indian and early colonial whaling are particularly interesting.

III — MANUSCRIPT MATERIAL

Original manuscripts of whaling merchants and masters obviously constitute the most authoritative and important source of information regarding many phases of the industry. Curiously enough, however, the great mass of this manuscript material has been little used, if not entirely overlooked. The best collections, consisting largely of log-books and account-books, are to be found in the New Bedford Public Library, in the Old Dartmouth Historical Society Museum at New Bedford, in the Fearing Collection at Harvard University, and in the Nantucket Whaling Museum. Other ports also yield individual manuscripts and smaller, informal collections. The New Bedford Public Library has incomparably the best collection of account-books; while the Old Dartmouth Museum has perhaps a larger number of log-books. The Fearing Collection contains a smaller number of carefully chosen manuscripts, mainly log-books of more than ordinary interest. Because these various repositories contain hundreds of individual manuscripts, reference to them is made in terms of groups, or types, of material.

A—*Material in the New Bedford Free Public Library*

New Bedford Free Public Library, "Collection of Books, Log-Books, Pamphlets, etc., Illustrating the Whale Fishery." Second Edition, April, 1920. A convenient guide to the manuscripts available in this one collection.

Wood, Dennis, "Whaling Vessels From the United States." 4 Volumes. These four volumes, each containing from 600 to 700 pages of legible notes closely written in long-hand, and carefully indexed, constitute a veritable Domesday Book of American whaling. They include a summarized record of hundreds of whaling vessels which sailed between 1831 and 1876. Dates of voyages, when spoken at sea, sizes of cargoes, accounts of mutinies, wrecks, or other unusual events, the final disposition of each vessel, when known,—all these facts, and others as well, the indefatigable author set down with meticulous care and a fine pen.

——, Crew-Accounts of Some Sixty-five Voyages Made Between 1836 and 1879 by the following New Bedford Vessels: Ships *James Maury, Fabius, Canton, Wm. C. Nye, Montreal, Brighton, Benjamin Tucker, Charles W. Morgan, Adeline, Triton, Alice Mandell, Isaac Howland, Samuel Robertson, Three Brothers,* and *Hibernia;* and barks *Marcella, Minerva, Endeavour, Mars, Arab,* and *Gypsy.*

——, A Collection of More than One Hundred Account-Books of Whaling Merchants and Insurance Companies, Including Journals, Day-Books, Waste-Books, Invoice-Books, Cash-Books, Ledgers, Slop-Accounts, Receipt-Books, Letter-Books, Memorandum-Books, Labor-Accounts, Insurance Policies, and Registers of Ships and Voyages. Most of these manuscripts came from the whaling firms of Charles R. Tucker & Company, Charles W. Morgan, and George Hussey, with a few early volumes belonging to William Rotch; while the insurance forms were those of the Bedford Commercial Insurance Company and of the Mutual Marine Insurance Company.

——, A Large Collection of Whaling Log-Books, Dating for the Most Part From the Middle Decades of the Nineteenth Century. Because of the endless monotony which characterized whaling voyages, log-books possess an air of sameness which renders them far less valuable for research purposes than account-books. There is both interest and significance, however, in their unstudied, mechanical portrayal of the atmosphere which pervaded a whaler at sea, and in their brief, sententious, matter-of-fact records of many terrifying and tragic experiences.

B—*Manuscripts in the Daniel B. Fearing Collection, Widener Library, Harvard University*

——, A Collection of Whaling Account-Books and Log-Books, Relatively Small But Well-Chosen. The Log-Books Include Those of Voyages Made by the Ships *Heroine, Orozimbo, Martha, Lucas, Herald, Acushnet,* and *Archer;* and of Other Voyages Made by the Barks *Charles Drew, Chase, Awashonks,*

Canton, Charleston Packet, Triton, Platina, Ohio, Osprey, Marcella, Kathleen, and *Laconias.*

C—*Material in the Old Dartmouth Historical Society Museum*

——, A Large and Comprehensive Collection of Log-Books and Whaling Equipment, Together with Other Documentary Material. (Noteworthy amongst the last are the crew-lists of the ship *Lydia,* beginning a "Whaling Voyage to Delago Bay and Other Places" during the 8th Month, 1795; and of the ship *Acushnet,* of Fairhaven, sailing on December 30, 1840, with the name of Herman Melville inscribed as a member of the crew.) Easily the best and largest collection of material pertaining to American whaling. The complete exhibits of whaling equipment, illustrating every phase of the industry, are particularly valuable. Relatively weak, however, in the field of whaling accounts.

IV — Further Source Material Dealing With Nineteenth Century American Whaling

Goode, G. B., (Editor, With the Coöperation of the Commissioner of Fisheries and the Superintendent of the Tenth Census), "The Fisheries and Fishery Industries of the United States." Five Sections, in Seven Volumes. Washington, 1884 and 1887. Printed as Senate Miscellaneous Document No. 124, 47th Congress, 1st Session. A mine of information on whaling and allied subjects. Especially valuable in furnishing detailed and authoritative descriptions of the apparatus, methods, and technique of the industry, as well as much historical and narrative material. Section V contains the bulk of the significant matter relating to whaling.

Wilkes, Charles, (Lieutenant, United States Navy), "Narrative of the United States Exploring Expedition During the Years 1838, 1839, 1840, 1841, and 1842." Five Volumes. London and Philadelphia, 1845. (A second edition, in two volumes, appeared in 1852; and there were several later editions.) Contains Maps, Diagrams, Plates, and Portraits in addition to the text. References to whaling largely in Vol. V. Valuable because the author had unusual opportunities for close observation of the activities of whalemen in the Pacific.

——, Whalemen's Shipping List and Merchant's Transcript. Published weekly at New Bedford from March 17, 1843, to December 29, 1914. Complete file in the New Bedford Public Library. The recognized and authoritative organ of American whaling. Contains a wealth of carefully gathered statistical material, and presents an invaluable summary of the business aspects of the industry.

Ringgold, Fayette M., (United States Consul at Paita, Peru), Consular Report to the State Department, Washington. Dated September 1, 1858. Original manuscript now amongst the Consular Letters, in the Library of the Department of State. Reference to these Letters is by year and port. See Appendix A for a reproduction of this report and for an explanation of its significance.

Reynolds, J. N., "Report on Islands discovered by Whalers in the Pacific." Transmitted to the Secretary of the Navy on September 24, 1828, but not ordered to be printed until January 27, 1835. House Executive Document, III, No. 105, 23rd Congress, 2nd Session.

Jefferson, Thomas, "Report of the Secretary of State on the Subject of the Cod and Whale Fisheries, February 1, 1791." Philadelphia, 1791. Also printed as part of House Miscellaneous Document No. 32, 42nd Congress, 2nd Session.

Sabine, Lorenzo, "Report on the Principal Fisheries of the American Seas."

Completed December 6, 1852. Printed as part of House Executive Document No. 23, 32nd Congress, 2nd Session.

GRINNELL, JOSEPH, "Speech on the Tariff, With Statistical Tables of the Whale Fishery." Washington, 1844. A sixteen-page pamphlet which provides a cross-section statistical view of the whaling industry on January 1, 1844. Originally in the form of a speech delivered in the House of Representatives on May 1, 1844, and reported in the Congressional Globe, 28th Congress, 1st Session, Vol. XIII, p. 570. See Appendix D for several tables and for further explanatory matter.

BROWN, J. T., "The Whale Fishery and Its Appliances." Bulletin No. 27, U. S. National Museum. Washington, 1884.

SEWARD, WILLIAM H., "The Whale Fishery, and American Commerce in the Pacific Ocean." Speech delivered in the U. S. Senate on July 29, 1852, and reported in the Congressional Globe, 32nd Congress, 1st Session, Vol. XXIV, Part III, pp. 1973–1976. Also reprinted at Washington in 1852.

MAURY, M. F., (Lieutenant, United States Navy), "The Winds and Currents of the Sea." National Observatory, Washington, 1851. Contains material regarding the migratory habits of whales.

HUSSEY AND ROBINSON (Printers and Publishers), "Catalogue of Nantucket Whalers, and Their Voyages From 1815 to 1870." Nantucket, 1876.

TRUSTEES OF THE PEABODY ACADEMY OF SCIENCE, "The Whaling Industry. Exhibition of Objects Illustrating the Whaling Industry and the Natural History of Whales." Salem, Mass., 1908. Contains list of Essex County whaling vessels, list of whaling log-books in the Essex Institute, and list of works on whales and whaling in the Salem Public Library.

ENDERBY, CHARLES, "Proposal for Re-Establishing the British Southern Whale Fishery, Through the Medium of a Chartered Company, and in Combination with the Colonization of the Auckland Islands." — Revision and enlargement of a letter written to a Mr. T. R. Preston on October 31, 1846. London, 1847. Contains some valuable comparisons between British and American whaling.

——, "New Bedford and Fairhaven Signal Book." New Bedford, 1850. A sample of the flags and semaphore telegraph codes which were used in this and other ports.

CRAPO, H. H., "The New Bedford Directory. — With List of Vessels in the Whale Fishery." New Bedford, 1841.

MARTIN, J. S., "Photographs of Old Whaleships and Wharf Scenes. — One Hundred Photographs." New Bedford Studio of J. S. Martin.

KIRBY, H. S., "Outfits for a Whaling Voyage." New Bedford, not dated. A pamphlet list furnished by a New Bedford ship-chandler to his customers. See Appendix G for a summary of the articles listed.

V — SECONDARY HISTORICAL ACCOUNTS

STARBUCK, ALEXANDER, "History of the American Whale Fishery from Its Earliest Inception to the Year 1876." Waltham, Massachusetts, 1878. (Also printed as Volume IV of the Report of the U. S. Commission of Fish and Fisheries for 1875–1876).

MACY, OBED, "History of Nantucket and of the Whale Fishery." Boston, 1835.

STARBUCK, ALEXANDER, "History of Nantucket, County, Island, and Town." Boston, 1924.

RICKETSON, DANIEL, "History of New Bedford." New Bedford, 1858.

SCAMMON, C. M., "Marine Mammals of the Northwestern Coast of North America, with an Account of the American Whale Fishery." San Francisco and New York, 1874. A valuable work written by a trained observer.

JENKINS, J. T., "History of the Whale Fisheries: From the Basque Fisheries of the 10th Century to the Finner Whale at the Present Time." London,

1921. The best single work which covers the development of whaling both in Europe and in America.

STARBUCK, ALEXANDER, "The Nantucket Whale Fishery." Pages 329–341 in Godfrey, E. K., "Island of Nantucket," 1882.

——, Old Dartmouth Historical Sketches. Forty-six numbers. Published quarterly from June 30, 1903, to March 29, 1917, as the Proceedings of the Old Dartmouth Historical Society, New Bedford. Numbers 2, 14, 43, 44, and 45 contain especially significant material on whaling.

TOWER, W. S., "A History of the American Whale Fishery." Publications of the University of Pennsylvania, Series in Political Economy and Public Law, No. 20. Philadelphia, 1907. A short and uninspired summary of the more obvious facts concerning the development of American whaling.

ELLIS, L. B., "History of New Bedford and Its Vicinity, 1602–1892." Syracuse, New York, 1892. Only one chapter on whaling proper; but the remainder of the volume contains much material bearing on the background of whaling development.

MORISON, S. E., "The Maritime History of Massachusetts, 1783–1860." Boston and New York, 1921. Combines a fascinating style and sound scholarship; and presents an invaluable picture of the maritime career of that state which was almost synonymous with American whaling. Chapter XX is devoted entirely to the whalers.

ASHLEY, C. W., "The Yankee Whaler." Boston, 1926. Because of the extinction of the whaling fleet, probably the best pictorial record of American whaling which can be published. Emphasizes the artistic and picturesque phases of the industry.

JENKS, S. H., "Compendium of the American Whale Fishery." In Hunt's Merchants' Magazine, 1840, Vol. 3, pp. 172–182.

DOUGLAS-LITHGOW, R. A., "Nantucket: A History." New York and London, 1914.

BANKS, C. E., "The History of Martha's Vineyard." Two Volumes. Boston, 1911. Chap. 28, Vol. I deals especially with whaling.

LANMAN, J. H., "The American Whale Fishery." In Hunt's Merchants' Magazine, 1840. Vol. 3, pp. 361–394.

WILLIAMS, W. F., "Destruction of the Whaling Fleet in the Arctic Ocean in 1871." New Bedford, 1902.

WILLIAMS, E. C., "History of the Whale Fisheries, Habits of the Whale, Perils of the Chase, and Method of Capture." New York, 1862.

WILLIAMS, J. R., "The Whale Fishery." Article in the North American Review for January, 1834. Vol. 38, p. 105.

——, "The Whale and Dangers of Whaling." New Haven, 1839.

——, "The Whale and Perils of the Whale Fishery." New Haven, 1840.

——, "New Bedford in 1845, and Aspects of the Whale Fishery." Article in the National Magazine for September, 1845.

BIRGE, W. S., "The Whale and the Whaleman." In New England Magazine, 1903. Vol. 40, pp. 265–272.

SPEARS, J. R., "The Story of the New England Whalers." New York, 1908.

VERRILL, A. H., "The Real Story of the Whaler." New York, 1916. A popular account. Interestingly written, but hardly trustworthy in all particulars.

CLARK, A. H., "The American Whale Fishery, 1877–1886." In Science, 1887. Vol. 9, pp. 321–324.

TODD, C. B., "Whale Boat Privateersmen of the Revolution." In Magazine of American History, 1882. Vol. 8, pp. 168–181.

WEAVER, R. M., "Herman Melville, Mariner and Mystic." New York, 1921. In tracing the connections between whaling and Melville's biography, the author devotes Chapters VII and VIII to a brief résumé of the outstanding facts of the industry.

TRIPP, GEORGE H., "Whaling Ventures and Adventures." Included in the 108th Report of the New England Society of the City of New York.

HAWES, CHARLES BOARDMAN, "Whaling." New York, 1924.

CHATTERTON, E. KEBLE, "Whalers and Whaling." New York, 1926.

VI — HISTORICAL MATERIAL CONTAINING INCIDENTAL REFERENCES TO WHALING

——, The New Bedford Mercury. Published weekly from August 7, 1807, until February 28, 1831. Thereafter daily. Published in New Bedford, the greatest whaling port in the world, this newspaper inevitably contained countless references to whaling.

HARRIS, JOHN, "Zebulon; or, The Moral Claims of Seamen Stated and Enforced." Boston, 1837. (This first American edition was adapted and revised from the third London edition.)

JEWELL, J. G., "Among Our Sailors." New York, 1874. The author was for some years an American consul in the Orient; and in this book he describes the shocking treatment of American seamen, — a subject which his official duties forced him to observe at close range.

——, Hunt's Merchants' Magazine and Commercial Review. Contains frequent references to whaling, especially in Volumes III, VI, VIII, IX, X, XIV, XVI, XXVI, and XLI.

——, The Sailors' Magazine. The organ of the American Seamen's Friend Society. Published monthly at New York. Volumes I–XLV, 1828–1873.

PEASE, Z. W., and HOUGH, G. A., "New Bedford, Mass.: Its History, Industry, Institutions, and Attractions." Published by the New Bedford Board of Trade, 1889. Chapter II is devoted particularly to the whale fishery.

THOMPSON, B. F., "The History of Long Island." New York, 1839. (2nd Edition, revised and enlarged, New York, 1843).

BLISS, W. R., "Quaint Nantucket." Boston and New York, 1896.

HOWLAND, FRANKLYN, "A History of the Town of Acushnet, Bristol Co., Mass." New Bedford, 1907.

SPOONER, NATHANIEL (Compiler), "Gleanings From the Records of the Boston Marine Society, Through Its First Century, 1742 to 1842." Boston, 1879.

PITKIN, TIMOTHY, "Statistical View of the Commerce of the United States." New Haven, 1835.

——, "Accounts of Shipwreck and of Other Disasters at Sea, designed to be Interesting and Useful to Mariners." Compiled by a "Friend of Seamen." Brunswick, Maine, 1823.

——, "The Mariner's Chronicle: Containing Narratives of the Most Remarkable Disasters at Sea, such as Storms, Shipwrecks, Fires, and Famines. Also Naval Engagements, Piratical Adventures, etc." New Haven, 1834.

——, "The Mariner's Library, or Voyager's Companion. Containing Narratives of the Most Popular Voyages, From the Time of Columbus to the Present Day, with Accounts of Remarkable Shipwrecks, Naval Adventures, the Whale Fishery, etc." Boston, 1840.

MCFARLAND, RAYMOND, "A History of the New England Fisheries." Publications of the University of Pennsylvania, Series in Political Economy and Public Law. Philadelphia, 1911. Devoted largely to the inshore fisheries.

VII — NARRATIVE AND DESCRIPTIVE ACCOUNTS OF WHALING LIFE

BROWNE, J. ROSS, "Etchings of a Whaling Cruise, and History of the Whale Fishery, with Notes on a Sojourn on the Island of Zanzibar." London, 1846, and New York, 1850. One of the best of the accounts of whaling life written by men who actually shipped as foremast hands. Before going to sea in a whaler, the author had been a reporter in the United States Senate.

SCORESBY, WILLIAM, "An Account of the Arctic Regions, with a History and Description of the Northern Whale-Fishery." Edinburgh, 1820. Two Volumes. Despite its early date, Scoresby's work remained the classic account of Arctic and Greenland right whaling throughout the century. Captain Scoresby, who made repeated voyages to the North Atlantic, wrote with the authority of one who was thoroughly conversant with every detail of his subject.

SCORESBY, WILLIAM, "Journal of Voyage to the Northern Whale Fishery." Edinburgh, 1823.

SCORESBY, WILLIAM (REVEREND), "The Adventurous Life of Captain Scoresby." London, 1851. A biography of Captain Scoresby, written by his son.

COLNETT, "A Voyage to the South Atlantic and Round Cape Horn into the Pacific Ocean for the Purpose of Extending the Spermaceti Whale Fisheries, and other Objects of Commerce." London, 1798.

BEALE, THOMAS, "A Few Observations on the Natural History of the Sperm Whale, with an Account of the Rise and Progress of the Fishery, and of the Modes of Pursuing, Killing, and 'Cutting-in' that Animal," etc. London, 1835.

BEALE, THOMAS, "The Natural History of the Sperm Whale. . . . To Which Is Added a Sketch of a South-Sea Whaling Voyage." London, 1839. The author sailed as surgeon of a British sperm whaler, and used his scientific training and developed powers of observation to good effect.

BENNETT, F. D., (Fellow of the Royal Geographical Society), "Narrative of a Whaling Voyage Round the Globe, From the Year 1833 to 1836." London, 1840. Two Volumes. Another account of whaling life by a man of scientific training.

LAING, JOHN, "An Account of a Voyage to Spitzbergen," etc. London, 1815. Written by the surgeon of the whale-ship *Resolution,* of Whitby, 1806–1807.

DAVIS, W. M., "Nimrod of the Sea; or, The American Whaleman." New York, 1874. A good, trustworthy account of whaling life.

WHITECAR, W. B., JR., "Four Years Aboard the Whaleship. Embracing Cruises in the Pacific, Atlantic, Indian, and Antarctic Oceans, in the Years 1855, '6, '7, '8, '9." London and Philadelphia, 1860.

OLMSTED, F. A., "Incidents of a Whaling Voyage." New York, 1841.

NORDHOFF, CHARLES, "Life on the Ocean: Being Sketches of Personal Experience in the United States Naval Service, the American and British Merchant Marine, and the Whaling Service." Cincinnati, 1874.

HAZEN, JACOB A., "Five Years Before the Mast, or, Life in the Forecastle, Aboard of a Whaler and Man-of-War." Philadelphia, 1854.

CHEEVER, H. T., "The Whale and His Captors; or, The Whaleman's Adventures, and the Whale's Biography, as Gathered on the Homeward Cruise of the *Commodore Preble.*" New York, 1850.

HOLMES, LEWIS, "The Arctic Whaleman; or, Winter in the Arctic Ocean; Being a Narrative of the Wreck of the Whale Ship *Citizen,* New Bedford, . . . Sept. 25, 1852 . . . Together With a Brief History of Whaling." Boston, 1857.

CHASE, OWEN, "Narrative of the Most Extraordinary and Distressing Shipwreck of the Whale-ship *Essex,* of Nantucket; . . . with an Account of the Unparalleled Sufferings of the Captain and Crew During a Space of Ninety-three Days at Sea in Open Boats, in the Years 1819 and 1820." New York, 1821. A plain, unadorned narrative of one of the most harrowing maritime experiences on record. The author was first mate of the *Essex.*

LAY, WILLIAM and HUSSEY, C. M., "A Narrative of the Mutiny on Board the Ship *Globe,* of Nantucket, in the Pacific Ocean, Jan. 1824, and the Journal of a Residence of Two Years on the Mulgrave Islands." New London, 1828. Another record of whaling pathology.

JARMAN, ROBERT, "Journal of a Voyage to the South Seas, in the *Japan,* Employed in the Sperm Whale Fishery, Under the Command of Capt. John May." London, 1838.

WAYLAND, FRANCIS, "The Claims of the Whaleman on Christian Benevolence." New Bedford, 1843.

CHASE, WORTHINGTON, "A Voyage from the U. S. to South America Performed During the Years 1821, '22, and '23." Newburyport, Massachusetts, 1823.

DELANO, REUBEN, "Wanderings and Adventures of Reuben Delano, Being a Narrative of Twelve Years Life in a Whale Ship!" Boston, 1846.

HOLDEN, HORACE, "A Narrative of the Shipwreck, Captivity, and Sufferings of Horace Holden and Benj. H. Nute; who were cast away in the American ship *Mentor,* on the Pelew Islands, in the year 1832." Boston, 1836.

TORREY, WILLIAM, "Torrey's Narrative; or, The Life and Adventures of William Torrey, who . . . was held a captive by the Cannibals of the Marquesas." Boston, 1848.

NEVENS, WILLIAM, "Forty Years at Sea; or, A Narrative of the Adventures of William Nevens." Portland, Maine, 1846.

LIGHTCRAFT, GEORGE (Pseudonym), "Scraps From the Log Book of George Lightcraft, Who Was More Than Twenty Years a Sailor; an Account of the Whale Fishery; with Many Thrilling Incidents in the Life of the Author." Detroit, 1850.

THOMES, W. H., "The Whaleman's Adventures in the Sandwich Islands and California." Boston, 1873.

HALL, D. W., "Arctic Rovings: or, The Adventures of a New Bedford Boy on Sea and Land." Boston, 1861.

FAULKNER, J. P., "Eighteen Months on a Greenland Whaler." New York, 1878.

COMSTOCK, WILLIAM, "Life of Samuel Comstock, The Terrible Whaleman," etc. Boston, 1840.

SPEARS, J. R., (Editor), "Log of the Bark *Emily* as Kept by L. R. Hale, Third Mate, 1857–1860." In Harper's Monthly Magazine, 1903. Vol. 107, pp. 242–251.

MACY, W. H., "The Whales We Caught and How We Did It." Boston, 1889.

DODGE, G. A., "Narrative of a Whaling Voyage in the Pacific Ocean, and Its Incidents." Salem, Massachusetts, 1882.

MONTGOMERY, JAMES (Compiler), "Journal of Voyages and Travels by Daniel Tyerman and George Bennett, Sent Out by the London Missionary Society to Visit Their Stations in the South Pacific, China, India, etc., 1821 to 1829." Boston and New York, 1832. Three Volumes.

MULLETT, J. C., "A Five Year Whaling Voyage." Cleveland, 1859.

MUNGER, J. F., "Two Years in the Pacific and Arctic Oceans and China. Journal and Letters while on a Whaling Voyage on the Ship *St. George,* New Bedford, and on the Merchant Vessel *Annie Buckman,* Hong Kong to New York." Vernon, New York, 1852. Only a few copies printed.

MARKHAM, A. H., "A Whaling Cruise to Baffin's Bay and the Gulf of Boothia." London, 1875.

ALDRICH, H. L., "Arctic Alaska and Siberia, or, Eight Months With the Arctic Whalemen." Chicago and New York, 1889.

BULLEN, F. T., "The Cruise of the *Cachalot,* Round the World After Sperm Whales." New York, 1899.

JENKINS, T. H., "Bark Kathleen Sunk by a Whale. To Which Is Added an Account of Two Like Occurrences, the Loss of the Ships *Ann Alexander* and *Essex.*" New Bedford, 1902.

SMITH, C. E., "From the Deep of the Sea. An Epic of the Arctic. The Diary of C. E. Smith, Surgeon of the Whaleship *Diana* of Hull, on a Voyage in 1866 to the Arctic." New York, 1923. A document of maritime misfortune which might have formed a companion tale to the "Ancient Mariner"! Edited by the author's son.

BEANE, J. F., "From Forecastle to Cabin." New York, 1905.

PADDACK, W. C., "Life on the Ocean; or, Thirty-five Years at Sea." Cambridge, Massachusetts, 1904.

ASHLEY, C. W., "The Blubber Hunters." In Harper's Monthly Magazine, 1906. Vol. 112, pp. 670–682, 832–844.

LINDSAY, D. M., "A Voyage to the Arctic in the Whaler *Aurora*." Boston, 1911.

WHEELER, J. C., "There She Blows!" New York, 1909.

BURNS, W. N., "A Year With a Whaler." New York, 1913.

COOK, J. A., "Pursuing the Whale." Boston, 1926.

VILLIERS, A. J., "Whaling in the Frozen South." Indianapolis, 1926.

VIII — Works on the Natural History of Whaling

GOODE, G. B., (Editor), "The Fisheries and Fishery Industries of the United States." Five Sections, in Seven Volumes. Washington, 1884 and 1887. Printed as Senate Miscellaneous Document No. 124, 47th Congress, 1st Session. Section I deals with "The Natural History of Useful Aquatic Animals." One volume of text and one volume of 227 plates. The scope of the work includes valuable material with regard to whales.

SCAMMON, C. M., "Marine Mammals of the Northwestern Coast of North America, With an Account of the American Whale Fishery." San Francisco and New York, 1874.

FLOWER, W. H., and LYDEKKER, R., "An Introduction to the Study of Mammals, Living and Extinct." London, 1891.

LYDEKKER, RICHARD, Articles on the Whale, Sperm-whale, Whalebone, Whale-Oil, and Whale-Fishery. Encyclopedia Brittanica.

RACOUITZA, E. G., "Summary of General Observations on the Spouting and Movements of Whales." Included in the Annual Report of the Smithsonian Institution for 1903. Washington, 1903.

ALLEN, G. M., "The Whalebone Whales of New England." Boston Society of Natural History. Boston, 1916.

BEDDARD, F. E., "A Book of Whales." London, 1900. This work forms one volume in the Progressive Science Series.

SHALER, N. S., "Notes on the Right and Sperm Whale." In American Naturalist, 1873. Vol. 7, pp. 1–4.

TRUE, F. W., "The Whalebone Whales of the Western North Atlantic Compared with Those Occurring in European Waters, with Some Observations on the Species of the North Pacific." Smithsonian Institution Publication No. 1414. Washington, 1904.

FLOWER, W. H., "Paper on the Cachalot," read before the Royal College of Surgeons of England on November 14, 1867; and a "Paper on Ziphioid Whales," read before the same body on November 7, 1871. These two papers, bound together, form part of the Daniel B. Fearing Collection, Treasure Room, Widener Library, Harvard University.

FLOWER, W. H., "On Whales, Past and Present, and Their Probable Origin." In Nature, 1883. Vol. 28, pp. 199–202, 226–230.

BEALE, THOMAS, "The Natural History of the Sperm Whale. . . . To Which Is Added a Sketch of a South-Sea Whaling Voyage." London, 1839.

GILL, T., "The Sperm Whales, Giant and Pigmy." In American Naturalist, 1871. Vol. 4, pp. 725–743.

JARDINE, WILLIAM (SIR), "Natural History of the Ordinary Cetacea or Whales." Vol. VI of a larger work on Mammalia, in the Naturalist's Library. Edinburgh, 1837. Although this volume was known during the middle of the nineteenth century as the work of a reputable scientist, later researches brought out the fact that it was often inaccurate and inadequate.

BUCHANAN, J. Y., "The Sperm Whale and Its Food." In Nature, 1896. Vol. 53, pp. 223–225.

DEWHURST, H. W., "Natural History of the Order Cetacea, and the Oceanic Inhabitants of the Arctic Regions." London, 1834.

SUCKOW, F. W. L., "Osteologische Beschreibung des Walls . . . Mit Fünf lythographirten Tafeln in Quertfolio, aufgenommen und gezeichnet von Anton Wüstner." Mannheim, 1837.

IX — Miscellaneous Material

A — The Whaleman in the Field of Literature

MELVILLE, HERMAN, "Moby-Dick; or, The Whale." New York, 1851. Unmistakably the masterpiece of whaling literature, as well as a significant contribution to the general field of American letters. Taken as a whole, the tale undoubtedly belongs to the realm of fiction; but the descriptions of whaling life, equipment, and activities which are scattered through its pages are essentially accurate. The author actually shipped as a foremast hand on a Pacific whaler, and was thoroughly familiar with the technique and the atmosphere of whaling.

HART, J. C., "Miriam Coffin; or, The Whale-Fishermen. A Tale." New York, 1835. The best tale of whaling life until "Moby-Dick" appeared.

ANTHONY, JOSEPH R., "Diary Written in 1823 and 1824." Published by the Old Dartmouth Historical Society under the title, "Life in New Bedford A Hundred Years Ago." New Bedford, 1922. An intimate portrayal of the social life of the wealthier Quaker whaling families of New Bedford during the decade of the twenties.

TRIPP, HOWLAND, "In Whaling Days." Boston, 1909. A collection of short stories dealing with whaling.

MACY, W. F., and HUSSEY, R. B., "The Nantucket Scrap Basket." Nantucket, 1916. A collection of yarns, anecdotes, and traditions concerning Nantucket life and people. Carries a quaint tang of harpoons and salt water.

ROBBINS, C. H., "The Gam: Being a Group of Whaling Stories." Boston, 1899.

KINGSTON, W. H. G., "The Two Whalers; or, Adventures in the Pacific." London and New York, 1879.

MACY, W. H., "There She Blows! or, The Log of the *Arethusa*." Boston and New York, 1877.

FROTHINGHAM, ROBERT (Compiler), "Songs of the Sea and Sailors' Chanteys." Boston, 1924.

B — Technical Works Dealing with the Construction, Equipment, and Handling of the Various Types of Sailing Vessels

DANA, RICHARD H., "The Seaman's Friend." Boston, 1857. A clear exposition of the mysteries of sailing-ship terminology and the intricacies of hull and rigging.

CHATTERTON, E. KEBLE, "Sailing Ships: The Story of Their Development." London, 1909. An elaborate history of the evolution of the various types of sailing vessels.

CHATTERTON, E. KEBLE, "Fore and Aft: The Story of the Fore and Aft Rig." London, 1912.

BRADY, WILLIAM N., "The Kedge-Anchor; or, Young Sailors' Assistant." New York, 1863.

——, "The Mate and His Duties." Liverpool, 1855. The anonymous author of this treatise performed a valuable service for aspiring and ambitious young officers, both actual and potential.

C — Documentary Material Concerning the Legal Status and Treatment of American Seamen

——, Consular Letters. Official Correspondence between American Consuls and the Department of State. Original Manuscripts, in Bound Volumes, Arranged Chronologically by Ports. State Department Library, Washington. The reports from the consulates at Liverpool, Calcutta, Canton, and Rio de Janeiro are particularly valuable in affording an insight into the conditions of forecastle life in the sailing-ship era.

MACARTHUR, WALTER (Compiler), "The Seaman's Contract 1790–1918. A Complete Reprint of the Laws Relating to American Seamen Enacted, Amended, and Repealed by the Congress of the United States." San Francisco, 1919.

MACARTHUR, WALTER (Compiler), "Handbook Navigation Laws of the United States Sections and Acts Applicable to Shipment and Discharge of Seamen, Qualifications of Licensed Officers, Manning Scale, etc." San Francisco, 1918. (Third Edition.)

D — Works on Whaling in the Twentieth Century

JENKINS, J. T., "History of the Whale Fisheries; From the Basque Fisheries of the 10th Century to the Finner Whale at the Present Time." London, 1921. The closing chapters of this comprehensive work present a summary view of the present-day state of whaling in various corners of the world.

RABOT, CHARLES, "The Whale Fisheries of the World." Published in the Annual Report of the Smithsonian Institution for 1913, pp. 481–489. Washington, 1914. Translation of an article which appeared in the issue of La Nature, Paris, for September 14, 1912.

STEVENSON, C. H., "Whalebone — Its Production and Utilization." U. S. Bureau of Fisheries, Document No. 626. Washington, 1907.

MURDOCH, W. G., "Modern Whaling and Bear-Hunting; a Record of Present-Day Whaling With Up-to-date Appliances in Many Parts of the World," etc. London, 1917.

ANDREWS, ROY CHAPMAN, "Whale Hunting With Gun and Camera; a Naturalist's Account of the Modern Shore-Whaling Industry, of Whales and Their Habits, and of Hunting Experiences in Various Parts of the World." New York, 1916.

MORLEY, F. V., and HODGSON, J. S., "Whaling North and South." New York, 1926.

——, "Cutting In a Whale. A Series of Twenty-five Photographs Taken on Board Bark *California*." Published by H. S. Hutchinson & Co., New Bedford, Mass., 1903. No Text. A graphic presentation of the processes involved in cutting-in sperm and right whales. These photographs, when added to the far more complete and pretentious works by Dow, G. F., "Whale Ships and Whaling," and Ashley, C. W., "The Yankee Whaler," present a pictorial record of the whaling industry which is both fascinating and revealing.

WATERMAN, T. T., "The Whaling Equipment of the Makah Indians." University of Washington Publications in Political and Social Science, Vol. I, No. 1. Seattle, 1920. A detailed account of primitive whaling methods, equipment, and superstitions.

INDEX

Names of vessels are in italics